Applied Molecular Biology

Chao-Hung Lee
Professor
Department of Pathology and Laboratory Medicine
Indiana University School of Medicine
Indianapolis, Indiana, USA

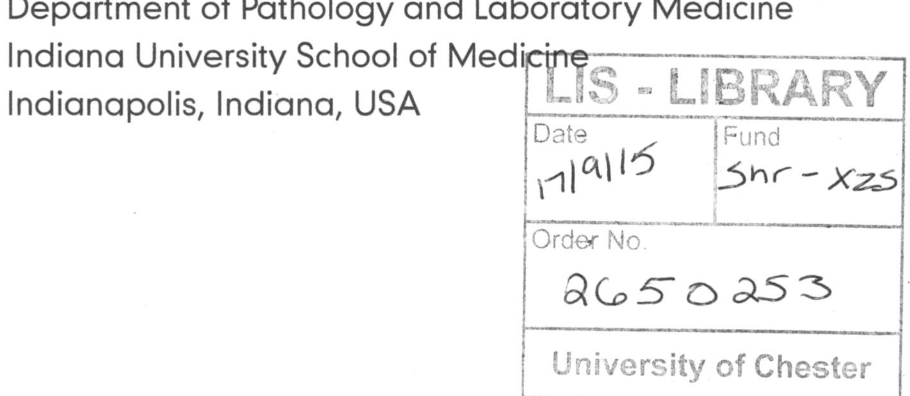

KOMPASS BOOKS KompassBooks Ltd.

Applied Molecular Biology
By **Chao-Hung Lee**
ISBN 978-988-18284-1-5

KOMPASS BOOKS is an academic publisher for leading scientists in medicine, physics, and chemistry.

Kompass Books Ltd.
Address : Suite 1907, 19/F, Office Tower Convention Plaza, 1 Harbour Road, Wanchai Hong Kong
TEL : 852 3102 8305
FAX : 852 2836 0813
Email : service@kompassbooks.com

Art Director, Cover Designer : Jump King Creativity Integration Ltd.
www.jumpking.com.tw

Chao-Hung Lee

Education:

- BS, Biology, Fu-Jen University, Taipei, Taiwan
- MS, Microbiology and Immunology, Graduate Institute of Microbiology and Immunology, College of Medicine, Taiwan University, Taipei, Taiwan
- Ph.D. Department of Microbiology and Immunology, Indiana University School of Medicine, Indiana University School of Medicine
- Postdoctoral fellow, Cold Spring Harbor Laboratory, New York, New York, USA

Current position:

- Professor, Department of Pathology and Laboratory Medicine, Indiana University School of Medicine, Indianapolis, Indiana, USA

Preface

It was in 1978 when I first heard the term EcoRI in a seminar. I had heard of *E. coli*, but not EcoRI. Since the speaker never explained what EcoRI was, I was totally lost during the seminar. I was also deeply troubled by my failure to understand the principle of DNA sequencing in the early 1980s even though I had heard talks from more than ten people and read almost everything that was published about it at that time. This incomplete understanding of important techniques in Molecular Biology really hampered my learning of the subject.

In 1987, I offered a Molecular Biology workshop with several colleagues for the first time, covering techniques such as cloning, restriction mapping, hybridization, and DNA sequencing. Although it was a graduate course, several faculty members from our school, Indiana University School of Medicine (IUSM), attended the class because they needed to use the techniques for their research. This workshop has become an annual event since then. As a firm believer of the importance of the basic principles of techniques, I tried to cover them as much as possible. I was astonished to learn that I was not the only person who had difficulties in understanding the principles of DNA sequencing. Numerous attendees commented in their evaluations that they had heard lectures on DNA sequencing principles many times but could never understand them until they attended my workshop. These comments were encouraging as they indicated that I was teaching in a way that effectively explained the basic principles of Molecular Biology. Hundreds of faculty members from IUSM and scientists from Eli Lilly and Co. are among the thousands of students who have since taken my workshop.

This book is intended to minimize the frustrations and difficulties that one may encounter in first learning Molecular Biology. It is focused on applications, not on theoretical or more advanced studies of Molecular Biology. Therefore, only the concepts related to techniques that are used in research are described. Efforts are devoted to making complicated concepts as simple as possible so that they can be easily understood. The book is presented in a manner similar to the way I teach, and contents of the book are based on the Molecular Biology workshop that I have taught for more than 20 years. The book starts with the structures of DNA and RNA followed by the processes of DNA replication, transcription, and translation, which are at the heart of Molecular Biology. The three major types of enzymes including degradation, synthesis, and modification enzymes that are routinely used in research are then introduced. Since the expression of a gene in cells is tightly regulated, the book describes in detail the mechanism of regulation and the concept of operons. Various methods for DNA and RNA isolation and electrophoresis are also covered.

Next, the book describes the process of cloning, which is the most basic technique in Molecular Biology. Vectors and different ways for selection of desired recombinant clones are discussed. Different types of nucleic acid hybridization and their applications are also introduced. Since the lambda bacteriophage and its use as a cloning vector for construction of genomic and cDNA libraries are the foundation of many studies in Molecular Biology, they are included even though they are no longer heavily used. The invention and evolution of DNA sequencing techniques are a major part of this book since DNA sequencing is a major breakthrough in the history of Molecular Biology. Polymerase chain reaction (PCR), which almost completely revolutionized Molecular Biology, is also discussed at length. The basic concept and technique of gene expression and its application in the production of therapeutic proteins are explained. Mutagenesis, homologous recombination, transgenic animals, viral vectors for gene therapy, and RNA interference which play a major role in modern Molecular Biology are also described. The book ends by describing various techniques such as DNA footprinting, chromatin immunoprecipitation, microarrays, various methods for global gene expression, and the second generation of DNA sequencing for more advanced and specific studies.

For me, writing a book is very time, energy, and labor intensive. I thank my family for their support throughout my efforts to write this book in the last several years. I am also grateful for the encouragement and assistance that I have received from many friends, colleagues, and former students. Without their help and support, this book would not be possible.

Chao-Hung Lee
Department of Pathology and Laboratory Medicine
Indiana University School of Medicine
June 1, 2009

Contents

Chapter 1

Nucleic Acid Structure and Properties

Outline

Nucleic acid
Ribose and Base
Nucleoside
Nucleotide
DNA and RNA structure
Properties of DNA and RNA

1.1: Nucleic acid

Nucleic acid is the genetic material of a cell. Two different types of nucleic acids exist in cells: ribonucleic acid (RNA) and deoxyribonucleic acid (DNA). DNA contains genes that control the functions of a cell. A gene is the basic unit of heredity. Before a gene is expressed, it is first transcribed into RNA which is then translated into protein.

Genes are represented by symbols (Table 1-1). Conventionally, a gene symbol consists of three to five italicized letters followed by a number or a capital letter, which is also italicized. The symbol of a protein is usually the same as that of the corresponding gene but is not italicized; however, the first letter of the symbol is usually capitalized. For example, *recA* is the symbol for a prokaryotic gene, and the product of the *recA* gene is the RecA protein. The letters for the symbol of a normal wild type (native) yeast gene are written in upper case, while its respective mutated gene is written in lower case letters. For example, *COX2* is the normal gene, and *cox2* is the mutated *COX2* gene. The yeast protein symbol is also written by capitalizing its first letter. For example, the protein product of the *COX2* gene is the Cox2 protein. For rat and mouse genes, all letters of a symbol are italicized, but only the first letter of the symbol is capitalized. The same symbol is used for the corresponding protein but is not italicized. For example, *Adh1* represents the alcohol dehydrogenase gene 1, whereas Adh1 is the protein encoded by the *Adh1* gene. Since mammals have two copies of most genes, one paternal and one maternal, a homozygous *Adh1* mutant is usually written as *Adh1-/-*, and a heterozygous mutant is written as *Adh1+/-*. Homozygous mutant means both paternal and maternal copies of the gene are mutated, and heterozygous mutant means either the paternal or maternal copy of the gene is mutated. For human genes and proteins, the symbols are the same as those of the rat and mouse systems except that all letters of the symbols are capitalized.

Table 1-1: Nomenclature of genes

Organism	Normal Gene	Mutated Gene	Protein
Bacteria	*recA*	*recA*	RecA
Yeasts	*COX2*	*cox2*	Cox2
Rats and mice	*Adhl*	*Homozygous: Adhl-/-* *Heterozygous: Adhl+/-*	Adhl
Humans	*ADHI*	*Homozygous: ADHI-/-* *Heterozygous: ADHI+/-*	Adhl

1.2: Ribose and Base

Both DNA and RNA are composed of three different molecules: ribose, base, and phosphate. Ribose is a 5-carbon sugar. The five carbons are numbered 1' to 5' (pronounced 1 prime to 5 prime) as shown in Fig. 1-1. A regular ribose has a hydroxyl (OH) group attached to its 2' carbon. However, the ribose in DNA has a hydrogen (H) group at this position. Since it is missing an oxygen atom at this position compared to a regular ribose, this ribose is called 2'-deoxyribose (Fig. 1-1).

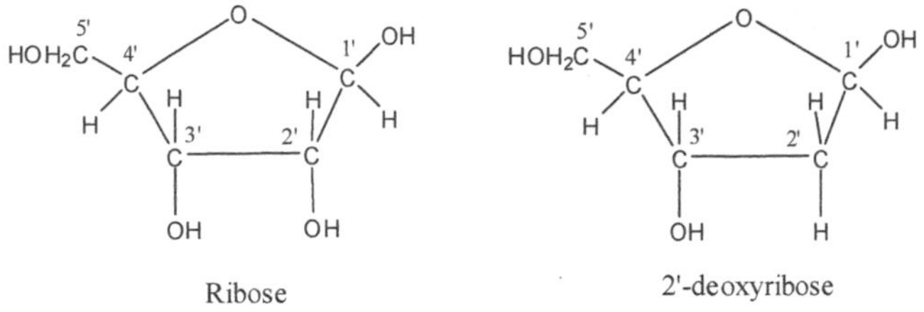

Ribose 2'-deoxyribose

Fig. 1-1: Ribose and deoxyribose.

Two types of bases exist: purine and pyrimidine. A purine is a heterocyclic aromatic compound consisting of a pyrimidine ring fused to an imidazole ring. The carbon and nitrogen atoms on the rings are numbered 1 - 9 with no prime symbol as shown in Fig. 1-2. Adenine (A) and guanine (G) are the purine bases.

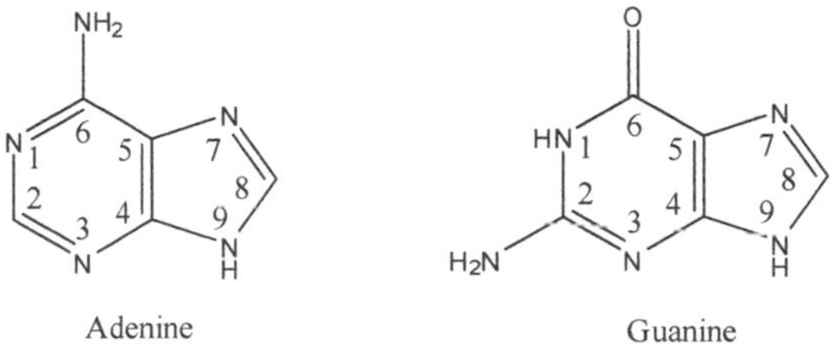

Adenine Guanine

Fig. 1-2: Adenine and guanine.

Pyrimidines are single ring aromatic molecules. As with purines, the carbon and nitrogen atoms of pyrimidines are numbered as shown in Fig. 1-3. The pyrimidines include cytosine (C), thymine (T), and uracil (U).

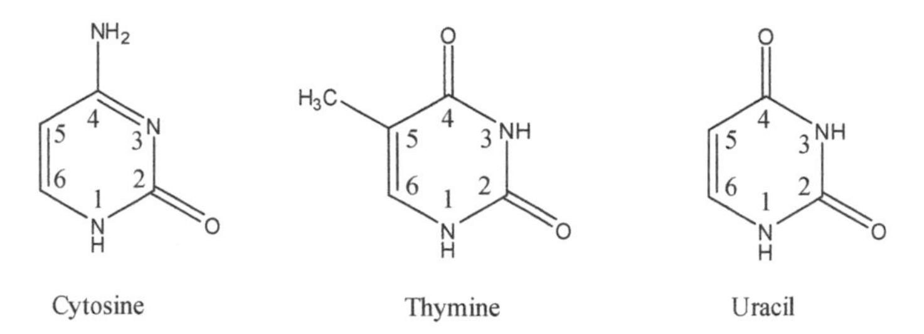

Fig. 1-3: Cytosine, thymidine and uracil.

1.3: Nucleoside

A nucleoside is formed when a base and a ribose are joined. Since there are two types of ribose molecules, ribose and 2'-deoxyribose, there are two types of nucleosides, ribonucleosides and deoxyribonucleosides. The ribonucleosides, adenosine, guanosine, cytidine, and uridine, are so called because a ribose is linked to adenine, guanine, cytosine, and uracil, respectively. If 2'-deoxyribose is the sugar component of a nucleoside, the resulting compounds are 2'-deoxyadenosine, 2'-deoxyguanosine, 2'-deoxycytidine, and 2'-deoxythymidine because the base is adenine, guanine, cytosine, or thymine, respectively. If the base is a purine (adenine or guanine), it is linked to a ribose between its number 9 nitrogen atom and the 1' carbon of the ribose or 2'-deoxyribose (Fig. 1-4). If the base is a pyrimidine (cytosine, thymine, or uracil), it is linked to a ribose or 2'-deoxyribose between its number 1 nitrogen atom and the 1' carbon of the ribose (Fig. 1-4).

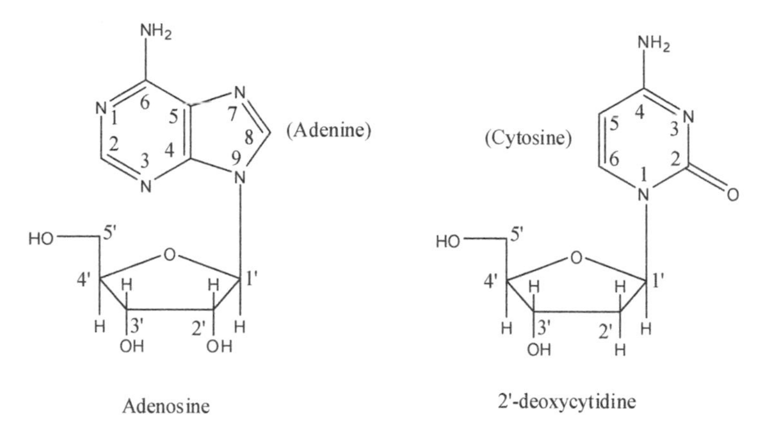

Fig. 1-4: Structure of adenosine and 2'-deoxycytidine.

1.4: Nucleotide

A nucleotide is formed when a phosphate group is linked to a nucleoside. This phosphate group is attached to the 5' carbon position of the ribose or the same position of the 2'-deoxyribose. A nucleotide may contain one, two, or three phosphate groups, so there can be many kinds of nucleotides, including monophosphate, diphosphate, and triphosphate nucleotides. If the base of a monophosphate nucleotide is adenine, guanine, cytosine, or uracil, the nucleotide is called adenosine monophosphate (AMP), guanosine monophosphate (GMP), cytidine monophosphate (CMP), or uridine monophosphate (UMP), respectively. Nucleotides with two phosphate groups include adenosine diphosphate (ADP), guanosine diphosphate (GDP), cytidine diphosphate (CDP), and uridine diphosphate (UDP). Those with three phosphate groups include adenosine triphosphate (ATP), guanosine triphosphate (GTP), cytidine triphosphate (CTP), and uridine triphosphate (UTP). These three phosphate groups are designated α, β, and γ phosphate, respectively; the α phosphate is linked to the 5' carbon of the ribose (Fig. 1-5). ATP, GTP, CTP, and UTP are used to make RNA.

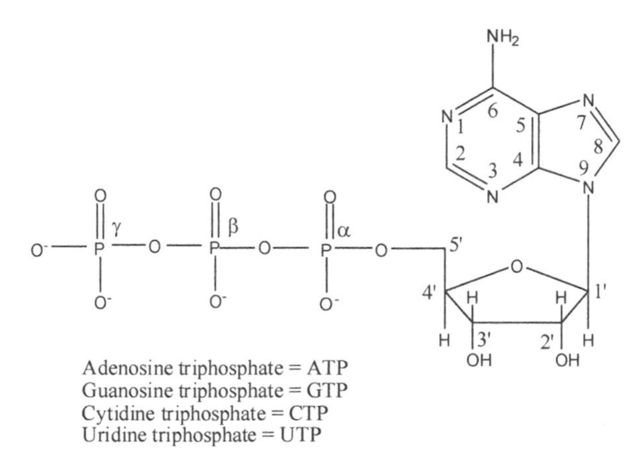

Fig. 1-5: Ribonucleotide triphosphate. The base can be adenine, guanine, cytosine, or uracil.

The following nucleotides may be formed with 2'-deoxyribose: 2'-deoxyadenosine monophosphate (dAMP), 2'-deoxyguanosine monophosphate (dGMP), 2'-deoxycytidine monophosphate (dCMP), 2'-deoxythymidine monophosphate (dTMP); 2'-deoxyadenosine diphosphate (dADP), 2'-deoxyguanosine diphosphate (dGDP), 2'-deoxycytidine diphosphate (dCDP), 2'-deoxythymidine diphosphate (dTDP); and 2'-deoxyadenosine triphosphate (dATP),

2'-deoxyguanosine triphosphate (dGTP), 2'-deoxycytidine triphosphate (dCTP), 2'-deoxythymidine triphosphate (dTTP). Among these, dATP, dCTP, dGTP, and dTTP are used to make DNA (Fig. 1-6).

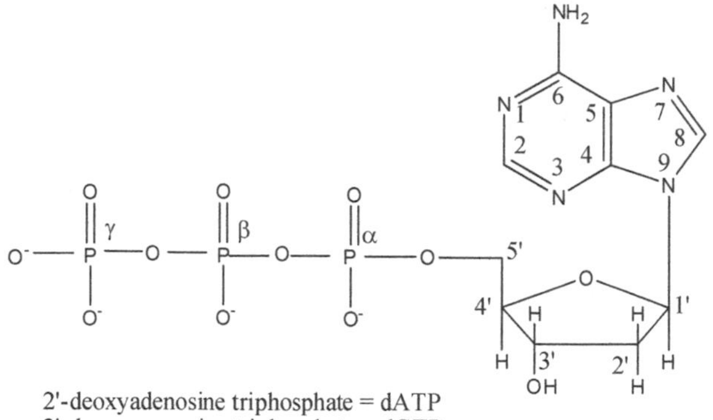

2'-deoxyadenosine triphosphate = dATP
2'-deoxyguanosine triphosphate = dGTP
2'-deoxycytidine triphosphate = dCTP
2'-deoxythymidine triphosphate = dTTP

Fig. 1-6: Deoxyribonucleotide triphosphate. The base can be adenine, guanosine, cytosine, or thymine.

1.5: Structure of DNA and RNA

DNA and RNA are polymers of nucleotides. When nucleotides are joined together to form DNA or RNA, the 3'-OH group of the first nucleotide is linked to the α phosphate of the second nucleotide to form a phosphodiester bond, and the β and γ phosphates of the second nucleotide are released as inorganic pyrophosphate (PPi).

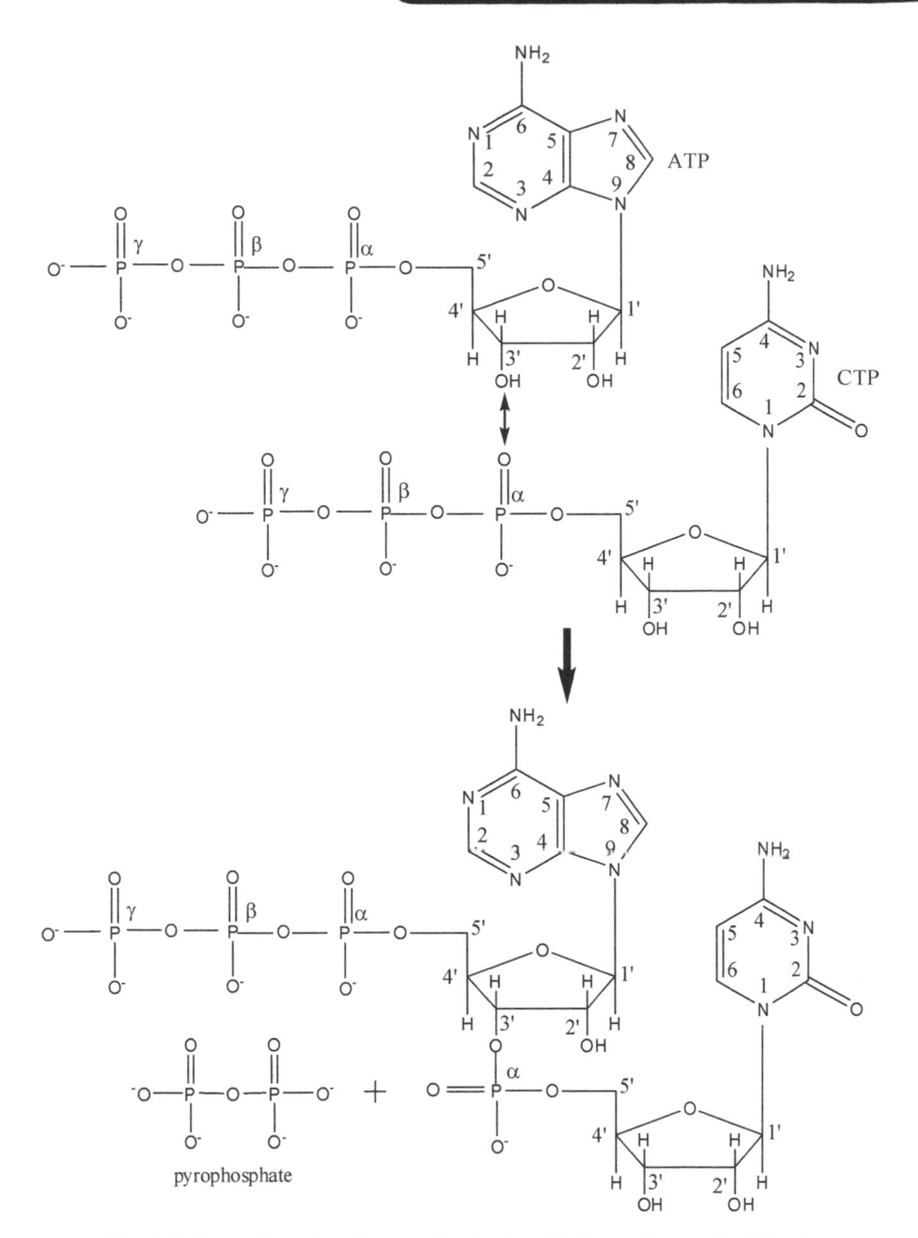

Fig. 1-7: Formation of a phosphodiester bond between two nucleotides to make RNA. The β and γ phosphates of the second nucleotide are released as pyrophosphate.

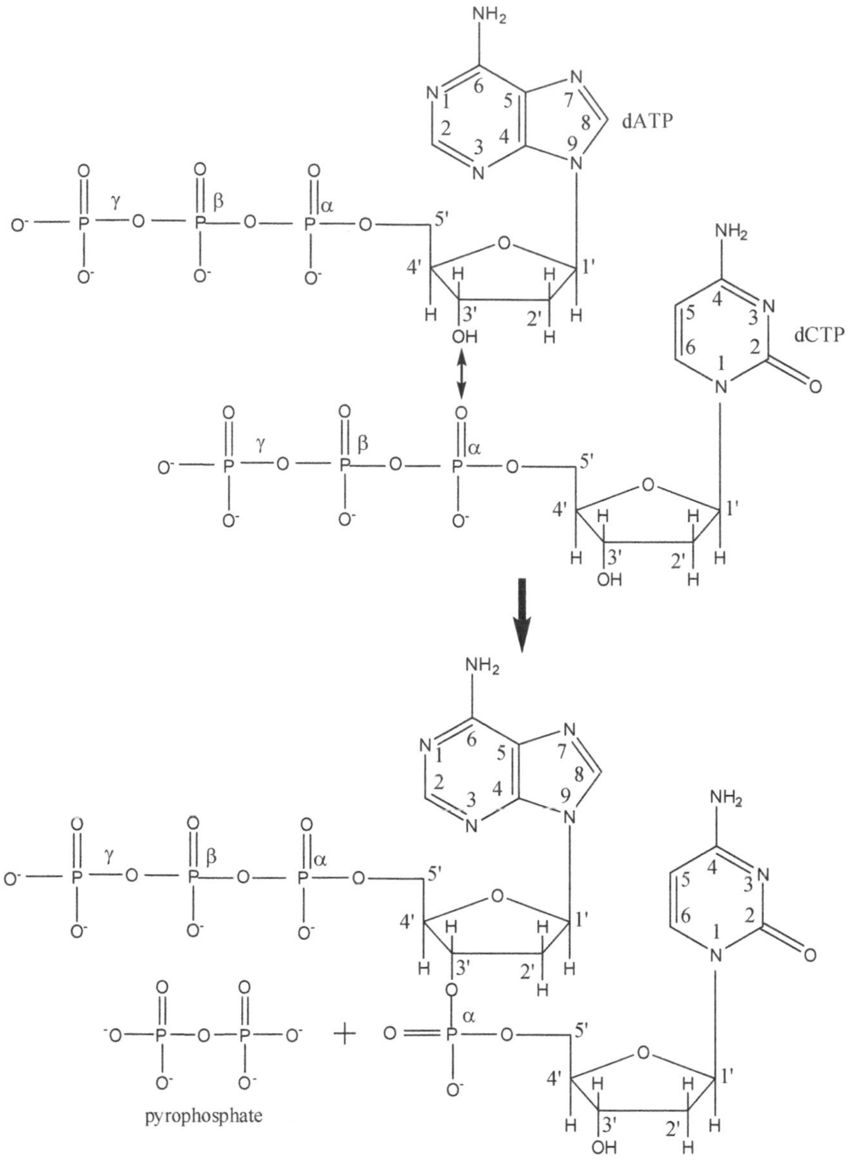

Fig. 1-8: Formation of a phosphodiester bond between two nucleotides to make DNA. The β and γ phosphates of the second nucleotide are released as pyrophosphate.

Linear DNA and RNA molecules consist of a phosphate group at one end of the molecule and a hydroxyl group at the other end (Fig. 1-7 and 1-8). Since the phosphate group is linked to the 5' position of the first nucleotide, this end is referred to as the 5' end. The hydroxyl group is linked to the 3' position of the last nucleotide; therefore, the end with the last nucleotide is referred to as the 3' end (Fig. 1-9). When drawing a line representing RNA or DNA, the 5' end is conventionally placed on the left and the 3' end is on the right. In the case of double-stranded DNA, the 5' end of the upper strand is on the left and the 3' end is on the right, while the 5' end of the lower strand is on the right and its 3' end is on the left (Fig. 1-10). Since these two strands are in opposite directions, they are commonly referred to as being anti-parallel.

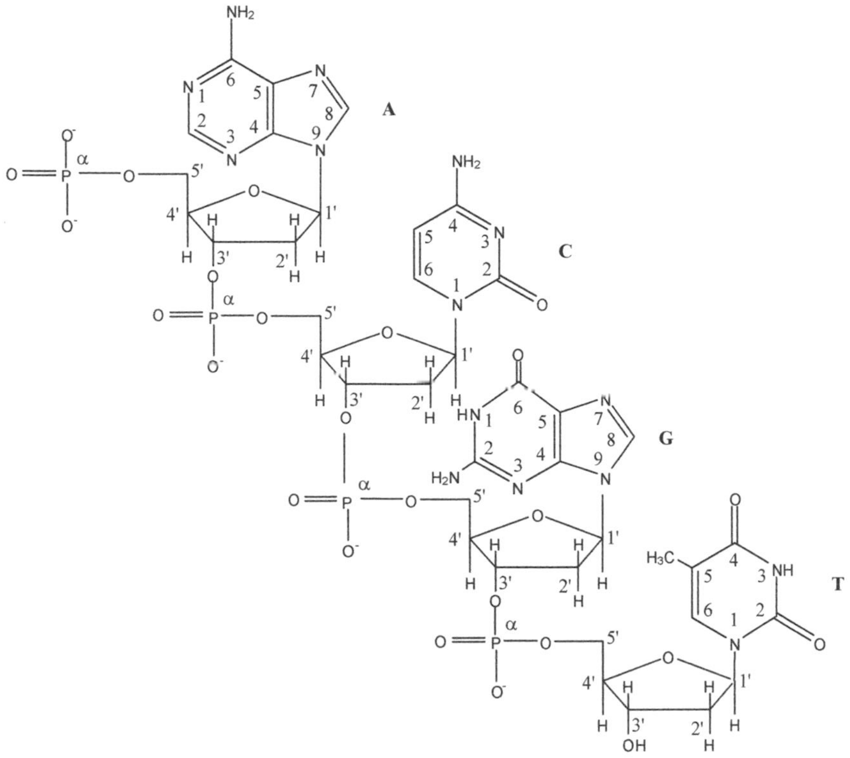

Fig. 1-9: A tetranucleotide DNA.

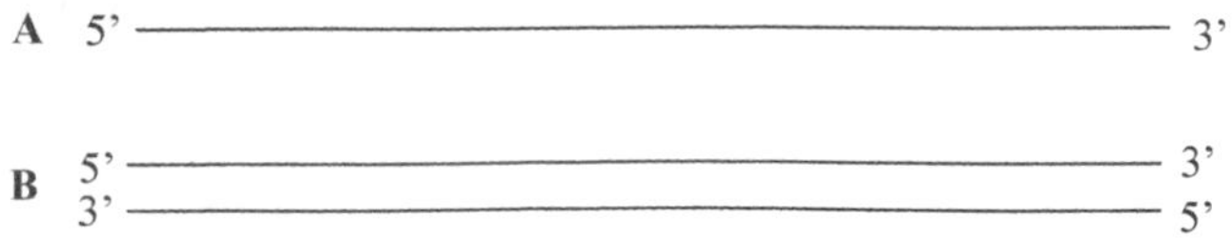

A 5' ———————————————— 3'

B 5' ———————————————— 3'
  3' ———————————————— 5'

Fig. 1-10: Naming convention of the ends of single (A) and double-stranded (B) forms of DNA or RNA.

The two DNA strands of a double-stranded DNA are held together by hydrogen bonds between complementary bases, referred to as base pairing, in which G is always paired with C, and A is always paired with T. Three hydrogen bonds are formed between G and C, and two of them are formed between A and T (Fig. 1-11).

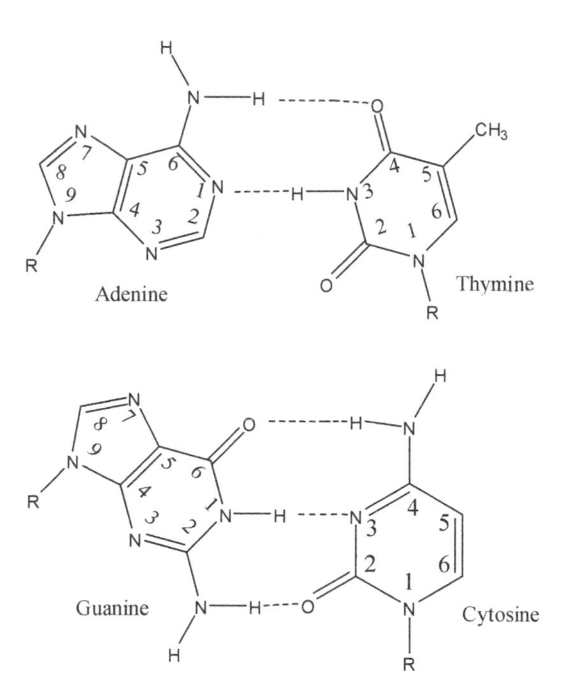

Fig. 1-11: Base pairing between DNA strands.

DNA has a helical structure. The helical conformation of a double-stranded DNA is called a double helix, which is a right-handed spiral. In this structure, the ribose and phosphate groups of DNA form the backbone of the helix, and the base pairs between the two strands hold the two backbones together. A characteristic feature of a double helix is that the spaces of alternating helical turns are not the same size (Fig. 1-12); each turn is composed of 10.4 base pairs. The bigger space is referred to as the major groove (22 Å wide), and the smaller one is referred to as the minor groove (12 Å wide). DNA may change its conformation under different conditions. The normal right-handed helical structural is B-form. In enzyme-DNA complexes or DNA and RNA hybrids, it may change to A-form which has a wider right-handed spiral, a shallower minor groove, and a deeper major groove. If the bases of a segment of DNA are chemically-modified such as methylation, it may become a left-handed spiral; this conformation is called Z-form.

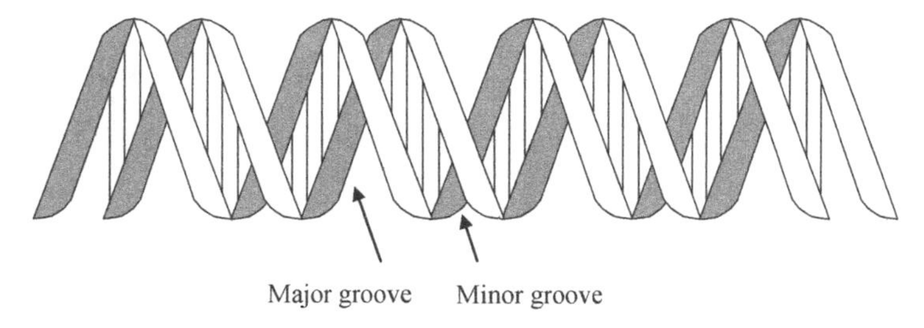

Fig. 1-12: Conformation of a DNA double helix.

1.6: Physical Properties of DNA and RNA

There are only two structural differences between naturally occurring DNA and RNA. The first is that thymine is present only in DNA and uracil is present only in RNA. The other difference is that the 2' position of the ribose is an H group in DNA but is an OH group in RNA. Although DNA and RNA are very similar structurally, their physical properties are very different. DNA is very susceptible to degradation in acidic solutions, whereas RNA is relatively resistant to acid but very sensitive to degradation in alkaline solutions. When an RNA molecule is placed in alkaline solution, the OH- ions in the solution will interact with the OH group at the 2' position, removing the hydrogen atom to form H_2O. The remaining oxygen then links to the phosphate group of the phosphodiester linkage, thus breaking the linkage between two adjacent nucleotides to form a nucleotide with a 2', 3'-cyclic phosphoric acid (Fig. 1-13). In this manner, RNA is broken down to nucleotides and is no longer able to carry out its function.

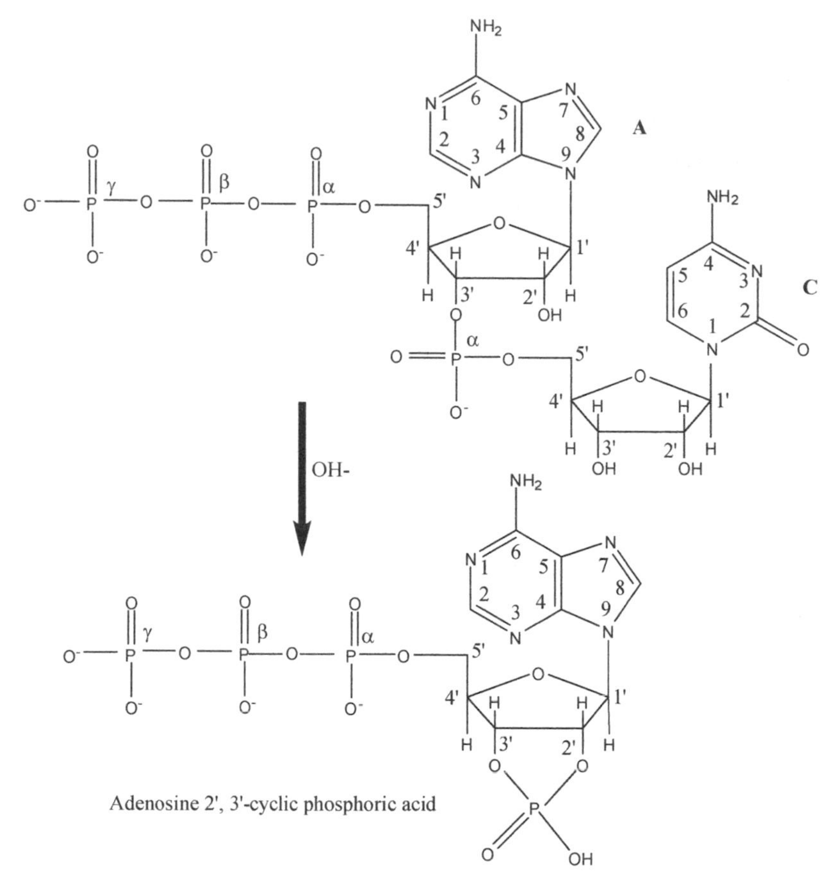

Fig. 1-13: Effect of alkali on RNA.

In contrast, DNA is not susceptible to degradation by alkali but is very sensitive to acid. In an acidic solution, the H^+ ions will attack the nitrogen atoms at position 1, 3, or 7 of a purine, resulting in an electron imbalance in the nucleotide and breakage of the link between the purine base and ribose (Fig. 1-14). A DNA with no purine residues is called apurinic DNA. The phosphodiester bonds in the sugar-phosphate backbone of an apurinic DNA are very susceptible to hydrolysis, leading to chain breakage of DNA molecules.

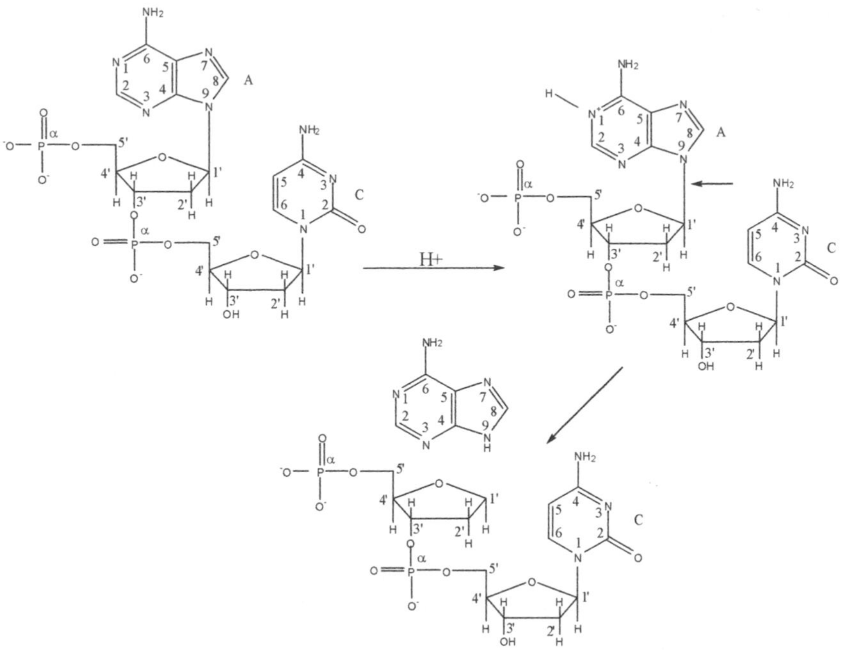

Fig. 1-14: Effect of acid on DNA.

RNA also has purine bases. However, in an acidic solution, the H^+ ions in the solution may be neutralized by the 2' OH group; therefore, RNA is not as susceptible as DNA to acid depurination. As shown in Table 1-2, in a pH 2.4 solution, 780 x 10^6 purine residues are released from a DNA molecule per minute, but only 1.2 x 10^6 purines are released from RNA. In a pH 2.0 or pH 1.8 solution, DNA is instantly degraded, but only 2.7 x 10^6 and 3.2 x 10^6 purine residues, respectively, are released from an RNA molecule.

Table 1-2: Sensitivity of DNA and RNA to acid.

	DNA	RNA
pH	k_{pur} x 10^6/min	k_{pur} x 10^6/min
2.4	780	1.2
2.0	-	2.7
1.8	-	3.2

k = first-order velocity constant. Source: Organic Chemistry of Nucleic Acids. 1972. Plenum Press, Part B, p440.

Summary

DNA and RNA are polymers of nucleotides which are composed of base, ribose, and phosphate. There are two kinds of bases: purines and pyrimidines. Purines include adenine and guanine, whereas pyrimidines include thymine, cytosine, and uracil. When a ribose is linked to a base, the resulting compound is called nucleoside. When a base, ribose, and phosphate are joined together, a nucleotide is formed. Some nucleotides have only one phosphate group; others may have 2 or 3 phosphate groups. Major atoms in a nucleotide are assigned numbers or symbols. Nitrogen and carbon atoms in purines are numbered 1 - 9, and those in pyrimidines are designated 1 - 6. The five carbon atoms in ribose are named 1' - 5', and the three phosphate groups in a nucleotide are designated α, β, and γ phosphate. The α phosphate is linked to the 5' carbon of the ribose. Nucleotides with three phosphate groups are the building blocks of DNA or RNA.

When two or more nucleotides are joined together, a DNA or RNA molecule is formed. In this process, the 3' OH group of the first nucleotide and the α phosphate group of the second nucleotide form a phosphodiester bond. In a linear DNA or RNA molecule, one end of the molecule is a phosphate group, and the other end is a hydroxyl group. Since the phosphate group is linked to the 5' position of the first nucleotide, this end is called the 5' end. The other end is called the 3' end because the OH group is located at the 3' position of the last nucleotide.

Most DNA molecules are double stranded. The two strands of DNA are held together by base pairing between G and C or A and T with three hydrogen bonds formed between G and C and two formed between A and T. There are two major structural differences between DNA and RNA. The first difference is that thymine is present only in DNA and uracil is present only in RNA. The second difference is that the 2' position of the ribose is an OH group in RNA, while it is an H group in DNA. Although DNA and RNA are structurally similar, their physical properties are very different. DNA is very sensitive to degradation in acidic solutions, but RNA is resistant to acid. On the other hand, RNA is very sensitive to degradation in alkali solutions but is resistant in alkali solutions.

Sample Questions (There may be more than one correct answer for each question)

1. Which of the following are purines: a) adenine, b) guanine, c) cytosine, d) thymine, e) uracil
2. Which of the following are pyrimidines: a) adenine, b) guanine, c) cytosine, d) thymine, e) uracil
3. Which of the following are components of DNA: a) ribose, b) 2' deoxyribose, c) thymine, d) uracil
4. Which of the following are components of RNA: a) ribose, b) 2' deoxyribose, c) thymine, d) uracil
5. A nucleoside is composed of: a) glucose + galactose, b) purine + ribose, c) pyrimidine + deoxyribose, d) purine + ribose + phosphate
6. A nucleotide is composed of: a) glucose + galactose, b) purine + ribose, c) pyrimidine + deoxyribose, d) purine + ribose + phosphate
7. In nucleosides or nucleotides, the linkage between the purine and ribose occurs at: a) position 1 of purine and 1' position of ribose, b) position 9 of purine and 1' position of ribose, c) position 7 of purine and 5' position of ribose, d) position 9 of purine and 3' position of ribose
8. In nucleosides or nucleotides, the linkage between the pyrimidine and ribose occurs at: a) position 1 of pyrimidine and 1' position of ribose, b) position 9 of pyrimidine and 1' position of ribose, c) position 7 of pyrimidine and 5' position of ribose, d) position 9 of pyrimidine and 3' position of ribose
9. In nucleotides, the phosphate group which is linked to the 5' position of a ribose is: a) α phosphate, b) β phosphate, c) γ phosphate, d) none of above
10. RNA is very sensitive to: a) phenol, b) alcohol, c) acid, d) alkali
11. DNA is very sensitive to: a) phenol, b) alcohol, c) acid, d) alkali
12. In DNA or RNA synthesis, nucleotides are joined together between: a) 2' OH of 1st nucleotide and γ phosphate of 2nd nucleotide, b) 2' OH of 1st nucleotide and α phosphate of 2nd nucleotide, c) 3' OH of 1st nucleotide and 5' carbon of 2nd nucleotide, d) 3' OH of 1st nucleotide and α phosphate of 2nd nucleotide
13. The 5' end of nucleic acid is usually: a) H, b) OH, c) COOH, d) phosphate
14. The 3' end of nucleic acid is usually: a) H, b) OH, c) COOH, d) phosphate
15. In double-stranded DNA, the two strands of DNA are held together by base pairing between: a) A and C, b) G and C, c) A and T, d) C and T

Suggested Readings:

1. Kochetkov, N. K. and Budovskii, E. I. (1972). Organic Chemistry of Nucleic Acids. Plenum Press, Part B, p440.
2. Watson, J. D. and Crick, F. H. C. (1953). A structure for DNA. Nature 171: 737-738.
3. Wilkins, M. F. H., Stokes, A. R., and Wilson, H. R. (1953). Molecular structure of DNA. Nature 171: 738-740.

Chapter 2

Replication, Transcription and Translation

Outline

2.1: Introduction

Genetic information flows first from DNA to RNA and then from RNA to protein. Replication, transcription, and translation are the three major processes involved in the transmission of genetic information in a cell. The process in which a pre-existing DNA molecule is used as the template to make an identical piece of DNA is called replication. In transcription, DNA is used as the template to make RNA. RNA can also be used as the template to make DNA. Since this process is opposite to transcription, it is referred to as reverse transcription. Finally, translation is the process by which protein is formed using RNA as a template.

2.2: DNA Replication

When a cell divides, each daughter cell needs a copy of the genome. Therefore, the genome must duplicate. The process by which this occurs is DNA replication. DNA replication requires a template, primer, DNA polymerase, and dNTP (dATP, dCTP, dGTP, and dTTP). DNA is normally double-stranded, but a DNA template is single-stranded. To obtain a single-stranded DNA template, helicase unwinds a portion of a double-stranded DNA to form a replication fork (Fig. 2-1). A primer then anneals to the template to initiate DNA synthesis. Primers

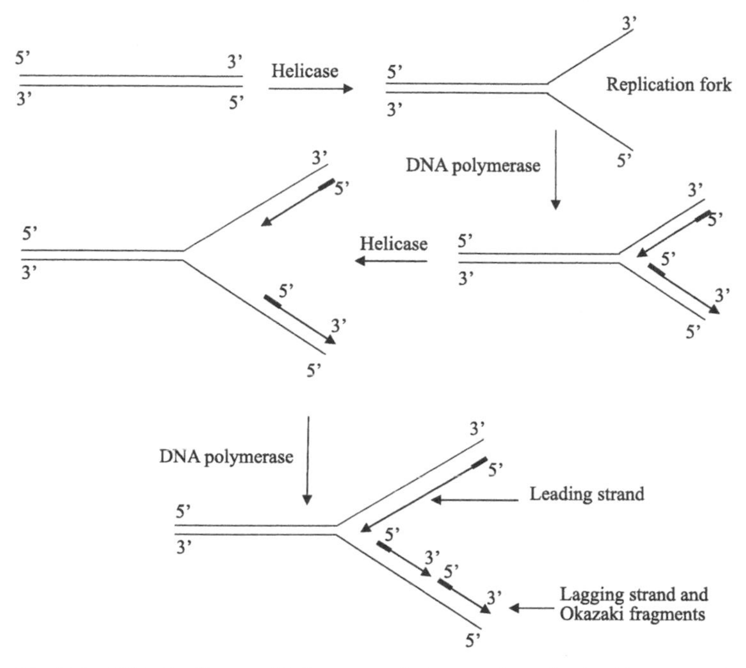

Fig. 2-1: DNA Replication. Helicase unwinds a portion of a double stranded DNA to form a replication fork. Primers indicated by thick bar anneal to the template and then initiate DNA synthesis in a 5' to 3' direction. Replication of the upper (plus) strand DNAoccurs in a continuous manner, whereas replication of the lower (minus) strand occurs in a discontinuous manner.

can be DNA, RNA, or protein. When DNA or RNA is used as the primer, it anneals to the template by base pairing; therefore, the sequence of the primer must be complementary to that of the region where it binds. DNA polymerase then starts to synthesize DNA by adding nucleotides to the 3' end of the primer. The nucleotide added to each position is complementary to the corresponding base in the template. If the base in the template is G, a C is added. In this manner, extension of the primer and replication of the DNA template occurs.

DNA synthesis by DNA polymerase always proceeds from the 5' to the 3' end of the strand being synthesized. To replicate the upper strand of the template, the primer anneals to the 3' end of the template and is extended toward the bottom of the replication fork (Fig. 2-1). The helicase then opens another portion of the DNA to use as the template, and the newly synthesized DNA continues to be extended. Replication of the lower strand (or minus strand) of the template is quite different from that of the upper strand (or plus strand). The primer first anneals to a region near the bottom of the replication fork and extends outward until reaching the end of the template. When additional template becomes available, another primer anneals to the bottom of the fork to initiate new DNA synthesis. Therefore, replication of the lower strand of the template is discontinuous, resulting in many DNA fragments. These fragments are then joined together by ligase. These DNA fragments, first discovered by Okazaki, are referred to as Okazaki fragments. Since continuous DNA synthesis is more efficient and faster, the newly synthesized upper strand DNA is called the leading strand, and that of the lower strand is referred to as the lagging strand.

DNA is a polymer of nucleotides. A second nucleotide is linked to the 3' OH group of the first nucleotide. If a protein is used as the primer, the protein must have an OH group to allow a nucleotide to join. Usually, proteins that are used as primers for DNA replication contain at least one nucleotide. For example, the reverse transcriptase of the hepatitis B virus (HBV) used as the primer to replicate the minus strand of HBV contains a dGTP, and that for the replication of the minus strand of duck hepatitis B virus has a tetranucleotide (5'-GTAA-3') associated with it. The pre-terminal protein of adenovirus is the primer for adenoviral DNA replication and has a dCMP linked to it.

2.3: Transcription

Transcription is the process through which DNA is used as a template to make RNA. Either strand of a double-stranded DNA can be used as the template. Like DNA synthesis, transcription also goes from 5' to 3'. In addition to a template, transcription requires an RNA polymerase and NTP (ATP, CTP, GTP, and UTP). In general, transcription does not require a primer. One exception is for transcription of the influenza virus which does require a primer. The influenza virus cuts a portion of the 5' end of the host mRNA and uses it as primer for transcription. This is one of the mechanisms for the pathogenesis of the influenza virus.

In transcription, only a portion of the template is copied to make RNA. Several kinds of RNAs are made including messenger RNA (mRNA), transfer RNA (tRNA), ribosomal RNA (rRNA), small nuclear RNA (snRNA), and small RNA (smRNA) including micro RNA (miRNA). Among these, mRNA is the template for protein synthesis; tRNA brings amino acids to the ribosome to make protein. rRNA is a component of the ribosome; snRNA regulates the modification of eukaryotic mRNA, and smRNA regulates translation. SmRNA usually inhibits translation; therefore, this phenomenon is referred to as RNA interference (RNAi). RNAi can also occur through the use of synthetic double-stranded RNA referred to as small interfering RNA (siRNA).

Only one RNA polymerase is present in prokaryotic cells, such as bacteria, and this RNA polymerase makes all of the different kinds of RNA in the cell. At least 3 different RNA polymerases are present in eukaryotic cells, RNA Polymerase I, II and III. RNA Polymerase I is responsible for making the 18S, 5.8S, and 28S rRNA, where S refers to the Svedberg constant or sedimentation coefficient. Svedberg is a measure of particle size with larger S values representing larger molecules. RNA Polymerase II makes mRNA, snRNA, and smRNA. RNA Polymerase III makes tRNA and 5S rRNA.

2.4: Capping of Pre-mRNA

In prokaryotic cells, the mRNA transcribed from the template is a functional mRNA and can be used directly to make protein without further modification. This is in contrast to the primary transcripts of eukaryotic mRNA which need to be modified to become functional. These primary transcripts are therefore referred to as pre-mRNA. Pre-mRNA processing takes place at both 5' and 3' ends as well as in the central portion of the molecule. The modification that occurs at the 5' end is the addition of a 7-methylguanosine to the first nucleotide of the RNA molecule. This process is referred to as capping. During capping, the α phosphate of a 7-methylguanosine triphosphate is covalently linked to the β phosphate of the first nucleotide of a pre-mRNA. After the addition of this 7-methylguanosine, the 2'-OH group of the first and the second nucleotide of the pre-mRNA may be methylated. Therefore, three types of caps exist: Cap 0, Cap 1, and Cap 2. Cap 0 has no methylation at the 2'-OH group of both the first and the second nucleotides. Cap 1 is methylated only at the 2'-OH group of the first nucleotide, and Cap 2 is methylated at the 2'-OH group of both the first and the second nucleotides (Fig. 2-2).

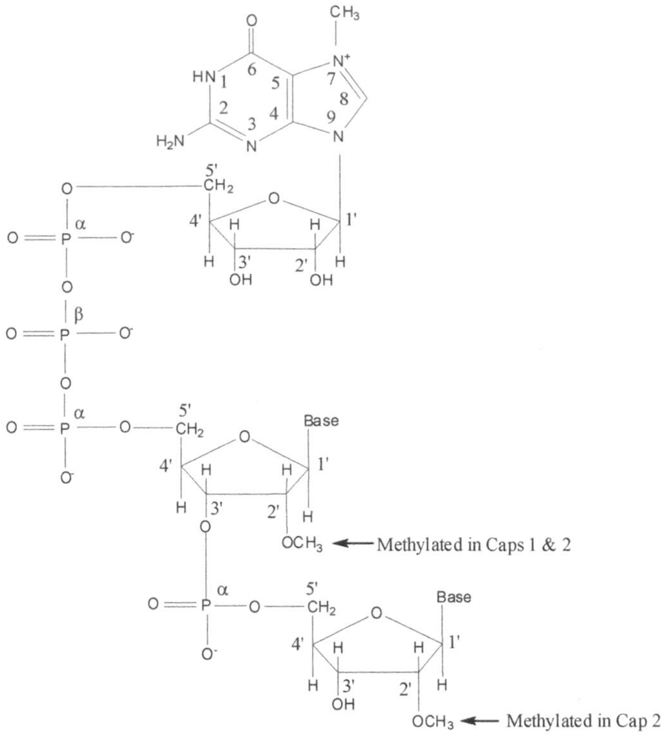

O CH3 HN 1 6 5 N+ 7 8 2 3 4 9 H2N N N
5' CH2 O
O α O=P—O⁻ 4' H H 1' H 3' 2' H OH OH
O β O=P—O⁻
O α O=P—O—5' CH2 O Base
O⁻ 4' H H 1' H 3' 2' H O OCH3 ← Methylated in Caps 1 & 2
α O=P—O—5' CH2 O Base
O⁻ 4' H H 1' H 3' 2' H OH OCH3 ← Methylated in Cap 2

Fig. 2-2: Capping of Pre-mRNA. A 7-methylguanosine is added to the 5' end of a pre-mRNA through the linkage between the α phosphate of an m7GTP and the β phosphate of the first nucleotide of the pre-mRNA. The 2'-OH group of the first and/or the second base may also be methylated. Cap 1 is methylated at the first base position only. Cap 2 is methylated at both the first and second base positions.

2.5: Splicing of Pre-mRNA

In the central portion of a pre-mRNA, some regions may or may not be translated into protein. The translated regions are referred to as expressed regions, or more commonly as "exons." Since the unexpressed regions are located between exons, they are referred to as intervening sequences or "introns." During pre-mRNA processing, introns are removed, and exons are joined together. This process is called splicing. The removal of introns is a highly regulated event. Pre-mRNA is cut at very specific places, and there are nucleotide sequences that facilitate the splicing process by signaling where splicing should occur. These signals are located at the 5' end, central region, and 3' end of every intron. As seen in Fig. 2-3, the first two nucleotides of the 5' end of an intron are GU, and those of the 3' end are AG. These are the most conserved splicing signals. When many intron sequences are compared, a consensus sequence of GURAGU at the 5' end and YYYYYNYAG at the 3' end becomes apparent; where R represents a purine (A or G), Y represents a pyrimidine (U or C), and N represents any base. The sequence YYYYYNYAG is commonly referred as PPT-N-YAG, where PPT stands for polypyrimidine tract. In addition to these signals, the sequence YURAY located 15 - 50 bases upstream from the 3' end of an intron is another splicing signal, referred to as the branch point (Fig. 2-4).

	5' Splice Site		3' Splice Site	
Ovalbumin	AUAAAUAAG	**GU**GAGCCUA	CAAUUAC**AG**	GUUGUUCGC
	GAAGCUCAG	**GU**ACAGAAA	UGUAUUC**AG**	UGUGGCACA
	AUCCUGCCA	**GU**AAGUUGC	GCUUUAC**AG**	GAAUACUUG
	AGACAAAUG	**GU**AAGGUAG	UUCUUAA**AG**	GAAUUAUCA
	GUGACUGAG	**GU**AUAUGGG	GUUCUCC**AG**	CAAGAAAGC
	CUUGAGCAG	**GU**AUGGCCC	UCCUUGC**AG**	CUUGAGAGU
Ovomucoid	UCCUCCCAG	**GU**GAGUAAC	UUCCCCC**AG**	AUGCUGCCU
	GGGGCUGAG	**GU**GAGAAAG	UUUGUCG**AG**	GUGGACUGC
	CUACAGCAU	**GU**GUGUACU	CCUCUUC**AG**	AGAAUUUGG
	ACUGUUCCU	**GU**AAGUGAA	CUUCCAC**AG**	AUGAACUGC
Human ß-globin	CCUGGGCAG	**GU**UGGUAUC	CACCCUU**AG**	GCUGCUGGU
Rabbit ß-globin	AACUUCAGG	**GU**GAGUUUG	UUCCUAC**AG**	CUCCUGGGC
Mouse ß-globin	AACUUCAGG	**GU**GAGUCUG	UUCCCAC**AG**	CUCCUGGGC
SV40 Large T	GCAACUGAG	**GU**AUUUGCU	GUAUUUU**AG**	AUUCCAACC
Rat Insulin	ACCCACAAG	**GU**AAGCUCU	CCCUGGC**AG**	UGGCACAAC
Ig λI L-VI	UCAGCUCAG	**GU**CAGCAGC	UGUUUGC**AG**	GGGCCAUUU
Ig λI J1-C1	CUGUCCUAG	**GU**GAGUCAC	CAUCCUG**AG**	GCCAGCCCA
Ig λII L-VII	UCUGCUCAG	**GU**CAGCAGC	UGUUUGC**AG**	GAGCCAGUU
Ig κ J1-1	AGCUGAAAC	**GU**AAGUACA	CUUCCUC**AG**	GGGCUGAUG
Ig μ J H1-CH1	UCUCCUCAG	**GU**AAGCUGG	UGUCCUC**AG**	AGAGUCAG
Ig μ CH4-CM1	AGUCCACUG	**GU**AAACCCA	CCUUCAU**AG**	AGGGGGAGG
Ig μ CM1-CM2	CUGUUCAAG	**GU**AGUAUGG	CACCUGC**AG**	GUGAAAUGA
Ig γ1 CH1-Hinge	AGAAAAUUG	**GU**GAGAGGA	UCUCCAC**AG**	UGCCCAGGG
Ig γ1 Hinge-CH2	UAUGUACAG	**GU**AAGUCAG	CAUCCUU**AG**	UCCCAGAAG
Ig γ1 CH2-CH3	AAACCAAAG	**GU**GAGAGCU	CACCCAC**AG**	GCAGACCGA
Ig γ2b CH1-Hinge	AAAAACUUG	**GU**GAGAGGA	UCUCUGC**AG**	AGCCCAGCG
Ig γ2b Hinge-CH2	AAUGCCCAG	**GU**AAGUCAC	CCUCAUC**AG**	CUCCUAACC
Ig γ2b CH2-CH3	AAAUUAAAG	**GU**GGGACCU	ACCCCAC**AG**	GGCUAGUCA
BKV Large T	AGCUCAGAG	**GU**UUGUGCU	UUUUUAU**AG**	GUGCCAACC
PY Large T	GGCUUCCAG	**GU**AAGAAGG	UUCUUAC**AG**	GGCUCUCCC
PY Small T	AUAAUCCAA	**GU**AAGUAUC	UUCUUAC**AG**	GGCUCUCCC
Ad5 EIA	AUGAAGAGG	**GU**GAGGAGU	UUUUAAA**AG**	GUCCUGUGU
Ad5 EIA	UUGUCUACA	**GU**AAGUGAA	UUUUAAA**AG**	GUCCUGUGU
Consensus		**GU**RAGU	YYYYYNY**AG**	

Fig. 2-3: Consensus Sequences of the 5' and 3' Ends of Introns.

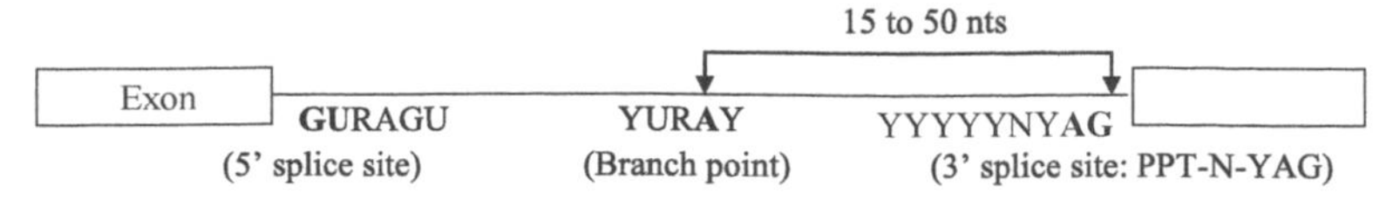

Fig. 2-4: Splicing Signals.

During splicing, the 2'-OH group of the A residue of the branch point (YURAY) is linked to the phosphate group located between the last base of an exon and the first base of the adjacent intron, thus breaking the 5' end of the intron. The resulting 3'-OH group of the last base of the first exon is then linked to the phosphate group located between the last base of the intron and the first base of the second exon, thus joining the two exons together and releasing the intron (Fig. 2-5).

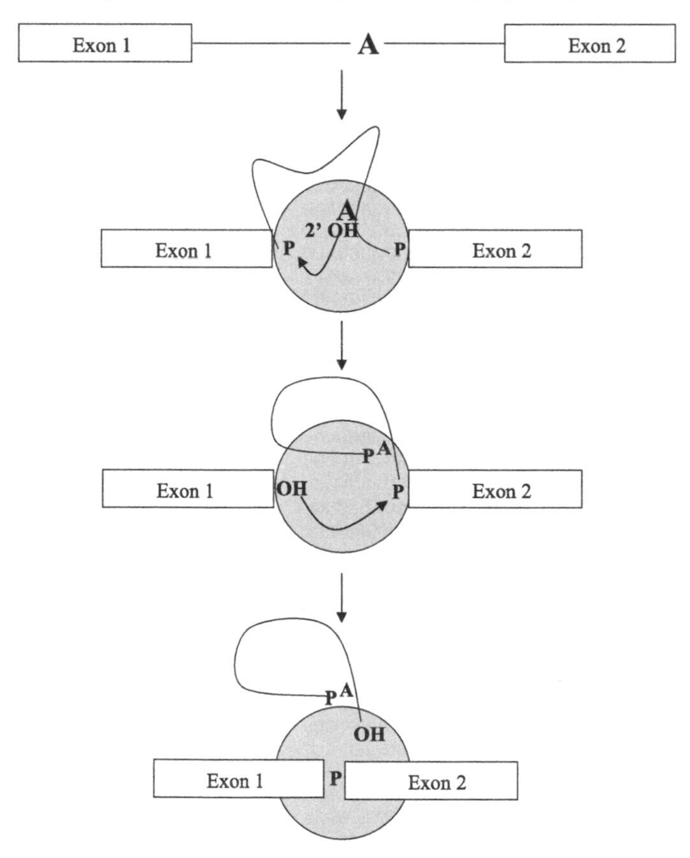

Fig. 2-5: An Overview of Pre-mRNA Splicing.

Many factors are involved in pre-mRNA splicing. The major ones are snRNA and protein complexes, referred to as snRNPs. Each snRNP is composed of one snRNA and at least 10 different proteins. More than 12 snRNAs have been identified. Since each one has a unique sequence, these snRNAs are named U1 - U12 snRNA; where U stands for unique. The length of snRNAs ranges from 56 to 217 nucleotides. During splicing, U1 snRNP first binds to the 5' end, and U2AF (U2 auxilliary factor) binds to the 3' end of the intron. U2AF is composed of two different proteins: p65 and p35. The p65 subunit interacts with PPT, and p35 binds to the AG sequence. Some introns do not have the typical 3' end sequence signal and therefore require a splicing enhancer to direct U2AF to bind to it. A splicing enhancer is usually located in the exon immediately downstream from the intron where the SR proteins bind. SR proteins are so named because they are rich in serine and arginine residues. Approximately 40 different SR proteins have been found. SF2/ASF (splicing factor 2/alternative splicing factor) and SC35 (splicing component 35) are the more common ones. SR proteins usually contain an RNA-binding domain and a protein-binding domain and therefore can bind to both RNA and proteins. When an SR protein binds to a splicing enhancer, it recruits U2AF to bind to the 3' end of the intron. SF1 (splicing factor 1) then binds to the branch point and U2AF, forming the E complex (Fig. 2-6). U2 snRNP then replaces SF1 and binds to the branch point. The same SR protein which binds to the splicing enhancer, or a different SR protein, now binds to U2 snRNP. This splicing complex is called the A complex.

At the 5' end of the intron, the binding of U1 snRNP promotes the binding of U4, U5, and U6 snRNPs. The SR protein which binds to U2 snRNP now also binds to the 70 kDa protein of the U1 snRNP, linking the 5' end of the intron to the branch point to form the B complex, commonly referred to as the spliceosome. U1, U4, and U5 snRNPs then dissociate forming the new complex called C complex. Finally, the snRNPs in the C complex remove the intron and join the two exons together.

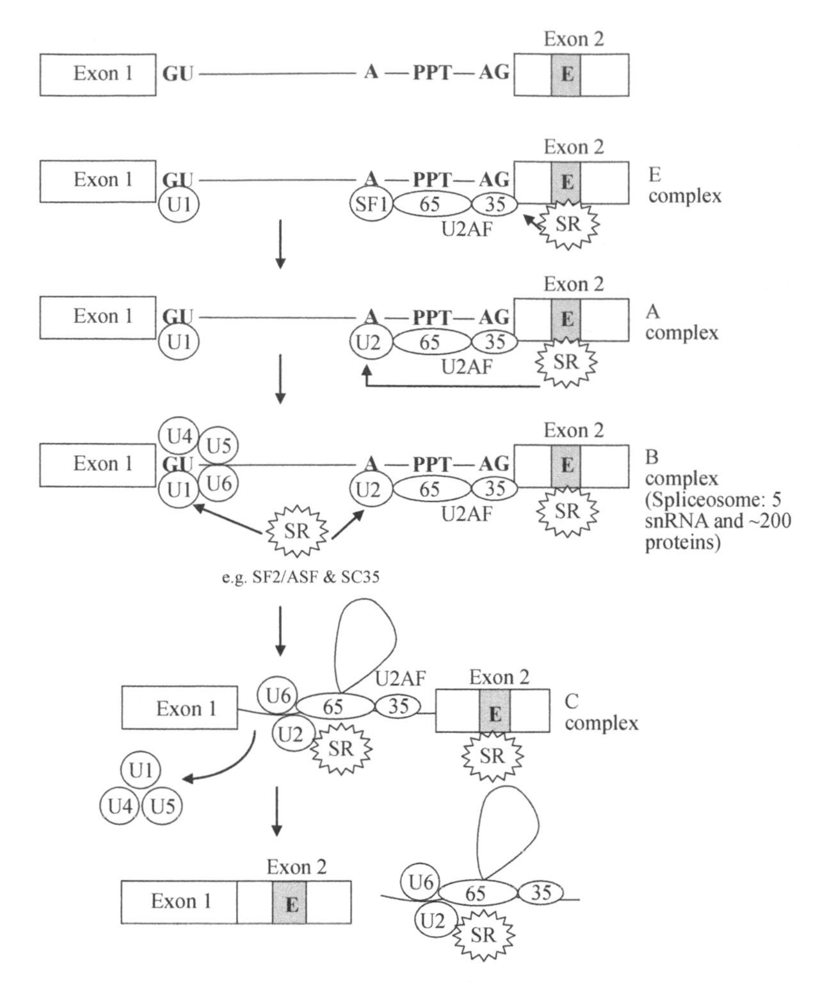

Fig. 2-6: Mechanisms of Pre-mRNA Splicing. "U" represents an snRNP composed of an snRNA and at least 10 different proteins. Other abbreviations or symbols used in this diagram include U2AF, U2 snRNP auxiliary factor; GU, 5' splicing signal; AG, 3' splicing signal; A, branch point; E, splicing enhancer; and SR, serine/arginine-rich proteins.

A pre-mRNA may produce more than one mRNA by splicing different combinations of exons. This process is referred to as alternative splicing. It has been estimated that at least 75% of human pre-mRNAs undergo alternative splicing. An example is the calcitonin pre-mRNA (Fig. 2-7A) which has 6 exons. In thyroid C cells, the first 4 exons are spliced to produce the calcitonin mRNA. Calcitonin regulates the blood calcium levels and is used to treat post menopausal osteoporosis. In nerve cells, exons 1, 2, 3, 5, and 6 are joined to form an mRNA encoding the calcitonin gene-related peptide (CGRP) which is a potent vasodilator. A similar example is the pre-mRNA of the Drosophila Double Sex gene (Fig. 2-7B). This pre-mRNA also has 6 exons. If the first 4 exons are joined during the embryonic stage, the protein produced by the resulting mRNA will cause the Drosophila to become female, while the joining of exons 1, 2, 3, 5, and 6 will result in a male Drosophila.

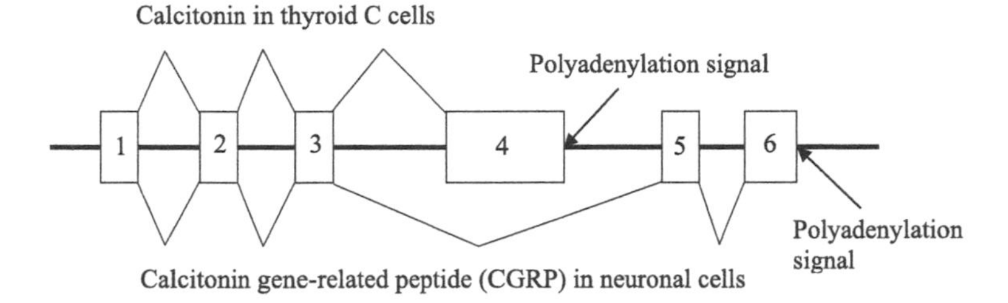

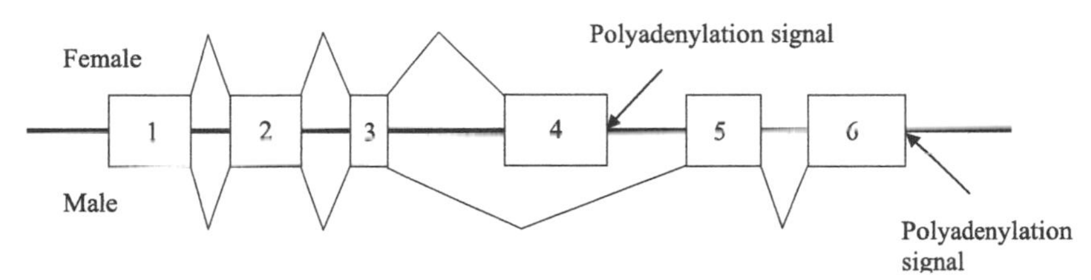

Fig. 2-7: Alternative Splicing of Calcitonin and Drosophila Double Sex Pre-mRNAs.

2.6: Polyadenylation of Pre-mRNA

The modification that takes place at the 3' end of pre-mRNA is the addition of many adenosine nucleotides. This process is referred to as polyadenylation or simply poly-A addition. The sequence signal AAUAAA located near the 3' end of pre-mRNA mediates polyadenylation. At lease 15 different proteins are involved in polyadenylation. An endonuclease that specifically recognizes the AAUAAA signal cleaves the pre-mRNA 10 - 30 bases downstream from this signal. Poly-A polymerase then adds 50 - 200 adenosine nucleotides to the cut end forming a poly-A tail (Fig. 2-8). After capping, splicing, and polyadenylation, a pre-mRNA becomes a functional and mature mRNA that is used to make protein.

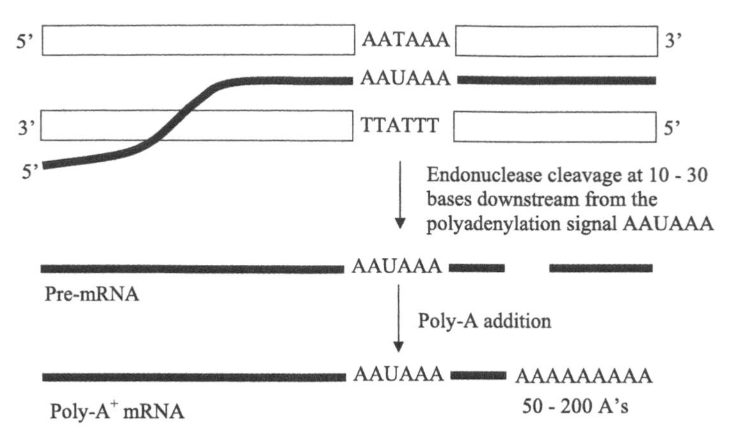

Fig. 2-8: Polyadenylation of Pre-mRNA. A pre-mRNA is cleaved at 10-30 nucleotides downstream from the polyadenylation signal AAUAAA. 50 - 200 adenosine nucleotides are then added to the 3' end to form a poly-A tail.

2.7: Translation

Translation is the process by which mRNA is used as a template to make protein. During translation, only a portion of an mRNA is translated. This region is therefore termed the translated, expressed, or coding region. The translated region is located in the central portion of an mRNA. The untranslated region located upstream from the translated region is called the 5' untranslated region (5'UTR), and that located downstream from the translated region is the 3' untranslated region (3'UTR). Both 5'UTR and 3'UTR vary in length ranging from a few to thousands of nucleotides (Table 2-1).

Table 2-1: Length of 5'UTR and 3' UTR.

	5' UTR				3' UTR			
	No.	Ave. length	Max. length	Min. length	No.	Ave. length	Max. length	Min. length
Human	1203	210.2	2803	18	1247	1027.7	8555	21
Other mammal	142	141.3	936	20	148	441.1	3324	37
Rodent	638	186.3	1786	16	457	607.3	3354	19
Other vertebrate	105	164.0	1154	15	111	446.5	2858	31
Invertebrate	5464	221.9	4498	14	3736	444.5	9142	15
Fungi	388	134.0	1088	16	326	237.1	1142	25

Source: Gene 276:73-81, 2001; No: number of genes examined.

In the coding region, three nucleotides together are translated into one amino acid. For example, AUG is translated into methionine. This 3-base unit code that forms one amino acid is referred to as codon. Different amino acids are encoded by different codons (Table 2-2). The codon at which translation begins is known as the initiation codon and is usually AUG. Translation terminates at stop codons. There are 3 different stop codons including UAG, UGA, and UAA.

Table 2-2: Amino Acid Codons.

3-letter code	1-letter code	No. of codons	Codon	Property
Ala	A	4	GCN	Non-polar, hydrophobic
Arg	R	6	CGN, AG(A/G)	Basic, very polar, hydrophilic
Asn	N	2	AA(T/C)	Noncharged polar, hydrophilic
Asp	D	2	GA(T/C)	Acidic, very polar, hydrophilic
Cys	C	2	TG(T/C)	Non-polar, hydrophobic
Gln	Q	2	CA(A/G)	Noncharged polar, hydrophilic
Glu	E	2	GA(A/G)	Acidic, very polar, hydrophilic
Gly	G	4	GGN	Non-polar, hydrophobic
His	H	2	CA(T/C)	Basic, very polar, hydrophilic
Ile	I	3	AT(T/C/A)	Non-polar, hydrophobic
Leu	L	6	CTN, TT(A/G)	Non-polar, hydrophobic
Lys	K	2	AA(A/G)	Basic, very polar, hydrophilic
Met	M	1	ATG	Non-polar, hydrophobic
Phe	F	2	TT(T/C)	Non-polar, hydrophobic
Pro	P	4	CCN	Non-polar, hydrophobic
Ser	S	6	TCN, AG(T/C)	Noncharged polar, hydrophilic
Thr	T	4	ACN	Noncharged polar, hydrophilic
Trp	W	1	TGG	Non-polar, hydrophobic
Tyr	Y	2	TA(T/C)	Noncharged polar, hydrophilic
Val	V	4	GTN	Non-polar, hydrophobic

Most amino acids have more than one codon. For example, arginine, leucine, and serine each have 6 codons. The six codons for arginine are CGU, CGC, CGA, CGG, AGA, and AGG. Among these, the first 2 bases of the first 4 codons are the same; only the third base is different. This third base is called the wobble base. Although some amino acids have more than one codon, the codons are not used in equal frequency in a cell. For example, 43% of the arginine codons in *Escherichia coli* (*E. coli*) are CGU, but only 2% are AGG. In contrast, only 9% of arginine codons in human cells are CGU, whereas 21% are AGG (Table 2-3). This difference in codon usage among the

Table 2-3: Codon Preference.

Amino acid	Codons	Codon frequency (%)			
		E. coli	*Sacchromyces cerevisiae*	*Drosophila melanogaster*	Primates
Leu	UUA	11	27	6	6
	UUG	11	36	18	11
	CUU	10	11	9	11
	CUC	10	5	15	21
	CUA	3	13	8	7
	CUG	55	9	44	45
Ile	AUU	47	50	33	33
	AUC	46	30	53	53
	AUA	7	20	14	13
Val	GUU	29	44	19	16
	GUC	20	25	26	25
	GUA	17	16	9	9
	GUG	34	15	46	49
Ser	UCU	18	31	8	18
	UCC	17	18	26	24
	UCA	12	19	8	13
	UCG	14	8	22	6
	AGU	13	15	12	13
	AGC	26	9	24	26
Pro	CCU	15	29	11	27
	CCC	10	13	36	35
	CCA	19	49	23	26
	CCG	56	9	30	11
Thr	ACU	20	38	16	23
	ACC	45	24	42	40
	ACA	12	26	17	25
	ACG	23	11	25	12
Ala	GCU	19	44	19	28
	GCC	24	24	19	42
	GCA	22	24	15	20
	GCG	35	8	17	11
Arg	CGU	43	17	19	9
	CGC	37	4	33	22
	CGA	5	5	13	10
	CGG	8	2	14	20
	AGA	4	54	9	19
	AGG	2	17	12	21
Gly	GGU	38	60	22	15
	GGC	40	15	43	36
	GGA	9	15	28	24
	GGG	13	9	7	25

different organisms is called codon preference or codon bias. This is an important concept for genetic engineering in which *E. coli* is often used to produce human proteins for therapeutic purposes. In this process, a human gene must be introduced into *E. coli*. While AGG is the most abundant arginine codon in humans, it is a rare arginine codon in *E. coli*, and very few transfer RNAs recognizing the AGG codon are present in *E. coli*. Therefore, the AGG codon in a human gene is usually changed to CGU in order to achieve optimal expression of the gene in *E. coli*. Amino acids are brought to ribosomes by tRNAs during protein synthesis. Many different tRNAs exist. Each tRNA carries a specific amino acid. A tRNA carrying its cognated amino acid is called aminoacyl-tRNA or charged tRNA.

2.8: Regulation of Translation

Although the 5'UTR and 3'UTR of an mRNA are not translated, they have important regulatory functions. The regulatory elements that exist in the 5'UTR include the m7G cap, upstream open reading frame (uORF), and internal ribosome entry site (IRES) (Fig. 2-9). The m7G cap is the first place where ribosome binds to mRNA to start translation. uORF ranges from 3 to 60 codons and may decrease or increase the translation efficiency of the main gene by regulating mRNA stability. uORF may or may not be expressed. Examples of expressed uORF are those of yeast genes GCN4 and CPA1 and *Neurospora crassa* ARG1 gene. Approximately 10% of human genes process uORF (Table 2-4). IRES allows translation to initiate from where it is located.

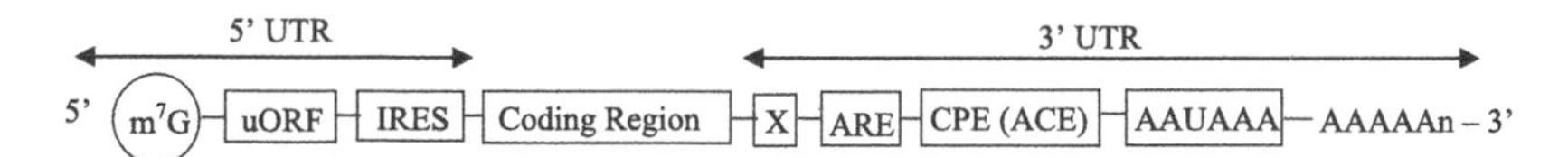

Fig. 2-9: Components of a Eukaryotic mRNA. Abbreviations or symbols: uORF, upstream open reading frame; IRES, internal ribosome entry site; ARE, AU-rich element; CPE, cytoplasmic polyadenylation element; ACE, adenylation control element; AAUAAA, the polyadenylation signal.

Table 2-4: Frequencies of uORF.

	No. of genes examined	uORF >10 codons (%)
Human	12633	10.90
Other mamma	3164	7.96
Rodent	10403	8.29
Other vertebrate	3370	5.31
Invertebrate	12910	7.39
Fungi	1463	4.99

Source: Gene 276:73-81, 2001.

Regulatory elements located in 3'UTR include the X region, AU-rich element (ARE), and cytoplasmic polyadenylation element (CPE), in addition to the polyadenylation signal AAUAAA. The X region regulates the export of mRNA from the nucleus to the cytoplasm. Transcription and pre-mRNA processing take place in the nucleus, whereas translation occurs in the cytoplasm. Therefore, a mature mRNA must be transported to the cytoplasm in order to make protein. Mutation in the X region may result in disease. One example of such is type 1 myotonic dystrophy due to mutations in the dystrophia myotonica protein kinase (DMPK) gene. A normal DMPK mRNA has 5 - 30 CUG repeats in the 3'UTR, but in patients with type 1 myotonic dystrophy, the DMPK mRNA has more than 1000 CUG repeats, leading to a defect in splicing and export of DMPK mRNA out of the nucleus and a decrease in the production of the DMPK protein.

The ARE region is usually 50 to 150 nucleotides in length and contains repeats of AUUUA or UUAUUUA(U/A)(U/A). This region regulates mRNA stability. The average half-life of eukaryotic mRNA is approximately 20 minutes, and that of prokaryotic mRNA is 2 minutes. mRNA with an increased or decreased half-life is abnormal and may cause problems. For example, two different sizes, 4.4 kb and 1.7 kb, of cyclin D1 (CCND1) mRNA are present in Mantle cell lymphoma. The 1.7 kb CCND1 mRNA is abnormal due to a deletion in the ARE region, making it more stable with an increased half-life from approximately 0.5 hours to more than 3 hours. This mutation results in production of more cyclin D1 leading to unregulated cell proliferation. CPE, also referred to as adenylation control element (ACE), regulates polyadenylation of mRNA.

2.9: Ribosome

Ribosomes are complexes of RNA and protein and are where proteins are made. A prokaryotic ribosome is 70S in size with a molecular weight of 2,800 kDa. It is composed of two different subunits: 30S and 50S (Fig. 2-10). The 30S subunit has one 16S rRNA of approximately 1500 bases and 21 different proteins. The 50S subunit contains one 5S rRNA, one 23S rRNA, and 33 different proteins. The 5S rRNA is approximately 120 bases, and the 23S is approximately 3000 bases in length.

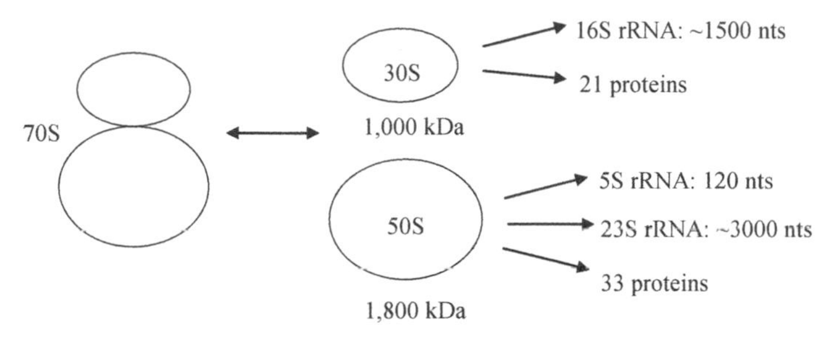

Fig. 2-10: Structure of Prokaryotic Ribosome.

Eukaryotic ribosome is 80S and also consists of two different subunits (Fig. 2-11). The 40S subunit contains one 18S rRNA of approximately 2000 bases and 33 different proteins. The 60S subunit is composed of one 5S rRNA of approximately 120 bases, one 5.8S rRNA of approximately 160 bases, one 28S rRNA of approximately 5000 bases, and 49 different proteins.

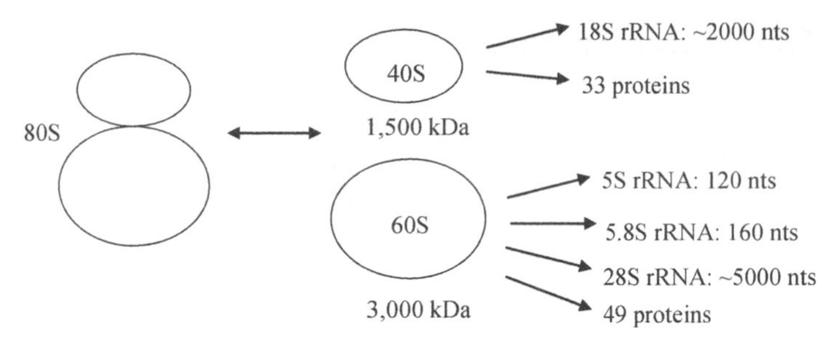

Fig. 2-11: Structure of Eukaryotic Ribosome.

In prokaryotic cells, mRNA binds to ribosome through an interaction with the 16S rRNA during translation. This interaction takes place between the 3' end of the 16S rRNA and the 5' end of the mRNA at the Shine-Dalgarno (SD) sequence which has a consensus sequence of GGAGG (Fig. 2-12). In eukaryotic cells, the Kozak sequence is functionally equivalent to the SD sequence of prokaryotic cells. The consensus of the Kozak sequence is CCG(A/G)CCATGG, of which ATG is the initiation codon. The Kozak sequence facilitates binding of mRNA to the 40S ribosomal subunit. The most important features are that the third base upstream from the ATG codon is an A or a G and that the base next to ATG is a G.

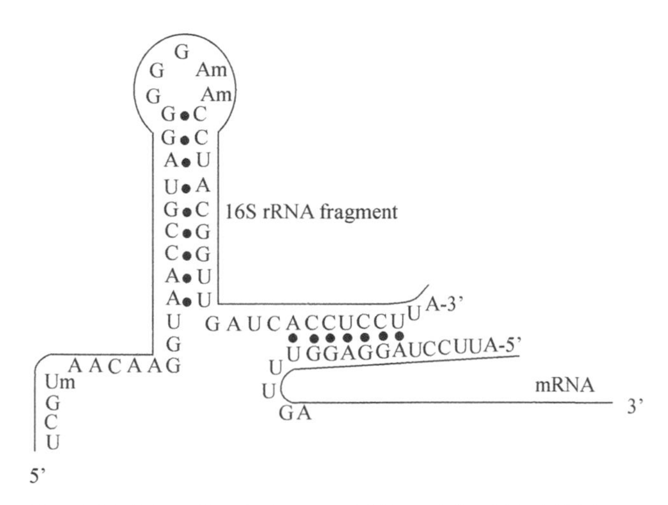

Fig. 2-12: Interaction between mRNA and the 16S rRNA in a prokaryotic ribosome.

2.10: Translation Factors

Translation is divided into three steps: initiation, elongation, and termination. Many proteins referred to as translation factors are involved in translation (Table 2-5). Initiation of prokaryotic translation requires initiation factors IF1, IF2, and IF3. The active form of IF2 is GTP bound (IF2-GTP). IF3, with the assistance of IF1, first binds to the 30S subunit and dissociate the 50S subunit from a complete 70S ribosome. IF1 then binds to the free 30S subunit and ensures that the initiation aminoacyl-tRNA, fMet-tRNAfmet, binds to the initiation codon. IF2-GTP binds to the initiator fMet-tRNAfmet and directs it to bind to the 30S subunit. The 50S subunit then joins the 30S subunit, and the GTP bound to IF2 is hydrolyzed, leading to disassociation of IF2-GDP and IF1 (Fig. 2-13). IF2-GDP is then reactivated by the 50S ribosomal subunit to become IF2-GTP and recycled.

Table 2-5: Translation Factors.

	Prokaryotes	**Eukaryotes**
Initiation	IF1 IF2 IF3	eIF1A, eIF1B eIF2A (eIF2α), eIF2B (eIF2β) eIF3 eIF4A, eIF4B, eIF4E, eIF4F, eIF4G eIF5, eIF5B
Elongation	EF-Tu EF-Ts EF-G	eEF1α eEF1βγ eEF2
Termination	RF1 (UAA, UAG) RF2 (UGA, UAA) RF3	eRF1 (UAG, UGA, UAA) eRF3

RF3: facilitates binding of RF1 and RF2 to ribosome.

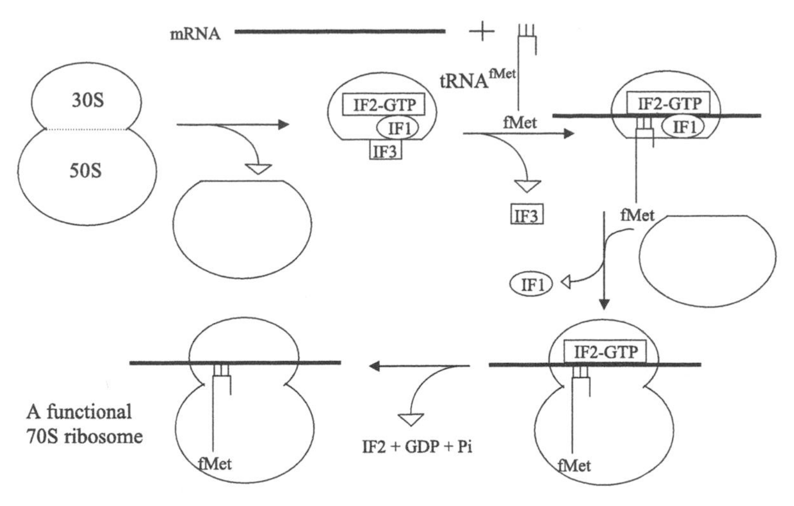

Fig. 2-13: Prokaryotic Translation Initiation.

When a ribosome binds to mRNA, it encompasses at least 3 codons referred to as the exit (E), peptide (P), and amino acid (A) sites. If the codon next to the initiation codon is UUU, tRNAphe will bring a phenylalanine to the ribosome to be linked to methionine. The ribosome then moves to the next codon and, at this point, covers the E, P, and A codons. The first codon is bound by a discharged tRNA which will soon exit the ribosome and is, therefore, called the E site. The second codon is bound by a tRNA carrying the growing peptide and is, therefore, called the P site. The third codon waits to accept another charged tRNA that is carrying an amino acid, and is, therefore, called the A site (Fig. 2-14).

Translation factors involved in the elongation step are elongation factors, including EF-Tu (EF1A), EF-Ts (EF1B), and EF-G (EF2). EF-Tu and EF-Ts are GTP-binding proteins. EF-Tu-GTP binds and delivers an aminoacyl-tRNA to the A site on the ribosome. Peptidyl transferase then links the amino acid on the A site to the peptide on the P site, with subsequent hydrolysis of GTP on EF-Tu to GDP and disassociation of EF-Tu from the complex. EF-G then moves the ribosome to the next codon to continue peptide synthesis. EF-Ts, which is a nucleotide exchange factor, converts EF-Tu to EF-Tu-GTP which is reused in the elongation process.

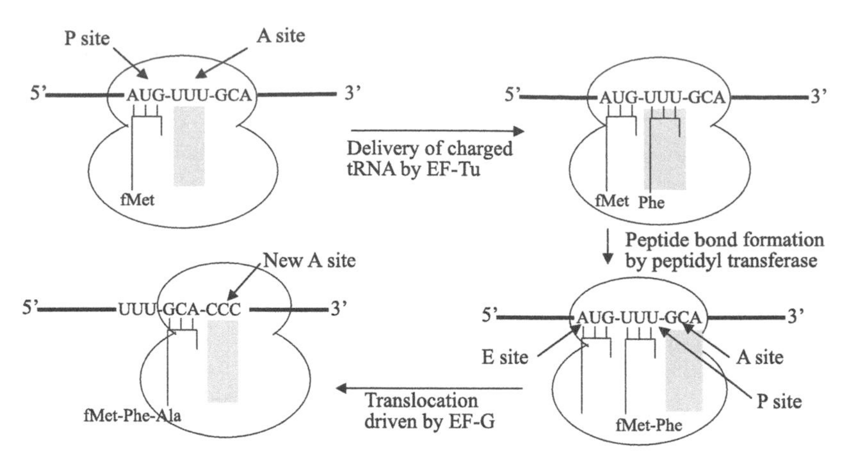

Fig. 2-14: Prokaryotic Translation Elongation.

During termination, release factors RF-1 and RF-2 bind to stop codons and terminate protein synthesis (Fig. 2-15). RF-1 recognizes UAA and UAG stop codons, and RF2 recognizes UGA and UAA. RF3-GTP facilitates the binding of RF-1 or RF-2 to the ribosome. A ribosome recycling factor (RRF) together with EF-G-GTP and IF3 release the uncharged tRNA from the P site. The ribosome then dissociates from the mRNA, and its two subunits separate.

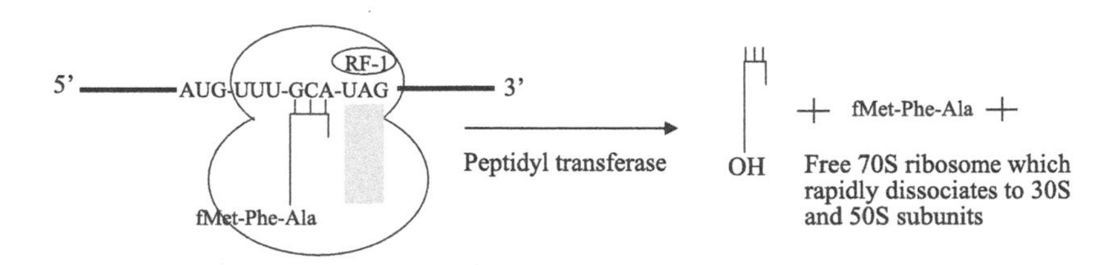

Fig. 2-15: Prokaryotic Translation Termination.

2.11: Eukaryotic Translation

Among the three steps in translation, elongation and termination of eukaryotic translation are very similar to that of prokaryotic translation. The Eukaryotic translation factors involved in elongation include eEF1α, eEF1βγ, and eEF2; where "e" represents eukaryotic. These 3 elongation factors are functionally equivalent to prokaryotic elongation factors EF-Tu, EF-Ts, and EF-G, respectively. In the termination step, eRF1 and eRF3 are required. eRF1 recognizes all three stop codons, and eRF3 is functionally equivalent to the prokaryotic termination factor RF3 and promotes binding of eRF1 to the ribosome.

Translation initiation in eukaryotic cells is very different from that in prokaryotic cells. The two ends of the mRNA are first joined to form a circle during translation. To achieve this, the cap-binding protein eIF4E first binds to the m7G cap and attracts eIF4G, eIF4A, and eIF4B to bind. At the same time, a poly-A-binding protein (PABP) binds to the poly-A tail. The PABP on the poly-A tail then binds to eIF4G, thus joining the two ends of the mRNA together. eIF4A is an RNA helicase and unwinds RNA, converting the secondary structure to a primary linear molecule to facilitate translation. eIF4B stimulates the helicase activity of eIF4A and is required for mRNA to bind to the ribosome.

Before binding to mRNA, the ribosome undergoes a series of changes. eIF2A-GTP first binds to tRNAmet and then attracts eIF1B, eIF3, and eIF5 to bind, forming a multifactorial complex (MFC). eIF1A then binds to the MFC and directs it to bind to the 40S ribosomal subunit to form a 43S complex. This complex then binds to the mRNA through the interaction between eIF3, eIF4G, eIF4A, and eIF4B. eIF4A, eIF4B, eIF4E, and eIF4G then dissociate from the mRNA, and the small ribosome subunit initiates a search for the AUG initiation codon. Once found, eIF2-GTP is converted to eIF2-GDP and dissociates together with eIF1B, eIF3, and eIF5 from the ribosome. Activated eIF5B (eIF5B-GTP) then binds to eIF1A, allowing the 60S subunit to join the small subunit to form a functional ribosome. The GTP on eIF5B is then hydrolyzed to GDP, and eIF5B and eIF1A dissociate from the ribosome. This ribosome is now ready to make protein. eIF2-GDP with the assistance of eIF2B is then reactivated to become eIF2-GDP and recycled (Fig. 2-16).

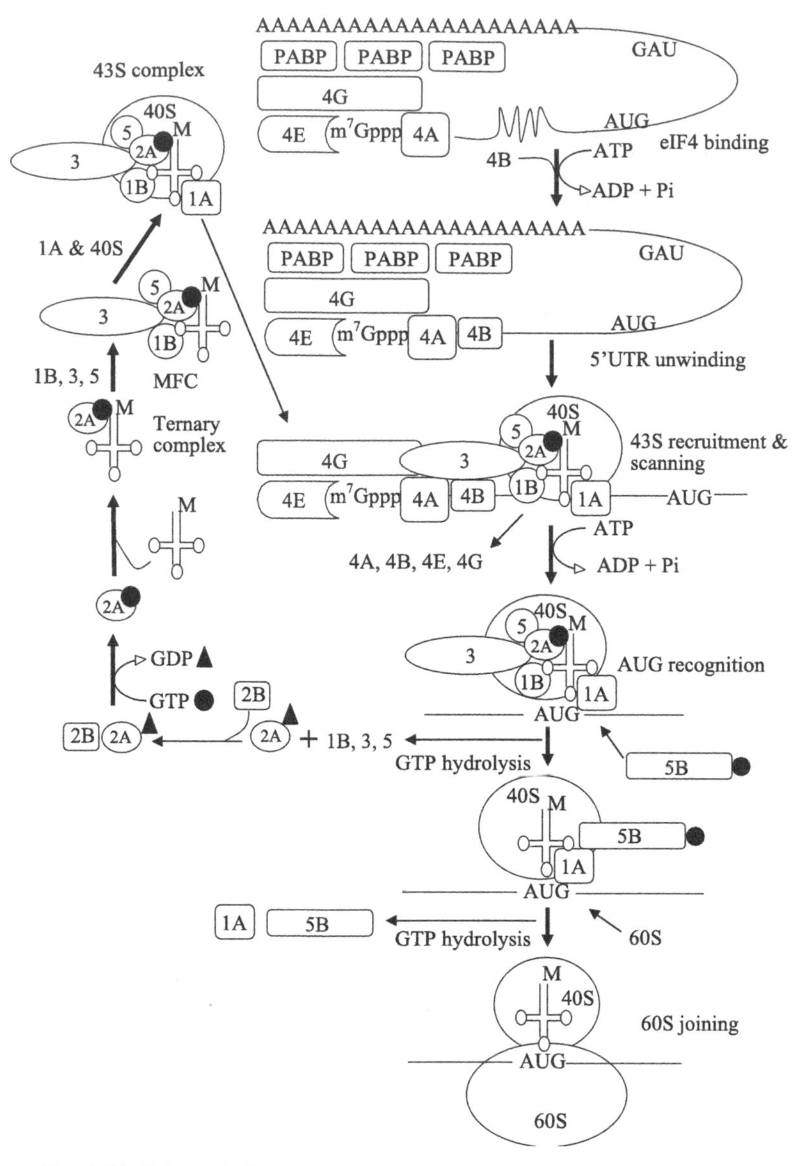

Fig. 2-16: Eukaryotic Translation Initiation.

2.12: Internal Ribosome Entry Site

Normally, a ribosome binds to the mRNA through the m7G cap to start translation. Some mRNAs have an internal ribosome entry site (IRES) where the ribosome initiates translation. IRES was first found in viruses. Since viruses do not have translation machinery, they must use the host translation system to make proteins and must compete with the host cells for ribosome and translation factors. To ensure successful translation, some viruses posses an IRES. It mediates translation initiation with reduced number of initiation factors or even without the need for initiation factors. As described above, normal cellular translation initiation requires eIF1, eIF1A, eIF2, eIF3, eIF4A, eIF4E, eIF4G, eIF4B, eIF5, and PABP. Hepatitis C virus (HCV) translation requires only eIF2 and eIF3, and that of Foot and Mouth Disease Virus (FMDV) requires only eIF4B, eIF4G, eIF2, and eIF3. Translation of Cricket Paralysis Virus mRNA does not require any initiation factors. An IRES is present in all picrornaviruses such as poliovirus, enterovirus, and rhinovirus; HCV; HIV; HHV8; and several RNA tumor viruses. Interestingly, some cellular mRNAs also have IRES (Table 2-6).

Table 2-6: Occurrence of IRES.

	No. of genes examined	IRES (%)
Human	12633	6.44
Other mammal	3164	3.45
Rodent	10403	5.49
Other vertebrate	3370	4.30
Invertebrate	12910	1.83
Fungi	1463	2.53

Source: Gene 276:73-81, 2001.

Summary

Before a cell divides, its genomic DNA must replicate. DNA replication requires a template. When helicase unwinds a double-stranded DNA, each of the two resulting single DNA strands can serve as the template for DNA replication. In addition to a template, DNA replication also requires primers, typically small RNA or DNA fragments. Some proteins can also serve as primers. DNA and RNA primers anneal to the template through base pairing; therefore, their sequences must be complementary to that of the region where they bind. Since a second nucleotide is joined to the 3'-OH group of the first nucleotide during DNA synthesis, primers must have a 3'-OH group. Therefore, proteins that are used as primers usually have at least one nucleotide attached to them to provide the 3'-OH group. The enzyme that carries out DNA synthesis is DNA polymerase which synthesizes DNA in a 5' to 3' direction. Both strands of DNA are replicated. To replicate the upper (or plus) strand of DNA, a primer anneals to the outside end of a replication fork. When this primer extends to the bottom of the fork, helicase opens more template, and DNA synthesis continues. To replicate the lower (or minus) strand, a primer anneals near the bottom of the replication fork and extends toward the end of the fork. When more template becomes available, another primer anneals to the bottom of the newly formed fork and reinitiates DNA synthesis. Therefore, this DNA synthesis is discontinuous and forms new DNA fragments, known as Okazaki fragments. These Okazaki fragments are then ligated together by ligase to form a contiguous strand of DNA.

The function of a gene is expressed by the protein it encodes. To make protein, a gene is first transcribed to produce RNA. This process is called transcription. Several kinds of RNAs exist, including mRNA, tRNA, rRNA, snRNA, and smRNA. During transcription, only a portion of a DNA template is transcribed. Transcription initiates from the promoter region and stops at the terminator. Unlike DNA replication, transcription does not require primers. The enzyme which carries out transcription is RNA polymerase. There is only one RNA polymerase in prokaryotic cells, but three different RNA polymerases are present in eukaryotic cells. Among these, RNA polymerase I makes 28S, 18S, and 5.8S rRNA; RNA polymerase II produces mRNA, snRNA, and smRNA, and RNA polymerase III is responsible for making tRNA and 5S rRNA. The manner in which rRNA and tRNA are made is similar in prokaryotic and eukaryotic cells. However, production of mRNA in these two types of cells is very different. In prokaryotic cells, the mRNA transcribed from the template is a functional and mature mRNA. In contrast, those that are produced in eukaryotic cells need to be modified to become functional and therefore are called pre-mRNA. Modifications of pre-mRNA include capping, splicing, and polyadenylation. Capping is the addition of a 7-methylguanosine to the 5' end of pre-mRNA. In polyadenylation, pre-mRNA is cleaved 10 - 30 bases downstream from the polyadenylation signal AAUAAA; 50 - 200 adenosine nucleotides are then added to the 3' end of the RNA.

Splicing is the removal of the regions that are not expressed in a protein, known as introns, and the joining of differentially expressed regions called exons. An intron usually contains 3 sequence signals that trigger splicing. The one located at the 5' end has a consensus sequence of GURAGU, and that located at the 3' end is PPT-N-YAG. The third signal is located 15 - 50 bases upstream from the 3' end and has a consensus sequence of YURAY. This location is called a branch point. The enzymes involved in splicing are snRNA and protein complexes called snRNP, including U1 - U6 snRNPs. These snRNPs form a spliceosome which is responsible for cutting out the intron and joining two exons together. A pre-mRNA may have several exons. Usually, all exons are joined together to form an mRNA. Sometimes, only some of the exons are joined. This phenomenon is called alternative splicing. Therefore, a pre-mRNA may be spliced in different ways to form several mRNAs.

Using mRNA as the template to make protein is the process called translation. During translation, only a portion of a certain mRNA is translated. Therefore, an mRNA has three different regions including the expressed region or coding region, 5' untranslated region (5'UTR), and 3' untranslated region (3'UTR). Translation starts from the initiation codon which is usually AUG and stops at stop codons. Three different stop codons exist including UAG, UGA, and UAA. Except for methionine and tryptophan, all other amino acids have more than one codon. For example, leucine has 6 different codons including CUA, CUG, CUC, CUU, UUA, and UUG. These six codons are not used in equal frequencies, and the usage frequency of a certain codon also varies in different organisms. This phenomenon is referred to as codon bias or codon preference. The complex which makes protein is called ribosome.

A prokaryotic ribosome is 70S in size and is composed of two different subunits, the 30S and 50S. The 30S subunit contains one 16S rRNA and 21 different proteins. The 50S subunit consists of one 5S rRNA, one 23S rRNA, and 33 different proteins. A eukaryotic ribosome is 80S in size and is composed of the 40S and 60S subunits. The 40S subunit contains one 18S rRNA and 33 different proteins. The 60S subunit contains one each of the 5S, 5.8S, and 28S rRNAs and 49 different proteins. Enzymes involved in translation are called translation factors. In prokaryotic cells, initiation factors IF1, IF2, and IF3 are required to enable the mRNA to bind to a ribosome to start protein synthesis. The mRNA binds to ribosome through the interaction between the Shine-Dalgarno (SD) sequence (GGAGG), located near the 5' end of mRNA, and the corresponding sequence (CCUCC) located near the 3' end of the 16S rRNA. Elongation factors EF-Tu, EF-Ts, and EF-G are required during the elongation step of translation, and releasing factors RF1, RF2, and RF3 are required for terminating translation.

Eukaryotic translation is similar to that of prokaryotic translation, except that many more initiation factors are required, including eIF1A, eIF1B, eIF2A, eIF2B, eIF3, eIF4A, eIF4B, eIF4E, eIF4G, eIF5, eIF5B, and poly-A binding protein (PABP). Among these, eIF4E recognizes the m7G cap, and eIF4G and PABP link the two ends of an mRNA together to form a circle to initiate translation. Eukaryotic elongation factors include eEF1α, eEF1βγ, and eEF2, and the two termination factors are eRF1 and eRF3. Although 5'UTR and 3'UTR are not translated, they have important regulatory functions. Some mRNAs have an internal ribosome entry site (IRES) that makes translation easier to start by reducing or even eliminating the requirement for initiation factors. An IRES allows translation to initiate from where it is located. In the 3'UTR, the X region mediates export of processed mRNA out of the nucleus to the cytoplasm for translation. The AU-rich element (ARE) controls the stability of mRNA, and the cytoplasmic polyadenylation element (CPE), also called adenylation control element (ACE) regulates polyadenylation. AAUAAA is the polyadenylation signal.

Sample Questions:

1. Which of the following is the polyadenylation signal of eukaryotic mRNA? a) YYYYYNYAG, b) GURAG, c) TGANNNNTGCT, d) AAUAAA
2. The 5' Cap of eukaryotic mRNA is: a) GTP, b) ATP, c) GMP, d) m7GTP
3. The first two nucleotides of an intron are usually: a) AT, b) GC, c) GU, d) AG
4. The last two nucleotides of an intron are usually: a) AT, b) GC, c) GU, d) AG
5. Which of the following RNA is involved in mRNA splicing? a) miRNA, b) siRNA, c) tRNA, d) snRNA
6. In eukaryotic cells, which of the following are made by RNA polymerase I: a) 28S, 18S, and 5.8S rRNA; b) mRNA and snRNA; c) tRNA and 5S rRNA; d) all of above
7. In eukaryotic cells, which of the following are made by RNA polymerase II: a) 28S, 18S, and 5.8S rRNA; b) mRNA and snRNA; c) tRNA and 5S rRNA; d) all of above
8. In eukaryotic cells, which of the following are made by RNA polymerase III: a) 28S, 18S, and 5.8S rRNA; b) mRNA and snRNA; c) tRNA and 5S rRNA; d) all of above
9. The X region in the 3' UTR of eukaryotic mRNA regulates: a) capping, b) polyadenylation, c) export of mRNA to the cytoplasm, d) splicing
10. The ARE region in the 3' UTR of eukaryotic mRNA regulates: a) capping, b) polyadenylation, c) export of mRNA to the cytoplasm, d) stability of mRNA
11. An IRES allows: a) DNA replication without primer, b) transcription without promoter, c) translation initiation with reduced requirements for initiation factors, d) alternative splicing
12. Arginine has the following 6 codons: CGU, CGC, CGA, CGG, AGA, AGG. The codon AGG is used 2% of the time in *E. coil* but 21% of the time in human cells. This phenomena is referred to as: a) codon degeneration, b) codon attenuation, c) codon preference, d) codon bias
13. During translation in a prokaryotic cell, mRNA binds to the 30S subunit of ribosome through interaction with: a) 5S rRNA, b) 5.8S rRNA, c) 16S rRNA, d) 28S rRNA
14. Which of the following is the initiation factor which first binds to the cap of mRNA during translation in eukaryotic cells? a) eIF1, b) eIF2, c) eIF3, d) eIF4E
15. Which of the following is an RNA helicase? a) eIF4A, b) eIF4B, c) eIF4E, d) eIF4G

Suggested Readings:

1. Ashley, C. T. and Warren, S. T. (1995). Trinucleotide repeat expansion and human disease. Annu. Rev. Genetics. 29: 703-728.
2. Bannerjee, A. K. (1980). 5'-terminal cap structure in eukaryotic mRNA. Microbiol. Rev. 44: 175-205.
3. Beaton, A. R. and Krug, R. M. (1981). Selected host cell capped RNA fragments prime influenza viral RNA transcription in vivo. Nucleic Acids Res. 9: 4423-4436.
4. Black, D. L. (2003). Mechanisms of alternative pre-messenger RNA splicing. Annu. Rev. Biochem. 72: 291-336.
5. Breathnach, R. and Chambon, P. (1981). Organization and expression of eukaryotic split genes coding for proteins. Annu. Rev. Biochem. 50: 349-383.

6. Conne, B., Stutz, A., and Vassalli, J.-D. (2000). The 3' untranslated region of messenger RNA: a molecular hotspot for pathology? Nat. Medicine 6: 637-641.

7. Crick, F. H. C. (1966). Codon-anticodon pairing: the wobble hypothesis. J. Mol. Biol. 19: 548-555.

8. Crick, F. H. C., Barnett, L., Brenner, S., and Watts-Tobin, R. J. (1961). General nature of the genetic code for proteins. Nature 192: 1227-1232.

9. Darnell, J. E., Philipson, L., Wall, R., and Adesnik, W. M. (1971). Poly(A) sequences: role in conversion of nuclear RNA into mRNA. Science 174: 507-510.

10. Dreyfuss, G., Matunis M. J., Pinol-Roma S., and Burd C. G. (1993). hnRNP proteins and the biogenesis of mRNA. Annu. Rev. Biochem. 62: 289-321.

11. Faustino, N. A. and Cooper, T. A. (2003). Pre-mRNA splicing and human diseases. Genes Dev. 17: 419-437.

12. Fire, A., Xu, S., Montgomery, M. K., Kostas, S. A., Driver, S. E., and Mello, C. C. (1998). Potent and specific genetic interference by double-stranded RNA in *C. elegans.* Nature 391: 806-811.

13. Gingras, A. C., Raught, B., and Sonenberg, N. (1999). eIF4 initiation factors: effectors of mRNA recruitment to ribosomes and regulators of translation. Annu. Rev. Biochem. 68: 913-963.

14. Gorlich, D. and Mattaj, I. W. (1996). Nucleo-cytoplasmic transport. Science 271: 1513-1518.

15. Grzybowska, E. A., Wilczynska, A., and Siedlecki, J. A. (2001). Regulatory functions of 3'UTR. Biochem. Biophy. Res. Commun. 288: 291-295.

16. Guhaniyogi, J. and Brewer, G. (2001). Regulation of mRNA stability in mammalian cells. Gene 265: 11-23.

17. Hellen, C. U. and Sarnow, P. (2001). Internal ribosome entry sites in eukaryotic mRNA molecules. Genes Dev. 15: 1593-1612.

18. Hertel, K. J. and Graveley, B. R. (2005). RS domains contact the pre-mRNA throughout spliceosome assembly. Trends Biochem. Sci. 30: 115 – 118.

19. Jansen, R. P. (2001). mRNA localization: message on the move. Nat. Rev. Mol. Cell. Biol. 2: 247-256.

20. Jurica, M. S. and Moore, M. J. (2003). Pre-mRNA splicing: awash in sea of proteins. Mol. Cell 12: 5-14.

21. Kloc, M., Zearfoss, N. R., and Etkin, L. D. (2002). Mechanisms of subcellular mRNA localization. Cell 108: 533-544.

22. Kozak, M. (1978). How do eukaryotic ribosomes select initiation regions in mRNA? Cell 15: 1109-1123.

23. Kozak, M. (1983). Comparison of initiation of protein synthesis in prokaryotes, eukaryotes, and organelles. Microbiol. Rev. 47: 1-45.

24. Lee, R. C., Feinbaum, R. L., and Ambros, V. (1993). The *C. elegans* heterochromic gene lin-4 encodes small RNAs with antisense complementarity to lin-14. Cell 75: 843-854.

25. Maniatis, T. and Reed, R. (1987). The role of small nuclear ribonucleoprotein particles in pre-mRNA splicing. Nature 325: 673-678.

26. Maitra, U., Stringer, E. A., and Chandhuri, A. (1982). Initiation factors in protein biosynthesis. Annu. Rev. Biochem. 51: 869-900.

27. Matlin, A., Clark, F., and Smith, C. W. J. (2005). Understanding alternative splicing toward a cellular code. Nature Reviews. (Mol. Cell. Biol.) 6: 386-397.

28. Meselson, M. and Stahl, F. W. (1958). The replication of DNA in *E. coli*. Proc. Natl. Acad. Sci. USA 44: 671-682.

29. Moore, M. J. (2000). Intron recognition comes of Age. Nature Structural Biology 7: 14-16.

30. Noller, H. F. (1984). Structure of ribosomal RNA. Annu. Rev. Biochem. 53: 119-162.

31. Palacios, I. M. and Johnston, D. S. (2001). Getting the message across: the intracellular localization of mRNA in higher eukaryotes. Annu. Rev. Cell Dev. Biol. 17: 569-614.

32. Pelletier, J. and Sonenberg, N. (1988). Internal initiation of translation of eukaryotic mRNA directed by a sequence derived from poliovirus RNA. Nature 334: 320-325.

33. Pesole, G., Mignone, F., Gissi, C., Grillo, G., Licciulli, F., and Liuni, S. (2001). Structural and functional features of eukaryotic mRNA untranslated regions. Gene 276: 73-81.

34. Ramakrishnan, V. (2002). Ribosome structure and the mechanism of translation. Cell 108: 557-572.

35. Reed, R. and Mamiatis, T. (1985). Intron sequences involved in lariat formation during pre-mRNA splicing. Cell 41: 95-105.

36. Sachs, A., Sarnow, P., and Hentze, M. W. (1997). Starting at the beginning, middle, and end: translation initiation in eukaryotes. Cell 89: 831-838.

37. Salas, M. (1991). Protein-priming of DNA replication. Annu. Rev. Biochem. 60: 39–71.

38. Sarnow, P. (2003). Viral internal ribosome entry site elements: novel ribosome-RNA complexes and roles in viral pathogenesis. J. Virol. 77: 2801-2806.

39. Chauvin, C., Salhi, S., Le Goff, C., Viranaicken, W., Diop, D., and Jean-jean, O. (2005). Involvement of human release factors eRF3a and eRF3b in translation termination and regulation of the termination complex formation. Mol. Cell Biol. 25: 5801-5811.

40. Sharp, P. A. (1987). Splicing of mRNA precursors. Science 235: 766-771.

41. Shine, J. and Dalgarno, L. (1975). Determinant of cistron specificity in bacterial ribosomes. Nature. 254:34-38.

42. Shine, J. and Dalgarno, L. (1975). Terminal-sequence analysis of bacterial ribosomal RNA. Correlation between the 3'-terminal-polypyrimidine sequence of 16S RNA and translational specificity of the ribosome. Eur. J. Biochem. 57: 221-230.

43. Stark, H., Rodnina, M. V., Wieden, H-J., van Heel, M., and Wintermeyer, W. (2000). Large-scale movement of EF-G and extensive conformational change of the ribosome during translocation. Cell 100: 301-309.

44. Staschke, K. and Colacino, J. (1994). Priming of duck hepatitis B virus reverse transcription in vitro: premature termination of primer DNA induced by the 5'-triphosphate of fialuridine. J. Virol. 68: 8265-8269.

45. Tazi, J., Durand, S., and Jeanteur, P. (2005). The spliceosome: a novel multi-faceted target for therapy. Trends Biochem. Sci. 30: 469-478.

46. van der Velden, A. W. and Thomas, A. A. M. (1999). The role of the 5' untranslated region of an mRNA in translation regulation during development. Intl. J. Biochem Cell Biol. 31: 87-106.

47. Vilela, C. and McCarthy, J. E. G. (2003). Regulation of fungal gene expression via

short open reading frames in the mRNA 5' untranslated region. Mol. Microbiol. 49: 859 – 867.

48. Wahle, E. and Keller, W. (1992). The biochemistry of 3'-end cleavage and polyadenylation of messenger RNA precursors. Annu. Rev. Biochem. 61: 419-440.

49. Wang, G. H. and Seeger, C. 1992. The reverse transcriptase of hepatitis B virus acts as a protein primer for viral DNA synthesis. Cell 71:663–670.

50. Webster, A., Leith, I. R., Nicholson, J., Hounsell, J., and Hay, R. T. (1997). Role of preterminal protein processing in adenovirus replication. J. Virol. 71: 6381-6389.

51. Wilson, K. S. and Noller, H. F. (1998). Molecular movement inside the translational engine. Cell 92: 337-349.

52. Zamore, P. D., Tuschl, T., Sharp, P. A., and Bartel, D. P. (2000). RNAi: double-stranded RNA directs the ATP-dependent cleavage of mRNA at 21 to 23 nucleotide intervals. Cell 101: 25-33.

53. Zhang, S., Zubay, G., and Goldman, E. (1991). Low-usage codons in *Escherichia coli*, yeast, fruit fly and primates. Gene 105: 61-72.

Chapter 3

Isolation of DNA and RNA

Outline

Plasmid DNA Isolation

Genomic DNA Isolation

Total RNA Isolation

Isolation of Poly-A$^+$ mRNA

Determination of Quality and Quantity of DNA

3.1: Plasmid DNA Isolation

A plasmid is an extra-chromosomal genetic element. It is most commonly found in bacteria. The size of a plasmid may range from 500 base pairs (bp) to 100 kilo base pairs (kb). Although plasmids are very small compared to chromosomal DNA, they have a replication origin as chromosomal DNA does, and can replicate autonomously in appropriate host cells. A host cell may contain more than one kind of plasmid, and each kind has a certain number. The number of a plasmid in host cell is referred to as copy number. Usually, the copy number of a plasmid is inversely proportional to its size. For example, the copy number of a 3-kb pUC plasmid is greater than 200, whereas that of the 100-kb F plasmid is one. The copy number of the 4.3-kb pBR322 is approximately 20. Plasmids are named for ease of identification. A typical example of plasmid nomenclature is pBR322; where "p" stands for plasmid, and "B" and "R" represent Bolivar and Rodriguez, respectively, who created the plasmid. The real meaning of the number 322 is known only to the creators. It may mean that the plasmid is constructed in their 322nd experiment or is the 322nd plasmid they have constructed. The "UC" in pUC refers to the University of California.

E. coli plasmids are double-stranded circular DNA. Any circular DNA has two major forms: relaxed circle and supercoiled. If a plasmid is cut by an endonuclease or broken by mechanical force, it becomes linear. These three forms of plasmid DNA, relaxed circle, supercoiled, and linear, exhibit different speeds of mobility in electrophoresis. The supercoiled DNA is more compact and therefore smaller in size; it migrates faster than the relaxed circle in electrophoresis. The linear form of the plasmid migrates at a speed between the other two forms. Some plasmid DNA may form a chain-like structure called a concatemer, which migrates much more slowly than the other three forms.

E. coli is the most commonly used host cell for recombinant and vector plasmids. The first step in plasmid DNA isolation is to grow *E. coli* cells containing the plasmid of interest. An *E. coli* colony is picked from an agar plate and inoculated in 1 ml of Luria-Bertani (LB) broth (10 g tryptone, 5 g yeast extract, 10 g NaCl in 1 liter of water, pH 7.0) containing a certain antibiotic to inhibit the growth of cells that have lost the plasmid. For example, if the plasmid in the *E. coli* has an ampicillin-resistance gene, 50 -100 μg/ml ampicillin is added to the LB broth. *E. coli* cells containing plasmids with tetracycline- or chloramphenicol-resistance gene are grown in LB broth with 12.5 μg/ml tetracycline or 25 μg/ml chloramphenicol, respectively. The culture is then grown overnight at 37°C after which it is transferred to a 10 ml LB broth containing the antibiotic. After 3-4 hours of incubation, this 10 ml culture is transferred to 1 liter of LB broth containing the antibiotic and incubated at 37°C with shaking to supply oxygen until the culture reaches an optical density at 600 nm light (OD_{600}) of 0.6 – 1.0 which is equivalent to the late log phase of growth. This stepwise culture method ensures rapid growth of *E. coli*. Direct inoculation of one colony to 1 liter of broth may work but will take several days for *E. coli* to grow to the required concentration. Chloramphenicol or spectinomycin is then added to the culture at a final concentration of 170 μg or 100 μg, respectively, to inhibit protein synthesis. Since chromosomal DNA replication requires continual protein synthesis in *E. coli*, the addition of chloramphenicol or spectinomycin inhibits its replication. Plasmids, on the other hand, can use pre-existing proteins to continually replicate. This process is called amplification; it not only increases the yield of plasmid DNA but also allows for more effective isolation of the plasmid DNA. If chromosomal DNA also continues to replicate, enormous amount of chromosomal DNA would be present. This would make plasmid DNA isolation very difficult.

After an overnight incubation, the cells are collected by centrifugation and resuspended in Solution I (50 mM glucose, 10 mM EDTA, 25 mM Tris-HCl, pH 8.0) containing 2 mg/ml lysozyme to digest the cell walls of *E. coli* cells. The glucose in this solution maintains the osmotic pressure so that the cells do not rupture immediately. EDTA (ethylene diamine tetraacetic acid) chelates divalent ions such as Mg^{++} and Ca^{++} and inhibits the activity of DNases because they require these ions to function. Tris-HCl assures that the pH of the solution does not change greatly during the procedure. The chemical name of Tris is 2-amino-2-hydroxymethylpropane-1,3-diol. The pH of solutions for DNA isolation is always slightly alkaline because DNA is very sensitive to acid. After the cell wall is digested, Solution II (0.2 N NaOH, 1% SDS) is added such that the final concentration of NaOH is 0.1 N and SDS is 0.5%. SDS (sodium dodecyl sulfate) is a detergent that dissolves the cell membrane and releases all cell contents to the solution. NaOH destroys hydrogen bonds that hold the two strands of DNA together and thus denatures DNA to become single-stranded. Solution III (3M NaOAc) is then added to a final NaOAc concentration of 1.5 M. Most proteins precipitate under high salt concentrations. Since chromosomal DNA is bound by numerous proteins, it precipitates together with proteins under high salt conditions. Plasmid DNA has very few proteins associated with it and therefore does not precipitate. Instead, it renatures back to the double-stranded form of DNA. The cell lysate is then centrifuged to remove the protein-DNA precipitate, thus removing the great majority of chromosomal DNA and proteins. Since DNA and RNA are not soluble in alcohol, they are then precipitated with alcohol. Two different kinds of alcohol, ethanol and isopropanol, can be used to precipitate DNA or RNA. When ethanol is used, 2.5 times the volume of the solution is required, whereas only 0.6 times the volume of the solution is required for isopropanol to precipitate DNA or RNA. Since the volume of cell lysate in this plasmid DNA isolation is quite large, isopropanol instead of ethanol is used to precipitate the plasmid DNA that is then dissolved in TE (10 mM Tris, 1 mM EDTA, pH 8.0) buffer. Ammonium acetate [$(NH_4)OAc$] is added to a final concentration of 2.5 M to perform another round of protein precipitation. After removal of the protein precipitate by centrifugation, RNase A is added to a final concentration of 10 μg/ml to digest the RNA.

To further purify plasmid DNA, a cesium chloride (CsCl) density gradient centrifugation is performed. CsCl is added to the plasmid DNA solution at 1 g/ml and allowed to dissolve by shaking the tube in a 70°C water bath. The CsCl-plasmid solution containing 200 μg/ml ethidium bromide (EtBr) is then sealed in an ultracentrifuge tube and centrifuged at 170,000 – 250,000 x g (Beckman Type 70.1 Ti rotor, 50,000 – 60,000 rpm) for 15 hours. Since EtBr causes DNA to break under fluorescent light, it is usually mixed with DNA solution in the dark. EtBr binds to the DNA and RNA and releases an orange color light upon UV irradiation, making DNA and RNA visible. During the ultracentrifugation, a CsCl gradient is formed with higher density in the lower part of the tube. Since RNA is most dense, it is pelleted to the bottom of the gradient. Protein, polysaccharide, and lipid float at the top of the gradient. DNA forms bands in the gradient at the equilibrium density (Fig. 3-1). Since the densities of linear and supercoiled DNA are different, they form bands at different positions. Super coiled DNA bands at a density (ρ) of 1.7, whereas linear DNA bands at $\rho = 1.59$. Relaxed circular form of the plasmid bands at $\rho = 1.5$. Since this density is very close to that of the linear DNA, the relaxed circular plasmids usually band together with the linear DNA molecules that are mostly contaminating chromosomal DNA fragments in this isolation procedure. A syringe needle is then inserted at the top of the gradient to release the vacuum, and the supercoiled plasmid band is withdrawn with another syringe. Since the band containing the relaxed circular plasmid also contains chromosomal

DNA fragments, this band is usually discarded. At this stage of plasmid purification, the solution contains plasmid DNA, EtBr, and CsCl. EtBr is then extracted with water-saturated butanol 3 to 4 times until the solution becomes clear, and CsCl is removed by dialysis. The resulting plasmid DNA is ultra pure and is stable for years.

Plasmid DNA can also be purified using commercial column chromatography kits. After removal of the chromosomal-protein precipitate, the clarified supernatant is passed through a column containing a certain type of resin. Plasmid DNA binds to the resin; other materials including contaminating chromosomal DNA fragments, proteins, polysaccharides, and lipids flow through the column. After washing, the plasmid DNA is eluted off the resin with water or other proprietary reagents. This method is rapid requiring less than two hours to complete, whereas the CsCl density gradient method requires two days. However, the quality of plasmid DNA purified by the CsCl gradient is much superior to that purified by the rapid column chromatographic method.

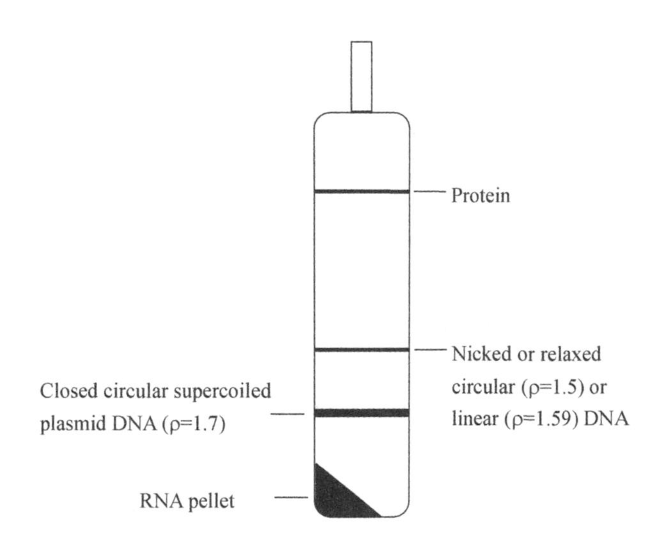

3.2: Genomic DNA Isolation.

To isolate genomic DNA from mammalian cells, cells are washed with phosphate-buffered saline (PBS) and resuspended in Tris-EDTA buffer (200 mM Tris-HCl, 100 mM EDTA, pH 8.0). SDS is then added to a final concentration of 0.5% to dissolve the cell membrane. After digestion of RNA at 37°C for one hour with 100 μg/ml RNase A, proteinase K is added to a final concentration of 100 μg/ml and incubated at 37 – 50°C overnight to digest proteins. Proteinase K is used because it is not sensitive to SDS and is relatively heat resistant. Therefore, protein digestion can be carried out at high temperature (e.g., 50°C) to minimize DNase activity. Phenol is then used to extract residual proteins and polysaccharides, followed by extraction with chloroform ($CHCl_3$) which removes lipids and phenol as well as proteins. The amount of phenol or $CHCl_3$ used is usually equal to the volume of the cell lysate. A small amount of isoamylalcohol is usually added to $CHCl_3$ at a ratio of 1:24 (v/v) because $CHCl_3$ causes surface denaturation of proteins making them prone to foaming; isoamylalcohol reduces this foaming during the extraction. Since $CHCl_3$ is not miscible with aqueous solution, two layers of solutions are formed. Isoamyl alcohol can stabilize the two layers so that they do not diffuse, making it easier to separate the aqueous layer that contains the DNA. DNA is then precipitated with alcohol. Since DNA is negatively charged, salt is added to neutralize the negative charges during alcohol precipitation; otherwise, the DNA molecules will repel each other and can not be precipitated by alcohol. Several different kinds of salt can be used for the neutralization, including NaOAc, KOAC, NaCl, and $(NH_4)OAc$. When NaOAc, KOAC, or NaCl is used, a final concentration of 0.3 M is required, while 2 M of $(NH_4)OAc$ is required. NaCl is rarely used because it is not very soluble in alcohol and precipitates with DNA in the presence of alcohol. The addition of salt and 2.5 volumes of ethanol (EtOH) results in a cotton-like genomic DNA precipitate, which is then fished out with a pipet tip or fine glass rod and washed with 70% EtOH to remove salt that may have precipitated with the DNA. After evaporation of EtOH, the DNA is dissolved in an appropriate amount of

TE (10 mM Tris, 1 mM EDTA, pH 8.0) buffer.

3.3: Total RNA Isolation

Total RNA includes pre-mRNA, mRNA, tRNA, rRNA, snRNA, and smRNA. Pre-mRNA and snRNA are located in the nucleus, while mRNA, tRNA, rRNA, and smRNA are present in the cytoplasm. To isolate RNA, the first step is to lyse the cells. Cells contain enormous amounts of various kinds of RNases. In an intact cell, RNases are tightly regulated and do not readily degrade RNA. When a cell is lysed, the RNases are no longer regulated and will degrade any RNA which they encounter. Therefore, the most important point in RNA isolation is to prevent RNA from being degraded by RNases. Microorganisms in the air as well as our hands, skin, sweat, and tears also contain enormous amounts of RNases. To prevent the RNases on our hands from contaminating the samples, gloves are always worn during RNA isolation. RNases are extremely stable and are not destroyed by autoclaving. To inactivate RNases that may be present in the water used to make reagents, the water is treated with 0.1% diethylpyrocarbonate (DEPC) for several hours and then autoclaved to inactivate the DEPC. Disposable plasticware is preferred for RNA isolation. If any glassware is used, it should be baked at 180°C overnight to destroy RNases.

The most commonly used method for total RNA isolation is that of Chomzynski and Sacchi (Anal. Biochem. 162: 156-159, 1987). Almost all commercially available RNA isolation kits are based on this method. This method uses a reagent called Solution D to isolate RNA. Solution D is composed of 4 M guanidinium thiocyanate, 25 mM sodium citrate, 0.5% N-lauryl sacrosine, and 0.1 M β-mercaptoethanol. Guanidinium thiocyanate is a potent RNase inhibitor. Sodium citrate is used to maintain the pH. N-lauryl sacrosine is a detergent and is used to lyse the cells. β-mercaptoethanol is a reducing agent that breaks disulfide bonds and decreases the activity of some RNases.

To isolate RNA, cells are lysed in Solution D to release the cell contents in solution. The most critical step in RNA isolation is to separate RNA from DNA. This is usually achieved by using acidic phenol (water-saturated phenol) because RNA stays in the aqueous layer, but DNA, protein, and polysaccharides are extracted into the phenol layer. After cells are lysed, sodium citrate is added to a final concentration of 100 mM to bring the pH of the cell lysate to approximately 4.0. Water-saturated acidic phenol is then used to extract RNA. The aqueous layer is then separated and extracted with $CHCl_3$-isoamylalcohol (24:1) to remove phenol, lipids, and residual proteins. The RNA in the aqueous layer is then precipitated with isopropanol and washed with 70% ethanol. After evaporating the ethanol, RNA is dissolved in an appropriate amount of DEPC-treated water.

3.4: Isolation of Poly-A$^+$ mRNA

Among various cellular components, only mRNAs have a poly-A tail. Taking advantage of this feature, poly-A$^+$ mRNAs can be easily isolated. Cells are lysed in Solution D and then centrifuged to remove cell debris. Oligo-dT-conjugated latex beads are added to bind poly-A$^+$ mRNAs. Since both oligo-dT and mRNA are negatively charged, this reaction is done in a solution containing a high concentration of salt (e.g., 500 mM NaCl, 5 mM Tris-HCl, pH 7.5). The oligo-dT-latex beads-poly-A$^+$ complexes are then pelleted and washed with a medium salt buffer (e.g., 100 mM NaCl, 5 mM Tris-HCl, pH 7.5). To remove the oligo-dT beads, the mRNAs and oligo-dT are allowed to regain their negative charges thus repelling each other. This is achieved by placing the mRNA-oligo-dT complexes in a solution containing no salt (e.g., 5 mM Tris-HCl, pH 7.5). The oligo-dT beads are then pelleted, and poly-A$^+$ mRNA in the supernatant is precipitated by salt and ethanol. The poly-A$^+$ mRNA precipitate is washed with 70% ethanol, dried, and then dissolved in an appropriate amount of DEPC-treated water.

3.5: Determination of the Quality and Quantity of DNA or RNA

The most common method for determining the quality and quantity of DNA or RNA is spectrophotometry, which determines the optical density (OD) of a certain molecule in solution. Optical density is the amount of light absorbed by the molecule. An optical density value at 260 nm light (OD_{260}) of 1 equals 50 μg/ml of double-stranded DNA, 33 μg/ml of single-stranded DNA including synthetic oligonucleotides, or 40 μg/ml of RNA. The quality of a DNA or RNA sample is determined by the ratio of its OD_{260} to OD_{280} values. A ratio of approximately 1.8 indicates good quality DNA. If the ratio is far less than 1.8, it means that the OD_{280} value is too big, most likely due to protein contamination since proteins absorb 280 nm light. On the other hand, if the OD_{260} to OD_{280} ratio is far greater than 1.8, DNA may be degraded to nucleotides which absorb more 260 nm light than DNA, resulting in a big OD_{260} value and a high OD_{260} to OD_{280} ratio. The quality of an RNA sample is considered good if its OD_{260} to OD_{280} ratio is approximately 2.0. Another value for quality determination is OD_{260} to OD_{230} ratio; the value should be greater than 1.5. Since carbohydrates, small peptides, phenols, and other aromatic compounds absorb 230 nm light, the OD_{260} to OD_{230} ratio would be smaller than 1.5 if a DNA or RNA sample is contaminated by any of these substances.

Summary

Two kinds of DNA, plasmid and genomic, are used in molecular biology research. Plasmid usually exists in bacteria such as *E. coli.* To isolate plasmid DNA, *E. coli* cells are grown and digested with lysozyme to remove cell walls. Cells are then lysed with 0.5% SDS, and the DNA is denatured with 0.1 N NaOH. Since chromosomal DNA is bound by many proteins, addition of high concentration of salt (150 mM NaOAc) will precipitate these proteins together with the chromosomal DNA.The precipitate is removed by centrifugation, thus eliminating the great majority of chromosomal DNA and proteins. The plasmid DNA is renatured under a high salt condition and is then precipitated by the addition of alcohol such as 0.6 volumes of isopropanol or 2.5 volumes of ethanol. After removal of RNA by RNase A digestion, the plasmid is purified by a cesium chloride density gradient or column chromatography. To perform a density gradient purification, CsCl is added to the DNA solution to 1 g/ml, mixed with 200 μg/ml ethidium bromide, and centrifuged at 170,000 – 250,000 x g for 15 hours. During the centrifugation, a gradient is formed, and proteins, polysaccharides, lipids, RNA, chromosomal DNA, and plasmid DNA band at different positions in the gradient. The band containing the supercoiled plasmid DNA is collected. Butanol is then used to remove ethidium bromide, and CsCl is removed by dialysis from the plasmid DNA solution.

To isolate genomic DNA from mammalian cells, the cells are lysed with 0.5% SDS, followed by digestion with 100 μg/ml RNase A to remove RNA and 100 μg/ml proteinase K to remove proteins. The cell lysate is then extracted with phenol to remove polysaccharide and residual proteins and with chloroform to remove phenol and lipids. The DNA in the aqueous solution is then precipitated by adding NaOAc to 0.3 M and 2.5 volumes of ethanol. The DNA precipitate is fished out, washed with 70% ethanol, dried, and then dissolved in TE (10 mM Tris-HCl, 1 mM EDTA, pH 8.0) buffer.

To isolate total cellular RNA, cells are lysed with a certain detergent such as SDS or N-lauryl sacrosine containing high concentrations (4 M) of RNase inhibitor guanidinium thiocyanate. RNA is then separated from DNA, polysaccharide, and lipid by extraction with acidic phenol (pH 4.0). The RNA in the aqueous layer is then precipitated with salt and alcohol.

Poly-A^+ mRNAs are usually isolated by using oligo-dT-conjugated latex beads. Cells are lysed, and oligo-dT latex beads are allowed to bind to the poly-A tail of mRNAs under high salt (500 mM NaCl) conditions. The RNA-oligo-dT complexes are pelleted, washed, and then suspended in a buffer containing no salt to allow mRNA and oligo-dT to regain their negative charges, thus repelling each other. Oligo-dT-latex beads are then pelleted, and the poly-A^+ mRNA in the supernatant is precipitated with salt and ethanol.

Quantification of DNA and RNA can be done by spectrophotometry. An OD_{260} value of 1 is equivalent to 50 μg of double-stranded DNA, 33 μg of single-stranded DNA, or 40 μg of RNA. A DNA solution with an OD_{260} to OD_{280} ratio of ~1.8 and RNA with a ratio of ~2 is considered good quality.

Sample Questions

1. If OD_{260} of a double-stranded DNA solution is 0.1, the concentration of this DNA is: a) 50 μg/ml, b) 5 μg/ml, c) 30 μg/ml, d) 3 μg/ml
2. The best way to determine the quality of DNA is by: a) isoelectric focusing, b) fluorometry alone, c) spectrophotometry alone, d) spectrophotometry and gel electrophoresis
3. The quality of a DNA is considered good if its OD_{260}/OD_{280} ratio is: a) 2.5, b) 3.0, c) 1.8, d) 2.0
4. The quality of an RNA is considered good if its OD_{260}/OD_{280} ratio is: a) 2.5, b) 3.0, c) 1.8, d) 2.0
5. Which of the following is commonly used to inhibit endogenous RNase activity during RNA isolation: a) phenol, b) chloroform, c) ethanol, d) guanidinium thiocyanate
6. To precipitate nucleic acids with ethanol or isopropanol, which of the following can be used to neutralize their negative charges: a) KCl, b) NaCl, c) NaOAc, d) $(NH_4)OAc$
7. Diethyl pyrocarbonate (DEPC) is used to inactivate which of the following in reagents and glassware: a) DNase, b) RNase, c) proteinase, c) endonuclease
8. When acidic phenol is used to extract nucleic acid, which of the following stays in the aqueous layer: a) protein, b) DNA, c) RNA, d) polysaccharide

Suggested Readings

1. Birnboim, H. C. and Doly, J. (1979). A rapid alkaline extraction procedure for screening recombinant plasmid DNA. Nucleic Acids Res.7: 1513-1523.
2. Castro, A. R., Morrill, W. E., and pope, V. (2000). Lipid removal from human serum sample. Clin. Diagn Lab. Immunol. 7: 197-199.
3. Chomczynski, P. and Sacchi, N. (1987). Single-step method of RNA isolation by acid guanidinium thiocyanate-phenol-chloroform extraction. Anal. Biochem. 162: 156-159.
4. Marmur, J. (1963). A procedure for isolation of deoxyribonucleic acid from microorganisms. Meth. Enzymol. 6: 726.

Chapter 4

Enzymes Used in Molecular Biology

Outline

Types of Enzymes Used in Molecular Biology
Degradation Enzymes
Discovery of Restriction Enzymes
Classification of Restriction-Modification Enzymes
Nomenclature of Restriction Enzymes
DNA Ends Generated by Restriction Enzymes
Factors Affecting the Activity of Restriction Enzymes
Common Restriction Enzymes
Isoschizomers
Methyltransferases
Methylation-Dependent Restriction Systems
Other Degradation Enzymes
RNA Polymerases
DNA Polymerases
Modification Enzymes

4.1: Types of Enzymes used in Molecular Biology

Enzymes used in Molecular Biology research can be classified into three major groups: degradation, synthesis, and modification enzymes. Degradation enzymes digest DNA or RNA, synthesis enzymes synthesize DNA or RNA, and modification enzymes change the structure of DNA or RNA.

4.2: Degradation Enzymes

Since DNA and RNA are nucleic acids, degradation enzymes are commonly referred to as nucleases. Two types of nucleases, exonucleases and endonucleases, exist. Exonucleases digest DNA or RNA from the ends of the molecules, whereas endonucleases cut DNA or RNA within the molecules. Some degradation enzymes only digest DNA or RNA and are, therefore, called DNases or RNases, respectively. There are degradation enzymes that can degrade both DNA and RNA. Some degradation enzymes only digest double-stranded; some only cut single-stranded; and some can cut both single- and double-stranded DNA or RNA. Degradation enzymes do not cut DNA or RNA randomly. They cut at the phosphodiester bond that links two nucleotides together. Depending on where the molecule is cut, one of two types of DNA or RNA molecules is generated. If the enzyme cuts at the ester bond between the phosphate and the first nucleotide, a DNA or RNA molecule with a phosphate group at the 5' end and an OH group at the 3' end is generated. On the other hand, a DNA or RNA molecule with a phosphate group at both ends is produced if the enzyme cuts at the ester bond between the phosphate group and the second nucleotide (Fig. 4-1). Most degradation enzymes generate fragments with a 5' phosphate and a 3' OH group.

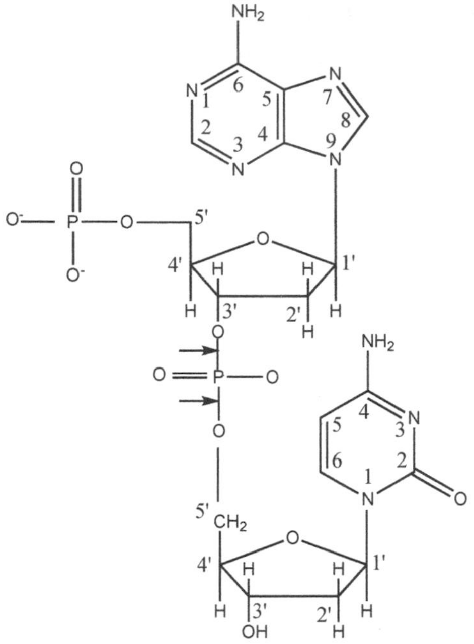

Fig. 4-1: Cutting Sites of Degradation Enzymes.

4.3: Discovery of Restriction Enzymes

The most important type of degradation enzymes is restriction enzymes that are involved in determining whether a bacteriophage can replicate in its host. A very well studied bacteriophage is the lambda (λ) phage which infects *E. coli*. Many different types of *E. coli* exist, including types K, B, C, R, etc. A λ phage usually produces at least 200 progenies if it successfully infects an *E. coli* cell. For ease of description, a λ progeny derived from *E. coli* type K is referred to as λ_k. When a λ_k is used to infect another type K *E. coli*, many λ progenies are produced. However, if a λ_k is used to infect other types of *E. coli* such as type B, very few or even no λ progenies are produced. This phenomenon is termed restriction. The replication of λ_k in *E. coli* B is restricted because the λ_k genomic DNA is degraded by enzymes present in *E. coli* B. These degradation enzymes are thus called restriction enzymes. The phenomenon that λ_k can infect *E. coli* K but not *E. coli* B is called host specificity. Since this host specificity is due to degradation of bacteriophage DNA by restriction enzymes, the genes encoding these restriction enzymes are called *hsd* (host specificity for DNA) genes. This type of restriction enzyme was the first type discovered and is, therefore, designated as Type I restriction enzyme. The first restriction enzyme was discovered in 1968 by Werner Arber, Daniel Nathans, and Hamilton Smith. Because of the extreme importance of restriction enzymes in Molecular Biology research, they were awarded the Nobel Prize in 1978.

Restriction enzymes do not cut DNA randomly; they recognize specific nucleotide sequences. Any DNA molecule with such a sequence can be cut by a certain restriction enzyme. To protect its own DNA from degradation by its own restriction enzyme, *E. coli* produces another type of enzyme, known as modification enzyme. This enzyme recognizes the same sequence as the restriction enzyme in the same cell. The modification enzyme modifies the genomic DNA so that it is not recognized and cut by the restriction enzyme. DNA modification by this type of enzyme is accomplished by the methylation of adenine or cytosine bases; thus, the modification enzyme is a methyltransferase. Adenine is usually methylated at position 6, and the product is called 6-methyladenine, while cytosine is methylated at position 5 to become 5-methylcytosine (Fig. 4-2). A λ_k bacteriophage can readily replicate in *E. coli* K because its genomic DNA is modified by the modification enzyme of *E. coli* K and is not recognized as foreign. However, λ_k DNA is recognized as foreign by *E. coli* B because it is not modified by the modification enzyme of *E. coli* B. Since the modification enzyme recognizes the same sequence as the restriction enzyme in the same cell, a foreign DNA, such as an invading λ DNA, may be modified and become resistant to the restriction enzyme if it is first detected by the modification enzyme. This λ then survives and replicates to produce progeny. On the other hand, if it is first detected by the restriction enzyme, it is degraded and produces no progeny. The modification enzyme of the Type I system is also encoded by the *hsd* genes.

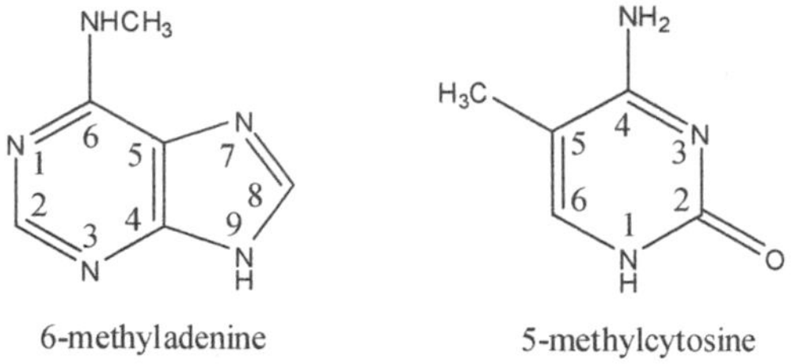

6-methyladenine 5-methylcytosine

Fig. 4-2: Methylated Adenine and Cytosine.

4.4: Classification of Restriction-Modification Enzymes

There are three major restriction-modification (R-M) systems (Table 4-1). As described above, the Type I R-M enzyme is encoded by the *hsd* genes. It is one protein with 3 different subunits (S, R, and M) and functions. The S subunit recognizes a certain DNA sequence and directs the M subunit to modify or the R subunit to cut the DNA containing the recognition sequence. All *E. coli* strains have the Type I R-M system. The Type II R-M system is usually encoded by plasmids, and only the bacteria containing such plasmids produce Type II enzymes. In the Type II system, restriction and modification functions are carried out by two separate enzymes encoded by two different genes. There are many kinds of Type II enzymes, including subtypes A, B, C, E, F, G, H, M, P, S, and T, but only types IIP and IIS (where P stands for palindrome and S for shifted) are useful for research purposes. The Type III R-M system is usually encoded by bacteriophages, such as SP1 and S15, and bacteria produce these enzymes only when they are infected with these bacteriophages. Type III enzymes are a single enzyme with both restriction and modification functions. However, unlike Type I enzymes, they do not have the S subunit. The major difference among these three types of enzymes is the specific sequence that they recognize. The Type I enzyme from *E. coli* B recognizes 5'-TGA-N8-TGCT-3', and that from *E. coli* K recognizes 5'-AAC-N6-GTGC-3'. The recognition sequence of a Type IIP enzyme is usually symmetrical such as GGATCC, in which the first two G's are complimentary to the last two C's and A is complimentary to

T. The Type III enzyme encoded by the SPI bacteriophage recognizes 5'-AGACC-3' and that by S15 recognizes 5'-CAGCAG-3'. Another major difference among the 3 types of enzymes is the location of their cutting sites. Type I enzymes cut at approximately 1,000 bp away from their recognition sequences, while Type III enzymes cut at 24-26 bp away from their recognition sequences. Type IIP enzymes cut exactly at their recognition sequences. Type IIS enzymes cut at specific sites several bases away from the recognition sequence and therefore are classified as subtype S (where S denotes a shifted cutting site). For example, FoKI recognizes 5'-GGATG-3'; its cutting site on the upper strand is 9 bases away from the 3' end of this sequence and that on the lower strand is 13 bases away. The modification enzymes of all three types methylate DNA at their respective recognition sequences. Once the recognition sequence of a certain restriction enzyme is methylated, the DNA cannot be cut by the restriction enzyme. Therefore, restriction and modification are mutually exclusive processes.

Table 4-1: Comparison of Restriction and Modification Enzymes.

	Type 1	Type IIP	Type IIS	Type III
Restriction and modification activities	Single multifunctional enzyme	Separate endonuclease and methylase	Single multifunctional enzyme	Single multifunctional enzyme
Protein structure	3 different subunits	Simple	2 different subunits	2 different subunits
Requirements for restriction	AdoMet, ATP, Mg^{++}	Mg^{++}	Mg^{++} (AdoMet)	ATP, Mg^{++}, AdoMet
Recognition sequence	sB:T-G-A-N8-T-G-C-T sK: A-A-C-N6-G-T-G-C	Two-fold symmetry	FokI: $GGATG(N)_9\downarrow$ $CCTAC(N)_{13}\downarrow$ AlwI: $GGATC(N)_4\downarrow$ $CCTAG(N)_5\downarrow$	SP1: A-G-A-C-C S15: C-A-G-C-A-G
Cleavage sites	Possibly random, at least 1,000 bp from host specificity site (recognition sequence)	Host specificity site (recognition sequence)		24-26 bp 3' of host specificity site (recognition sequence)
Requirements for methylation	AdoMet ATP, Mg^{++}	AdoMet ATP, Mg^{++}	AdoMet ATP, Mg^{++}	AdoMet ATP, Mg^{++}
Site of methylation	Recognition sequence	Recognition sequence	Recognition sequence	Recognition sequence

4.5: Nomenclature of Restriction Enzymes

Restriction enzymes are mostly produced by bacteria and are, therefore, named according to the scientific names of the bacteria. Typically, the name of a restriction enzyme is formed by the first letter of the genus name of the bacteria in the upper case, followed by the first two letters of the species name in the lower case. The serotype of the bacteria is denoted in an upper or lower case letter, followed by a roman numeral that represents the peak where the enzyme is found during isolation by chromatography. For example, EcoRI is the name of the restriction enzyme produced by a certain type of *E. coli* where "E" is the first letter of the genus name, *Escherichia*, and "co" is from the first two letters of the species name, *coli*. "R" represents the serotype of the *E. coli* which produces the enzyme, and the roman numeral "I" indicates that the enzyme is present in the first chromatography peak used to isolate the enzyme. The enzyme that is present in the second peak of the chromatography would be EcoRII. An exception to this convention is AccI in which "Ac" is from the genus name *Acinetobacter* and the second "c" is from the species name *calcoaceticus*. Previously, the letters derived from the scientific name of the bacteria were italicized or underlined. Therefore, EcoRI was written as *Eco*RI. This rule was changed in 2003 to better reflect the fact that restriction enzymes are proteins and their names do not need to be italicized. So, *Eco*RI is now written as EcoRI (Nucleic Acids Res. 31: 1805-1812, 2003).

4.6: DNA Ends Generated by Restriction Enzymes

Each restriction enzyme recognizes a specific nucleotide sequence. This sequence is always read from 5' to 3'. For example, the recognition sequence of HinfI is GANTC, not CTNAG. In fact, CTNAG is the recognition sequence of DdeI. Almost all restriction enzymes cut only double-stranded DNA. Three types of ends may be generated after cutting by restriction enzymes. For example, BamHI recognizes 5'-GGATCC-3' and cleaves the DNA between the two G's on both upper and lower strands. At the cut site, the 5' ends of the resulting two DNA fragments have four protruding bases (GATC). These types of ends are called 5' overhang ends. When a DNA fragment is cut with PstI which cuts at the sequence 5'-CTGCAG-3 between A and G on both strands, the newly generated ends have TGCA protruding from the 3' ends. These types of ends are called 3' overhang ends. SmaI recognizes 5'-CCCGGG-3 and cuts between C and G, resulting in DNA fragments with no protruding bases. These types of ends are referred to as blunt ends.

```
       ↓
5'-GGATCC-3'         5'-G          GATCC-3'
3'-CCTAGG-5'  ──►    3'-CCTAG          G-5'
            ↑
```

Fig. 4-3: 5' Overhang Ends Generated by BamHI Digestion.

```
            ↓
5'-CTGCAG-3'         5'-CTGCA          G-3'
3'-GACGTC-5'  ──►    3'-G          ACGTC-5'
     ↑
```

Fig. 4-4: 3' Overhang Ends Generated by PstI Digestion.

```
          ↓
5'-CCCGGG-3'         5'-CCC        GGG-3'
3'-GGGCCC-5'  ──►    3'-GGG        CCC-5'
          ↑
```

Fig. 4-5: Blunt Ends Generated by SmaI Digestion.

4.7: Factors Influencing the Activity of Restriction Enzymes

Many factors may affect the activity of restriction enzymes. Like all other enzymes, restriction enzymes require specific buffer solutions for optimal activity. A very important component in the buffer is salt. Restriction enzymes may require high (~ 100 mM NaCl), medium (~ 50 mM NaCl), or low (0 - 10 mM NaCl) salt conditions depending on the enzyme. EcoRI normally works under a high salt condition, recognizes the sequence 5'-GAATTC-3', and cuts at this sequence between G and A. However, if the same EcoRI is placed in a low salt buffer, it recognizes only 5'-AATT-3' and cuts in front of the first A. Therefore, a DNA molecule with the sequence 5'-GGCATACCGAATTCCCAATTCCCGG-3' is cut with EcoRI into two pieces with the sequences GGCATACCG and AATTCCCAATTCCCGG in a high salt buffer solution. But the same DNA molecule is cut with EcoRI into three pieces with the sequences GGCATACCG, AATTCCC, and AATTCCCGG in a low salt buffer solution. This phenomenon is called star activity, and EcoRI* indicates that the EcoRI is used under low salt conditions.

Temperature is another factor that can affect the activity of restriction enzymes. EcoRI is isolated from *E. coli* which normally grows at 37°C. Therefore, its optimal reaction temperature is 37°C. SalI is isolated from *Streptomyces albus* which lives in soil with an optimal growth temperature of 25°C, and therefore the optimal reaction temperature for SalI is 25°C. TaqI is isolated from *Thermus aquaticus* which lives in hot springs; thus, TaqI has an optimal reaction temperature of 72°C.

Reaction volume may also affect the activity of restriction enzymes. Since repeated freezing and thawing processes are harmful to restriction enzymes, they are usually stored in 50% glycerol to prevent them from becoming frozen in a freezer. However, a glycerol concentration greater than 5% inhibits restriction enzyme activity. Therefore, the amount of a certain restriction enzyme used in a reaction should not be greater than one-tenth of the total reaction volume. Restriction enzymes are normally stored at -20°C. Because temperature fluctuations adversely affect the stability of restriction enzymes, frost-free freezers are not appropriate for storing restriction enzymes. Although restriction enzymes are proteins, they are not measured by weight but by units. One unit of a certain restriction enzyme is defined as the amount of the restriction enzyme required to cut 1 μg of lambda bacteriophage DNA under optimal conditions.

4.8: Common Restriction Enzymes

Approximately 3,400 restriction enzymes have been discovered and approximately 500 of them are commercially available. Table 4-2 lists those that are more commonly used. In general, different

Table 4-2: Commonly Used Restriction Enzymes (arrow indicates cutting site)

Enzyme	Microorganism	Recognition Sequence
AccI	*Acinetobacter calcoaceticus*	GT↓(A/C)(T/G)AC
AluI	*Arthrobacter luteus*	AG↓CT
AvaI	*Anabaena variabilis*	C↓PyCGPuG
BalI	*Brevibacterium albidum*	TGG↓CCA
BamHI	*Bacillus amyloliquefaciens*	G↓GATCC
BclI	*Bacillus caldolyticus*	T↓GATCA
BglI	*Bacillus globigii*	GCCNNNN↓NGGC
BglII	*Bacillus globigii*	A↓GATCT
BstEII	*Bacillus stearothermophilus*	G↓GTNACC
BstNI	*Bacillus stearothermophilus*	CC↓(A/T)GG
ClaI	*Caryophanon latum*	AT↓CGAT
DdeI	*Desulfovibrio desulfuricans*	C↓TNAG
DpnI	*Diplococcus pneumoniae*	GA↓TC
EcoRI	*Escherichia coli*	G↓AATTC
EcoRV	*Escherichia coli*	GAT↓ATC
HaeII	*Haemophilus aegyptius*	PuGCGC↓Py
HaeIII	*Haemophilus aegyptius*	GG↓CC
HhaI	*Haemophilus haemolyticus*	GCG↓C
HincII	*Haemophilus influenzae*	GTPy↓PuAC
HindIII	*Haemophilus influenzae*	A↓AGCTT
HinfI	*Haemophilus influenzae*	G↓ANTC
HpaI	*Haemophilus parainfluenzae*	GTT↓AAC
HpaII	*Haemophilus parainfluenzae*	C↓CGG
KpnI	*Klebsiella pneumonia*	GGTAC↓C
MboI	*Moraxella bovis*	↓GATC
MspI	*Moraxella species*	C↓CGG
NcoI	*Nocardia coralline*	C↓CATGG
NdeI	*Neisseria denitrificans*	CA↓TATG
NlaIII	*Neisseria lactamica*	CATG↓
NotI	*Nocardia otitidis*	GC↓GGCCGC
PstI	*Providencia stuarti*	CTGCA↓G
PvuI	*Proteus vulgaris*	CGAT↓CG
PvuII	*Proteus vulgaris*	CAG↓CTG
SacI	*Streptomyces achromogenes*	GAGCT↓C
SalI	*Streptomyces albus*	G↓TCGAC
Sau3AI	*Staphylococcus aureus*	↓GATC
SmaI	*Serratia marcescens*	CCC↓GGG
SphI	*Strephtomyces phaeochromogwnes*	GCATG↓C
TaqI	*Thermus aquaticus*	T↓CGA
XbaI	*Xanthomonas badrii*	T↓CTAGA
XhoI	*Xanthomonas holcicola*	C↓TCGAG
XmaI	*Xanthomonas malvacaerum*	C↓CCGGG
XmnI	*Xanthomonoas manihotis*	GAANN↓NNTTC

restriction enzymes recognize different sequences that vary in length. AluI recognizes 4 base pairs, AGCT, and thus is referred to as a 4-base cutter. Since the probability for the AGCT sequence to appear in a DNA is 1/256, AluI may cut this DNA approximately every 256 bp, whereas a 6-base cutter such as BamHI may cut every 4,096 bp. NotI is an 8-base cutter and recognize GCGGCCGC. It cuts approximately every 65,536 bp.

4.9: Isoschizomers

Although different restriction enzymes recognize different sequences, some of them recognize the same sequences. These enzymes are referred to as isoschizomers. Isoschizomers are classified into three types. The first class of isoschizomers recognize the same sequence and cut at the same site (Table 4-3).

Table 4-3: Isoschizomers with the same recognition sequence and cut site

Restriction enzyme and its host bacteria		Recognition sequence and cut site
FspI (*Fischerella sp.*)	NsbI (*Neisseria subflava*)	-TGC↓GCA-
HincII (*Haemophilus influenzae,* Type c)	HindII (*Haemophilus influenzae,* Type d)	-GTPy↓PuAC-
MboI (*Moraxella bovis*)	DpnII (*Diplococcus pneumonia*) Sau3A (*Staphylococcus aureus*)	-↓GATC-
SstI (*Streptomyces stanford*)	SacI (*Streptomyces achromogenes*)	-GAGCT↓C-

The second class of isoschizomers, also called neoschizomers, recognize the same sequence, but cut at different sites as seen in Table 4-4.

Table 4-4: Isoschizomers with the same recognition sequence but different cut site

BstNI -CC↓(A/T)GG- (*Bacillus stearothermophilus*)	EcoRII -↓CC(A/T)GG- (*Escherichia coli*)
XmaI -C↓CCGGG- (*Xanthomonas malvacaerum*)	SmaI -CCC↓GGG- (*Serratia marcescens*)
DpnI -GA↓TC- (*Diplococcus pneumonia*)	MboI -↓GATC- (*Diplococcus pneumonia*) Sau3A -↓GATC- (*Staphylococcus aureus*)

The third class of isoschizomers are enzymes that recognize the same sequence and cut at the same site, but the cutting may or may not occur if there is methylation of the recognition sequence (Table 4-5). For example, both MspI and HpaII recognize CCGG and cut between the two C's in the sequence. If the first C of this sequence is methylated, MspI will not cut it. HpaII will not cut the sequence if either C residue is methylated.

Table 4-5: Effect of Methylation on the Activity of MspI and HpaII (boxed base is methylated)

	MspI (C↓CGG)	HpaII (C↓CGG)
- CCGG -	+	+
-C[C]GG-	+	-
-[C]CGG-	-	-

Other isoschizomers include MboI, DpnI, DpnII, and Sau3A that all recognize the sequence GATC. DpnI cuts between A and T of this sequence; the other three enzymes cut in front of G. MboI and DpnII can only cut unmethylated GATC sequences. Sau3A cuts both unmethylated and methylated GATC sequences, whereas DpnI cuts the sequence only when its A residues on both strands are methylated (Table 4-6).

Table 4-6: Effect of Methylation on the Activity of MboI, DpnII, Sau3A, and DpnI (boxed base is methylated)

	MboI and DpnII (↓GATC)	Sau3A (↓GATC)	DpnI (GA↓TC)
-GATC- -CTAG-	+	+	-
-G[A]TC- -CTAG-	-	+	-
-G[A]TC- -CT[A]G-	-	+	+

4.10: Methyltransferases

Any bacterium which produces a certain restriction enzyme must also produce a methyltransferase that recognizes and modifies the same sequence as that of the restriction enzyme. Otherwise, the bacterium will not survive. The methyltransferase methylates the recognition sequence to prevent the restriction enzyme from cutting the bacterial genome. For example, *Bacillus amyloliquefaciens* produces BamHI and also produces the BamHI methyltransferase (M.BamHI) which methylates the first C of the sequence GGATCC, thus preventing BamHI from cutting the sequence. Table 4-7 lists some of the common methyltransferases associated with their corresponding restriction enzymes.

Table 4-7: Common Methyltransferases and Their Recognition Sequences (bold and underlined letters represent methylated bases)

Methyltransferases	Recognition Sequence and Site of Methylation
M.AluI	5'-AG**C**T-3'
M.BamHI	5'-GGAT**C**C-3'
M.ClaI	5'-ATCG**A**T-3'
Dam	5'-G**A**TC-3'
Dcm	5'-C**C**(A/T)GG-3'
M.EcoRI	5'-GA**A**TTC-3'
M.FnuDII	5'-**C**GCG-3'
M.HaeIII	5'-GG**C**C-3'
M.HhaI	5'-G**C**GC-3'
M.HpaII	5'-C**C**GG-3'
M.MspI	5'-**C**CGG-3'
M.PstI	5'-CTGC**A**G-3'
M.TaqI	5'-TCG**A**-3'
M.SssI (CpG methytransferase)	5'-**C**G-3'

Most restriction enzymes cut unmethylated sequences, and the methylation of their recognition sequences prevent them from cutting those sequences. Therefore, DNA methylation is a mechanism by which bacteria prevent their DNA from cutting by their own restriction enzymes. Some bacteriophages can methylate their own DNA rendering them resistant to being cleaved by restriction enzymes. To defend against these bacteriophages, *E. coli* produces another type of restriction enzymes that cuts the DNA only if their recognition sequences on the DNA are methylated. This defense system is called the methylation-dependent restriction system (MDRS). Two types of restriction enzymes are produced by this system, including Mcr (modified cytosine restriction) and Mrr (methyladenine recognition and restriction) (Table 4-8). McrA recognizes CCGG and GCGC and cuts these sequences only when the second C of CCGG and the first C of GCGC are methylated. These sequences can be methylated by M.HpaII (HpaII methyltransferase) and M.HhaI, respectively. McrBC recognizes AGCT, GGCC, GCGC, and CAGCTG and cuts these sequences only when the underlined C residues of these sequences are methylated.

These sequences are methylated by M.AluI, M.HaeIII, M.HhaI, and M.PvuII, respectively. Mrr recognizes GANTC, CTGCAG, and TCGA and cuts the sequences only when the A residues in these sequences are methylated. These sequences are methylated by M.HinfI, M.PstI, and M.TaqI, respectively.

Since MDRS restriction enzymes digest DNA only when their recognition sequences on the DNA are methylated, *E. coli* simply does not produce any methyltransferases that can methylate MDRS recognition sequences. Therefore, MDRS enzymes in the *E. coli* do not cut its DNA. Although *E. coli* has methylation systems such as the Type I R-M system, deoxyadenosine methyltransferase (Dam), and deoxycytosine methyltransferase (Dcm), they do not methylate the recognition sequences of MDRS enzymes. Dam and Dcm recognize GATC and CC(A/T)GG, respectively and methylate the underlined residues of those sequences.

Table 4-8: Methylation-Dependent Restriction System (bold and underlined letters represent methylated bases)

MDRS Enzymes	Recognition Sequence
McrA	C**C**GG – M.HpaII G**C**GC – M.HhaI
McrBC	AG**C**T – M.AluI GG**C**C – M.HaeIII G**C**GC – M.HhaI CAG**C**TG – M.PvuII
Mrr	G**A**NTC – M.HinfI CTGC**A**G – M.PstI TCG**A** – M.TaqI

4.11: Other Degradation Enzymes

Other enzymes such as Bal 31, Exo I, Exo III, S1 nuclease, mung bean nuclease, DNase I, and RNase A are commonly used degradation enzymes. These enzymes are not restriction enzymes and do not recognize specific sequences. Bal 31, which is isolated from

Altermonas espeijiana, digests both DNA and RNA. It is an exonuclease for double-stranded but an endonuclease for single-stranded DNA or RNA. When cutting double-stranded DNA or RNA, it cuts both strands simultaneously and from both ends. Bal 31 requires Ca^{++} to function. It was mainly used to delete regions of DNA that are not needed. This enzyme is no longer used because other easier methods to delete unwanted DNA regions, such as the polymerase chain reaction (PCR), are now available.

Exo I (exonuclease I) cuts single-stranded DNA from 3' to 5'. It is used to remove unused primers after PCR. Exo III (exonuclease III) recognizes double-stranded DNA but cuts only one strand in the 3' to 5' direction. DNA molecules with 5' overhangs are preferred substrates of Exo III. It also cuts blunt-ended DNA, but does not cut DNA fragments with 3' overhangs. Exo III has RNaseH activity and can digest the RNA portion of an RNA-DNA hybrid. It also has 3' phosphatase activity which removes the phosphate group located at the 3' end of a DNA molecule. Exo III is also an AP (apurinic DNA) endonuclease and breaks DNA at where a purine base is missing.

S1 nuclease is isolated from *Aspergillus oryzae*. It is both an exo- and endonuclease for DNA and RNA and requires Zn^{++} for function. The most important feature of S1 is that it is single-strand specific. If a DNA or RNA molecule has both single- and double-stranded regions, S1 nuclease will cut only the single-stranded regions. Mung bean nuclease is another single-strand specific nuclease. It is isolated from the sprouts of *Vigna radiata*. Mung bean nuclease has a higher specificity for single-stranded DNA and RNA than S1 nuclease and therefore is more commonly used to remove unwanted single strand regions of a DNA or RNA molecule. Unlike S1 nuclease, mung bean nuclease does not cleave the intact strand of a nicked double-stranded DNA.

DNase I is an endonuclease and digests only DNA. In the presence of Mg^{++}, it cleaves either strand of a double-stranded DNA randomly, generating nicks at various places on the DNA. This process is called nicking. DNase I cuts both strands of DNA simultaneously resulting in double-stranded breaks in the presence of Mn^{++}.

RNase A is both an exo- and endonuclease. Its function is very similar to that of S1 and mung bean nuclease, except that it only digests RNA. RNase A cuts RNA at the 5' side of any pyrimidine (C or U) base. For example, it would cut the following sequence as follows: AG↓CGG↓U↓C↓UA↓C↓U.

4.12: RNA Polymerases

RNA polymerases synthesize RNA. Like most bacteria, *E. coli* has only one RNA polymerase responsible for making all RNAs, including mRNA, tRNA, and rRNA. Some bacteriophages such as SP6, T3, and T7 have their own RNA polymerases. As described in Chapter 2, eukaryotic cells have three different kinds of RNA polymerases that make different kinds of RNAs.

4.13: DNA Polymerases

Many kinds of DNA polymerases exist. An example of a DNA polymerase is *E. coli* DNA polymerase I. This enzyme is composed of two subunits of 76 kDa and 35 kDa. The large subunit (76 kDa) has the polymerase activity and synthesizes DNA in a 5' to 3' direction. It is also a 3' to 5' exonuclease which recognizes double-stranded DNA but cuts only one strand in a 3' to 5' direction. As mentioned in Chapter 2, DNA synthesis requires a template, dNTP, DNA polymerase, and primer. If the sequence of a template is 5'-GATCATCGTCGAGCA-3', a primer with the sequence 5'-TGCT-3' will anneal to the 3' end of the template. DNA polymerase then adds nucleotides to the primer. In this example, the first nucleotide to be added to the primer is C because the template has a G at the corresponding position. If the DNA polymerase makes a mistake and adds A instead of C to the primer, this nucleotide will fail to base pair with the template and will result in a hanging single-stranded DNA molecule. This molecule will trigger the 3' to 5' single-strand specific activity of the 76 kDa subunit to remove the hanging A residue, allowing DNA polymerase to add another nucleotide. If the DNA polymerase makes another mistake by placing the incorrect nucleotide, the same process is repeated until a C is added. This action is called DNA proofreading. It is an extremely important function of a cell.

The small (35 kDa) subunit has the 5' to 3' exonuclease activity. If a DNA molecule has a nick in one of the strands, it will cut this DNA from the nick in a 5' to 3' direction and expand the nick to a gap. The gap is then filled by the DNA polymerase activity of the large subunit. The large subunit of *E. coli* DNA polymerase I is fully functional when it is separated from the small subunit. This phenomenon was first discovered by H. Klenow; therefore, this subunit is called the Klenow fragment or Klenow enzyme. The small subunit loses it function when it is separated from the large subunit. Most DNA polymerases have 3' to 5' exonuclease activity in addition to their polymerase activity.

T7 DNA polymerase is another commonly used DNA polymerase. It is also composed of two different subunits. The large subunit is the 84 kDa T7 gene 5 protein, and the small subunit is the 12 kDa *E. coli* thioredoxin. T7 DNA polymerase is more processive than *E. coli* DNA polymerase I. In a DNA synthesis reaction, DNA polymerase may jump from one template to another before it completely replicates a template. If the DNA polymerase jumps too frequently, its DNA synthesis efficiency is lower. This efficiency is measured by processivity which reflects the number of nucleotides added by DNA polymerase per association with the template.

Sequenase is a modified T7 DNA polymerase lacking the 3' to 5' exonuclease activity. It is a DNA polymerase previously used for DNA sequencing. T4 DNA polymerase is another commonly used DNA polymerase. Its function is very similar to that of the Klenow fragment of *E. coli* DNA polymerase I with 5' to 3' DNA polymerase and 3' to 5' exonuclease activity. Its 3' to 5' exonuclease activity is approximately 200 times that of the Klenow fragment. T4 DNA polymerase is most commonly used to fill in the 5' overhang or chew up the 3' overhang of a DNA fragment making it blunt ended.

Taq DNA polymerase is isolated from *Thermus aquaticus*. It is heat resistant with an optimal reaction temperature of 72°C and is the major enzyme used in the polymerase chain reaction (PCR). Taq DNA polymerase has no 3' to 5' exonuclease activity and therefore is prone to making mistakes. It is estimated to mis-incorporate one in every 10,000 bases synthesized. Reverse transcriptase is another commonly used DNA polymerase. All retroviruses have this enzyme. It uses RNA as the template to make DNA and can also use DNA as template to synthesize DNA in vitro. Terminal deoxynucleotide transferase (TdT) adds nucleotides to the 3' end of a single or double-stranded DNA. It is commonly used to label oligonucleotides.

4.14: Modification Enzymes

Commonly used modification enzymes include methyltransferases, ligase, alkaline phosphatase, and kinase. Methyltransferases transfer a methyl group (CH_3) from S-adenosine methionine (AdoMet) to DNA as described above. Ligases link two pieces of DNA fragments together by forming a phosphodiester bond between the 3' OH group of the first DNA fragment and the 5' phosphate group of the second DNA fragment. Alkaline phosphatases remove the 5' phosphate group of a DNA fragment and converts it to an OH group. This enzyme is active only under alkaline conditions (e.g., pH 9.0). Calf intestinal and shrimp alkaline phosphatases are the two most commonly used. Kinase is used to transfer the γ phosphate of ATP to the 5' end of DNA fragments in which the phosphate group has been removed and converted to an OH group.

Summary

Enzymes used in Molecular Biology research include degradation, synthesis, and modification enzymes. Degradation enzymes are divided into endonucleases and exonucleases. Degradation enzymes may digest only DNA, RNA, or both DNA and RNA. In addition, they may cut only single-stranded, double-stranded, or both single- and double-stranded DNA or RNA. The most important degradation enzymes are restriction enzymes that are mostly produced by bacteria. The major function of restriction enzymes is to defend bacteria from infection by bacteriophages. To prevent their genomic DNA from digestion by their own restriction enzymes, bacteria also produce modification enzymes that methylate the recognition sites of restriction enzymes rendering the DNA in the bacteria not susceptible to their own restriction enzymes.

There are three major types of restriction enzymes. Type I restriction enzymes are present in all *E. coli* and are encoded by *hsd* genes. These enzymes recognize a certain sequence, but cut the DNA at approximately 1000 bp away from the recognition sequence. Type II enzymes recognize a certain sequence and cut at the sequence or at a specific site several bases from the sequence. Type III enzymes cut DNA 24 - 26 bp downstream from their recognition sequences. Type I and type III enzymes are a single enzyme with multiple functions. Type I enzymes are composed of three different subunits, each with a different function. The S subunit recognizes the recognition sequence and directs the R subunit to restrict (cut) or the M subunit to modify (methylate) the DNA. Type III enzymes are composed of R and M subunits responsible for cutting and modification, respectively. The restriction and modification functions of Type II enzymes are carried out by two independent enzymes although they recognize the same sequence.

Restriction enzymes are named according to the scientific name of the host bacteria. For example EcoRI is a restriction enzyme produced by *E. coli*; where "Eco" is derived from the first letter of the genus and first two letters of the species name. "R" represents the serotype of the *E. coli*, and "I" indicates that the restriction enzyme is present in the first protein peak of the chromatography used to isolate the enzyme. Restriction enzymes are measured in units. One unit of a restriction enzyme is the amount of the enzyme required to cut 1 μg of lambda bacteriophage DNA in one hour under optimal conditions. The activity of restriction enzymes may be affected by salt concentration, temperature, and reaction volume. Some enzymes such as EcoRI and BamHI require high salt (~ 100 mM NaCl) conditions. EcoRI recognizes the sequence GAATTC and normally cuts the sequence between G and A, but recognizes only AATT under low salt conditions in which it cuts in front of the first A. This activity is called star activity. EcoRI is isolated from *E. coli* which grows at 37°C; therefore, its optimal reaction temperature is 37°C. SalI is from *Streptomyces albus* which lives in soil and works the best at 25°C. TaqI has an optimal reaction temperature of 72°C because it is isolated from *Thermus aquaticus* which lives in hot springs. Restriction enzymes are usually stored in 50% glycerol at -20°C and need to be diluted at least 10 times to reduce the glycerol concentration to less than 5%; otherwise, the glycerol will inhibit their activity.

Restriction enzymes may generate DNA fragments with a 5' overhang, 3' overhang, or blunt ends. In general, different restriction enzymes recognize different sequences, but some restriction enzymes such as MboI and DpnII recognize the same sequence and cut at the same site. These restriction enzymes are called isoschizomers. Some bacteriophages have modification systems that methylate their DNA making them resistant to the restriction enzymes produced by the *hsd* system. To defend these bacteriophages, *E. coli* develops a methylation-dependent restriction system (MDRS). The MDRS enzymes also recognize certain specific sequences, but cut the DNA only when these recognition sequences on the DNA are methylated. MDRS enzymes include Mcr and Mrr. *E. coli* does not produce any enzymes that can methylate the recognition sequences of MDRS enzymes; thus its DNA is not susceptible to MDRS restriction enzymes.

Bal 31 is another degradation enzyme. It digests both DNA and RNA and requires Ca^{++} to function. It is an exonuclease for double-stranded and an endonuclease for single-stranded DNA or RNA. Exo I is a single-strand specific, 3' to 5' exonuclease. It is commonly used to remove primers after PCR. Exo III recognizes double-stranded DNA, but cuts only one strand in the 3' to 5' direction. S1 nuclease and mung bean nuclease are single-strand specific endo- and exonucleases. They both require Zn^{++} to function. DNase I makes single-strand nicks on DNA in the presence of Mg^{++} and double-strand breaks in the presence of Mn^{++}. RNase A cuts at the 5' side of any pyrimidine (C and U) nucleotide of single-stranded RNAs.

Synthesis enzymes include RNA polymerases and DNA polymerases. In addition to the *E. coli* RNA polymerase, commonly used RNA polymerases include SP6, T7, and T3 RNA polymerases. Many kinds of DNA polymerases are used in molecular biology research. *E. coli* DNA polymerase I is composed of two subunits of 76 kDa and 35 kDa. The 76 kDa subunit has both 5' to 3' DNA polymerase activity and 3' to 5' single-strand specific exonuclease activity. The former engages in DNA synthesis, and the latter is responsible for DNA proofreading in which a mistakenly incorporated nucleotide is removed because it does not base pair with the template and creates a single-stranded region. Most DNA polymerases have this function. The small subunit (35 kDa) of *E. coli* DNA polymerase I has the single-strand 5' to 3' exonuclease activity. The large subunit is called the Klenow fragment and is fully functional even when it is separated from the small subunit. T7 DNA polymerase is more processive than *E. coli* DNA polymerase I and was previously used in manual DNA sequencing. T4 DNA polymerase is functionally similar to the Klenow fragment. It is most commonly used to convert DNA fragments with 5' or 3' overhang ends to blunt ends. Taq DNA polymerase is heat resistant and is the major enzyme used for PCR. Reverse transcriptase uses RNA as template to make DNA. It can also use DNA as a template to make DNA in vitro. Terminal deoxynucleotide transferase adds nucleotides to the 3' end of a single- or double-stranded DNA.

Modification enzymes include methyltransferase, ligase, alkaline phosphatase, and kinase. Methyltransferases modify DNA by transferring a methyl group from S-adenosine methionine to DNA. Ligase joins two DNA fragments together by forming a phosphodiester bond between the 3' OH group of a DNA fragment and the 5' phosphate group of another DNA fragment. Alkaline phosphatase is used to remove the 5' phosphate group of DNA fragments, and kinase is used to transfer the γ phosphate of ATP to the 5' end of a dephosphorylated DNA fragment.

Sample Questions

1. The *hsd* genes encode: a) Type I restriction and modification system, b) MDRS system, c) Type II restriction and modification system, d) Type III restriction and modification system.
2. Which of the following mutations is lethal: a) *hsdS*, b) *hsdM*, c) *hsdR*, d) *dam*
3. Which of the following is a recognition sequence of a Type II restriction enzyme: a) T-G-A-N8-T-G-C-T, b) A-A-C-N6-G-T-G-C, c) G-A-A-T-T-C, d) A-G-A-C-C
4. Isoschizomers are: a) different restriction enzymes that recognize the same sequence, b) the same restriction enzyme which recognizes multiple sequences, c) MDRS restriction enzymes, d) MDRS methylases
5. Which of the following may be cut by MDRS restriction enzymes: a) CCGG, b) C$\underline{C}$GG, c) AGCT, d) AG$\underline{C}$T, in which the underlined bases are methylated.
6. *E. coli* protects its own DNA from degradation by MDRS restriction enzymes by a) producing certain enzymes to protect DNA, b) not producing methyltransferases that methylate the recognition sites of MRDS restriction enzymes, c) removing methyl groups from DNA, d) denaturing DNA
7. Restriction enzymes recognize different sequences under different salt concentrations; this activity is referred to as a) alternative activity, b) isozyme activity, c) star activity, d) nonsense activity
8. S1 nuclease cuts a) only single-stranded RNA, b) only single-stranded DNA, c) double-stranded DNA, d) single-stranded DNA and RNA
9. DNase I a) is an exonuclease, b) is an endonuclease, c) makes single-strand nicks in the presence of Mg^{++}, d) makes double-strand breaks in the presence of Mn^{++}
10. RNase A cuts a) double stranded RNA, b) only single-stranded RNA, c) at 5' side of pyrimidines, c) at 5' side of purines
11. *E. coli* DNA polymerase I has a) 5' - 3' polymerase activity, b) 3' - 5' single-strand specific exonuclease activity, c) 5' - 3' single-strand specific exonuclease activity, d) 3' - 5' polymerase activity
12. The Klenow fragment of *E. coli* DNA polymerase I has a) 5' - 3' polymerase activity, b) 3' - 5' single-strand specific exonuclease activity, c) 5' - 3' single-strand specific exonuclease activity, d) 3' - 5' polymerase activity

13. Reverse transcriptase can use which of the following as a template to make DNA: a) DNA, b) RNA, c) protein, d) polysaccharide
14. Alkaline phosphatase is used to remove which of the following from DNA: a) 3' OH group, b) 5' phosphate group, c) introns, d) exons
15. Kinase transfers which of the following to the 5' end of DNA: a) β phosphate of ATP, b) γ phosphate of ATP, c) α phosphate of ATP, d) γ phosphate of dATP

Suggested Readings

1. Bickle, T. A. and Kruger, D. H. (1993). The biology of DNA restriction. Microbiol. Rev. 57: 434-450.
2. Blumenthal, R. M. (1989). Cloning and restriction of methylated DNA in *Escherichia coli*. BRL Focus 11: 41-46.
3. Dryden, D. T. F., Murray, N. E., and Rao, D. N. (2001). Nucleoside triphosphate-dependent restriction enzymes. Nucleic Acids Res. 29: 3728-3741.
4. Klenow, H. and Henningsen, I. (1970). Selective elimination of the exonuclease activity of the deoxyribonucleic acid polymerase from *Escherichia coli* B by a limited proteolysis. Proc. Natl. Acad. Sci. USA 65: 168-175.
5. Murray, N. (2000). Type I restriction systems: sophisticated molecular machines (a legacy of Bertani and Weigle). Microbiol Mol. Biol. Rev. 64: 412-434.
6. Nathans, D. and Smith, H. O. (1975). Restriction endonucleases in the analysis and restructuring of DNA molecules. Annu. Rev. Biochem. 44: 273-293.
7. Pingoud, A. and Jeltsch, A. (1997). Recognition and DNA cleavage by type II restriction endonucleases. Eur. J. Biochem. 246: 1-22.
8. Raleigh, E. A. (1992). Organization and function of the mcrBC genes of *Escherichia coli* K-12. Mol. Microbiol. 6: 1079-1086.
9. Roberts R. J. et al. (47 authors) (2003). A nomenclature for restriction enzymes, DNA methytransferases, homing endonucleases and their genes. Nucleic Acids Research. 31:1805-1812.
10. Roberts R. J. and Macelis, D. (1991). Restriction enzymes and their isoschizomers. Nucleic Acids Research. 19 Suppl:2077-2109.
11. Szybalski, W., Kim, S. C., Hasan, N., and Podhajska, A. J. (1991). Class-IIS restriction enzymes - a review. Gene. 100:13-26.
12. Yuan, R. (1981). Structure and mechanism of multifunctional restriction endonucleases. Annu. Rev. Biochem. 50: 285-315

Chapter 5

Promoter and Operon

Outline

Promoter
Transcription terminators of prokaryotic genes
Eukaryotic gene transcription
Elements affecting eukaryotic gene transcription
Operon
Lac operon
Repression of *lac* operon
Activation of *lac* operon

5.1: Promoter

Promoter is a DNA element which initiates transcription. Although nucleotide sequences of genes differ, two regions containing consensus sequences in prokaryotic gene promoters are crucial for transcription. One of these regions is located approximately 10 bp upstream from the transcription start site and is therefore referred to as the "-10" box. The consensus sequence of the -10 box is $T_{77}A_{76}T_{60}A_{61}A_{56}T_{82}$, where the numbers in subscript represent percentages of occurrence (Fig. 5-1). Since the role of the -10 box in transcription is first determined by Pribnow, it is also called the Pribnow box. Another important region for transcription is located approximately 35 bp upstream from the transcription start site and is therefore called the -35 box; its consensus sequence is $T_{69}T_{79}G_{61}A_{56}C_{54}A_{54}$ (Fig. 5-1). If the sequence of any of these two regions of a gene is deleted or mutated, the gene is inactivated. The enzyme responsible for RNA synthesis is RNA polymerase. To transcribe a gene, the RNA polymerase must bind to a

DNA template. The -35 box is the first place where the RNA polymerase makes contact with the template, and the -10 box is where the template starts to open to become single-stranded to be transcribed (Fig. 5-2).

```
         -35                           -10
CCAGGCT|TTACAC|TTTATGCTTCCGGCTCG|TATGTT|GTGTGG|A|ATTG
CTTTTTG|ATGCAA|TTCGCTTTGCTTCTGAC|TATAAT|AGACAG|G|GTAA
GGCGGTG|TTGACA|TAAATACCACTGGCGGT|GATACT|GAGCAC|A|TCAG
GTGCGTG|TTGACT|ATTTTACCTCTGGCGGT|GATAAT|GGTTGC|A|TGTA
ATTGTTG|TTGTTA|ACTTGTTTATTGCAGCT|TATAAT|GGTTAC|A|AATA
CGTAACA|CTTTAC|AGCGGCGCGTCATTTGA|TATGAT|GCGCCC|G|CTTA
```

-35 sequence ($T_{69}T_{79}G_{61}A_{56}C_{54}A_{54}$) Pribnow box ($T_{77}A_{76}T_{60}A_{61}A_{56}T_{82}$) mRNA start

Fig. 5-1: Prokaryotic Gene Promoters.

E. coli RNA polymerase has two forms including holoenzyme and core enzyme. The holoenzyme is composed of two α subunits and one each of β, β', ω, and σ (sigma) subunits. The α subunit links other subunits together. The β subunit synthesizes RNA, and the β' subunit allows the RNA polymerase to bind to the DNA template. The ω subunit maintains the stability and activity of the RNA polymerase. The σ subunit is responsible for recognition of the -35 sequence. The only difference between the holoenzyme and core enzyme is that the core enzyme does not have a σ subunit. *E. coli* makes seven different σ subunits including σ^{70} (σ^{D}), σ^{54} (σ^{N}), σ^{38} (σ^{S}), σ^{32} (σ^{H}), σ^{28} (σ^{F}), σ^{24} (σ^{E}), and σ^{19} (σ^{FecI}). These σ factors are produced in *E. coli* under different conditions, and the core enzyme can associate with any of the 7 sigma factors to become a holoenzyme and transcribe different genes. σ^{70} is for transcription of housekeeping genes, σ^{54} is for expression of genes under a nitrogen-limitation condition, σ^{38} is for expression of genes during starvation or stationary phase, σ^{32} is for transcription of heat shock genes, σ^{28} is responsible for expression of flagellar genes, σ^{24} is responsive to extreme heat stress, and σ^{19} is the ferric citrate sigma factor responsible for transcription of genes involved in iron transport.

The holoenzyme is responsible for initiation of transcription because it has the σ factor that directs it to bind to the -35 region of a certain promoter. Once transcription is initiated, the σ factor is no longer needed and dissociate from the holoenzyme, converting it to the core enzyme that carries out the rest of RNA synthesis (Fig. 5-2). Like DNA synthesis, RNA synthesis also goes from 5' to 3'; thus the

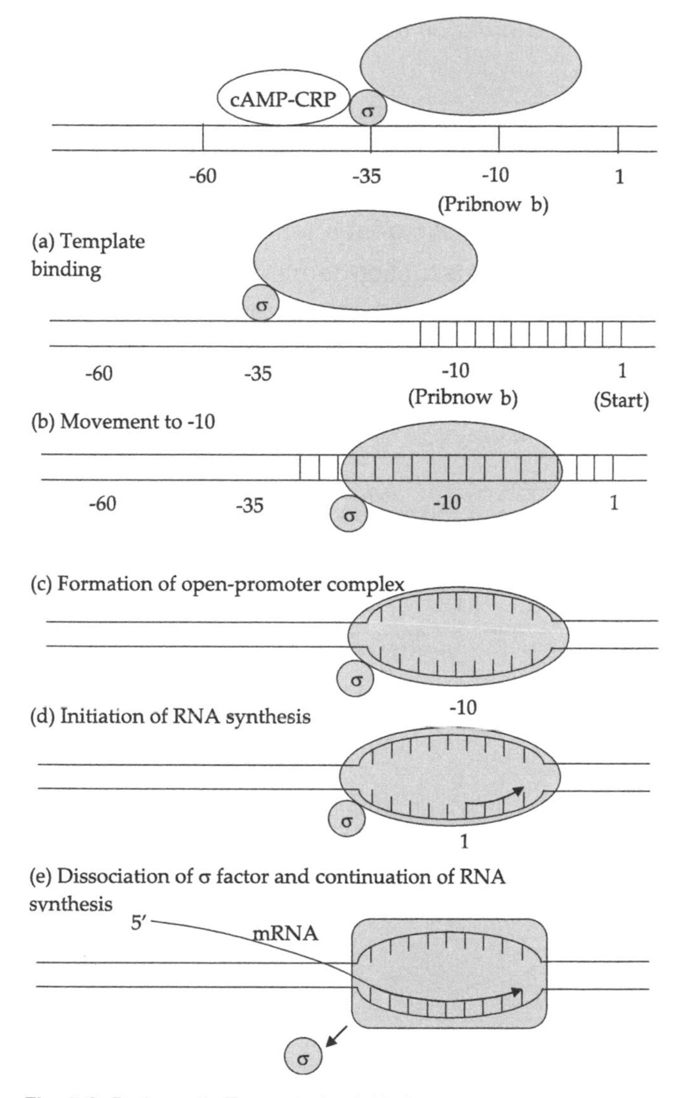

Fig. 5-2: Prokaryotic Transcription Initiation.

RNA polymerase reads the sequence on the DNA template in a 3' to 5' direction. During transcription, only one strand of the template is transcribed. Two different genes may be located on the same DNA fragment in opposite orientations. If the -35 and -10 sequences are in 5' to 3' order on the upper strand, the lower strand of the DNA is used as the template to synthesize RNA. On the other hand, the upper strand of the DNA is used as the template to make RNA if the -35 and -10 sequences are in 5' to 3' order on the lower strand.

5.2: Transcription Terminators of Prokaryotic Genes

Transcription of prokaryotic genes terminates at terminators. A characteristic feature of transcription terminators is the presence of inverted repeat sequences (Fig. 5-3). When this region is transcribed into RNA, the RNA may form a stem-and-loop structure. Two different

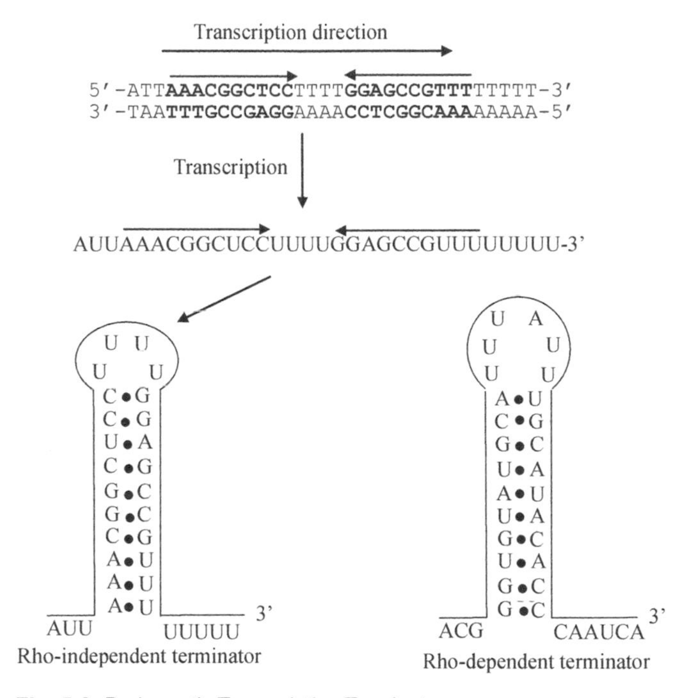

Fig. 5-3: Prokaryotic Transcription Terminators.

types of terminators exist. One requires a protein called Rho (ρ) to help terminate transcription. Rho protein binds to newly synthesized RNA and travels along the RNA. When the terminator region is transcribed and the RNA forms a stem-and-loop structure, the movement of the

RNA polymerase on the template is stalled, allowing the Rho protein to catch up and bind to it (Fig. 5-4). This binding inactivates the RNA polymerase, thus terminating transcription. The other type of terminator has a run of A's on the template; therefore, the newly synthesized RNA has a run of U's at its 3' end (Fig. 5-3). Since the hydrogen bonds between A and U are weak, this RNA will automatically fall off the template. Since this type of terminator does not require the Rho protein to help stop transcription, it is referred to as Rho-independent terminator. The other type of terminator is called Rho-dependent terminator.

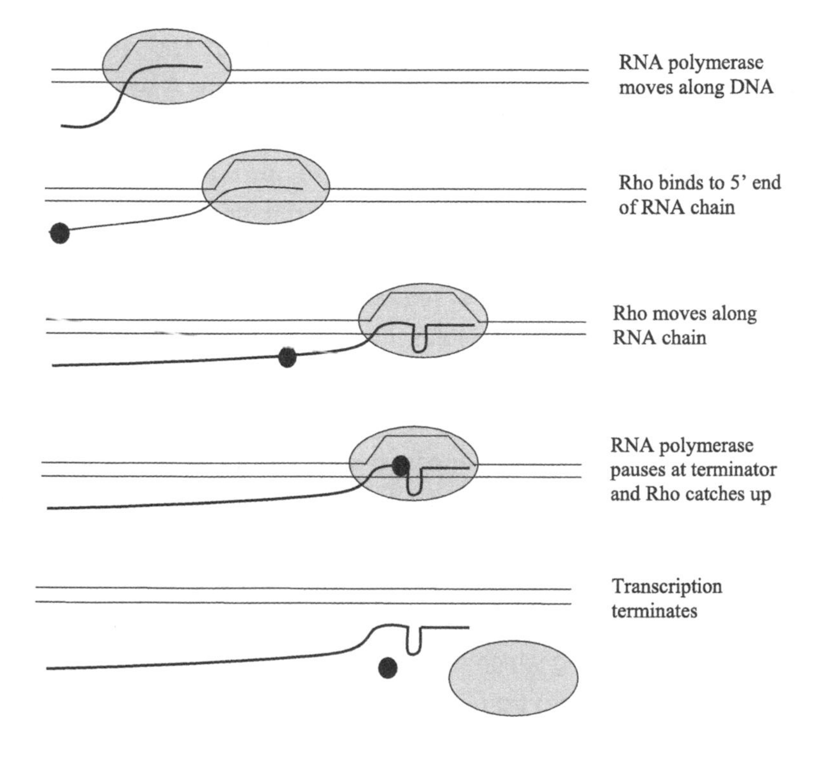

Fig. 5-4: Rho-dependent Transcription Termination.

5.3: Eukaryotic Gene Transcription

As described previously, there are 3 different RNA polymerases in eukaryotic cells. RNA polymerase I is responsible for making 28S, 18S, and 5.8S rRNA. RNA polymerase II synthesizes mRNA, snRNA, and smRNA. RNA polymerase III makes tRNA. Each RNA polymerase recognizes a different type of promoter. Since genes transcribed by RNA polymerase II are more important in practice, only promoters recognized by RNA polymerase II are described in this chapter. At least two types of eukaryotic promoters exist. The most common one has a consensus sequences GC(C/T)CAATCT located at 70 - 80 bp and TATA(A/T)A(T/A) located at 25 - 30 bp upstream from the transcription start site (Fig. 5-5). Since CAAT is pronounced "cat", the GC(C/T) CAATCT sequence is referred to as the "CAT" box. The other region is called the TATA box. The CAT and TATA boxes are functionally equivalent to the -35 and -10 boxes of prokaryotic promoters, respectively. The other type of eukaryotic promoter does not have CAT and TATA boxes but has several GC-rich sequences (Fig. 5-6). Since deletion or mutation of these sequences located upstream from the coding region of a gene disables the gene, the GC-rich regions are considered promoter elements of eukaryotic genes. In general, GC-rich promoters are found in genes that are always expressed, commonly referred to as constitutively expressed genes. Genes that are expressed only when they are induced usually have CAT and TATA boxes.

In contrast to prokaryotic genes, eukaryotic genes do not have transcription terminators. Their transcription termination is coupled to polyadenylation during which the newly synthesize mRNA is cleaved 10 - 30 bases downstream from the polyadenylation signal AAUAAA, thus terminating transcription.

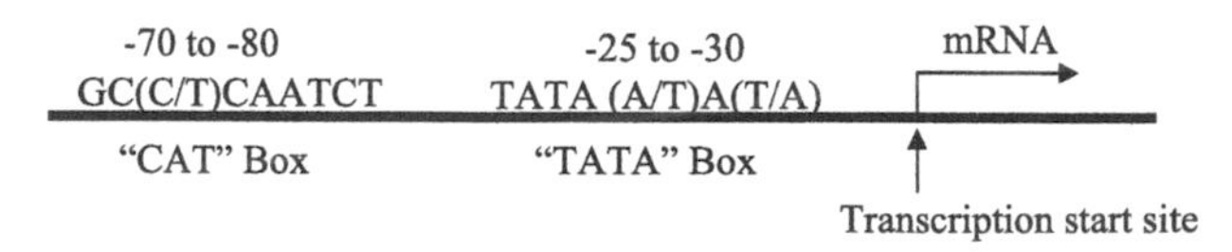

Fig. 5-5: Eukaryotic Promoter with CAT and TATA Boxes.

```
CGGAGGAAGATGGGTAAGAGAGTGAACCCTGTGGCGGCCGCAGCCGGAGAGGAACAGGA
      -420                    -400                    -380
ACCGCAGTGGACGCCCCTGGTCCCCGGGACGACGCGCAGTAGCGCCGGGAACTGGGTAC
      -360                    -340                    -320
CAGGGCGGGATGGGTGAGAGGCTCTAAGGGACAAGGCAGGGAGAAGCGCACGGGTGCGG
        -300                    -280                    -260
GGAACCAGCCCTCCCTTTGCCTCTGCTTCCCACCCCGAGGCGGCAGGGCGGGCGGGCAG
          -240                    -220                    -200
GTTCCGGGGGTGGGCGGGCTGGCGGGGCGGACGCGGGGGCCGACAGCTGGCTTCACCAG
            -180                    -160                    -140
CCTTCCGCCGTAGCAGAGCGAGCCGCGGCCAGCTCCGGCGGGCAGGGGGGCGCTGGAG
            -120                    -100                    -80
CGCAGCGCAGCGCAGCCCCATCAGTCCGCAAAGCGGACCGAGCTGGAAGTCGAGCGCTG
             -60                     -40                     -20
                 Met Gly Ala Gly Ala Thr Gly Arg Ala Met Asp
CCGCGGAGGCGGGCG ATG GGG GCA GGT GCC ACC GGC CGC GCC ATG GAC
                +1
Gly Pro Arg Leu Leu Leu Leu Leu Leu Leu Gly
GGG CCG CGC CTG CTG CTG TTG CTG CTT CTG GGG
```

Fig. 5-6: Eukaryotic Promoter with GC-rich Sequences. The boxed base is the transcription start site.

5.4: Elements Affecting Eukaryotic Gene Transcription

In addition to promoter, some eukaryotic genes require other regulatory elements to become fully functional. One such element is enhancer which is a short DNA sequence of approximately 15 bp (Table 5-1). The major function of an enhancer is to allow RNA polymerase to find the promoter more readily. A certain enhancer only affects the activity of the promoter located on the same DNA fragment. This phenomenon is called cis-acting. The function of an enhancer is position and orientation independent because it has the same function when it is placed upstream or downstream from the gene or in opposite orientations. However, enhancer by itself has no function. Enhancers express their function only when they are bound by transcription factors which attract the RNA polymerase to come and bind to the promoter to initiate transcription. Transcription factors are proteins and can bind to any gene that has the same or similar enhancer. Therefore, they are trans-acting and hence are commonly referred to as trans-acting factors. Some genes have other regulatory elements, including silencers whose function is opposite to that of enhancers.

Some enhancers are inducible. For example, the mouse mammary tumor virus (MMTV) enhancer is induced by glucocorticoid, and the metallothionine gene enhancer is induced by heavy metals such as Cd^{++}, Cu^{++}, Fe^{++}, and Zn^{++}. The human β-interferon gene enhancer becomes functional only when cells are infected by viruses or treated with poly-IC, and the heat shock gene enhancer is induced by fever. As mentioned above, enhancers themselves have no function. The inducers actually stimulate the production of trans-acting factors that bind to the enhancer to carry out their functions.

Table 5-1: Enhancer Sequences.

SV40	GGTGTGGAAAGTCC
Mo-MSV	TCTGTGGTAAGCAG
BKV	GTCATGGTTTGGCT
BPV	GGAGTGGTGTGTAC
Py (1)	CGTGTGGTTTTGCA
Py (2)	GGCCTGGAATGTTT
JC	AGTATGGATCCCTC
RSV	AAGGTGGTACGATC
Fe-SV	TCTGTGGTTAAGCA
MMTV	GACTTGGTTTGGTA
SNV	TTTGTGGGCTTGCT
SSV	CATGTGGGAGTTCT
Mouse Ig heavy chain	GCTGTGGTTTGAAG
HSV-tk	GGGCGGGTTTGTGT
Ad2-MP	AGGAAGGTGATTGG
Rabbit ß-globin	ACCCTGGTGTTGGC

Py: polyoma virus
SNV: Spleen necrosis virus
SSV: Simian sarcoma virus

5.5: Operon

In cells, proteins are made only when they are needed. Therefore, there must be mechanisms to turn genes on and off. Genes encoding proteins that are required for the function of a cell are called structural genes, and those that regulate the expression of structural genes are referred to as regulatory genes. Since a structural gene and its regulatory gene together is an operation unit in expression, this operational unit is called operon.

5.6: Lac Operon

Many operons exit in a cell. The first operon that was fully characterized is the *lac* operon. Lac stands for lactose which is a disaccharide composed of glucose and galactose (Fig. 5-7). The

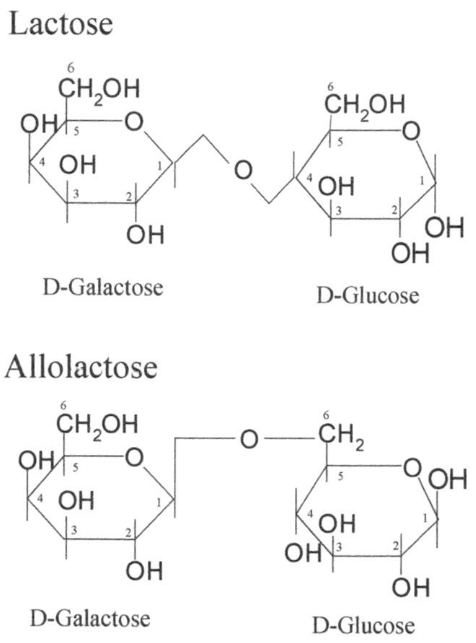

Lactose

D-Galactose D-Glucose

Allolactose

D-Galactose D-Glucose

Fig. 5-7: Lactose and Allolactose.

lac operon contains three structural genes designated Z, Y, and A (*lacZYA*), encoding β-galactosidase, β-galactoside permease, and β-thiogalactoside acetyl transferase, respectively (Fig. 5-8). β-galactosidase converts lactose to galactose and glucose. β-galactoside permease transports lactose from culture medium into the cell. β-thiogalactoside acetyl transferase transfers the acetyl group from acetyl-CoA to a β-galactoside; this enzyme, however, is not involved in lactose metabolism. These three different proteins are translated from the same mRNA. This type of mRNA is therefore called polycistronic mRNA. Some mRNAs encode only one protein and are called monocistronic mRNA. Most prokaryotic mRNAs are polycistronic, whereas most eukaryotic mRNAs are monocistronic. In *E. coli*, β-galactosidase is normally produced at a very low level. Its production is greatly increased when the cell is grown in a culture medium containing lactose as the only carbon source. If the culture medium contains glucose, *E. coli* turns off β-galactosidase production.

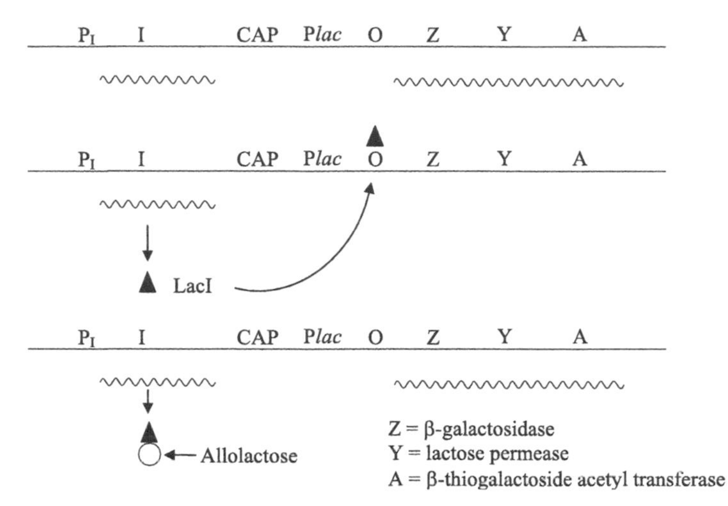

Fig. 5-8: *Lac* Operon

5.7: Suppression of Lac Operon

The protein that turns off lac operon is the *lac* repressor. It is encoded by the *lacI* gene; therefore, it is also called the LacI protein. The *lacI* gene is located immediately upstream from the *lacZYA* genes (Fig. 5-8). The *lac* repressor binds to a DNA element called *lac* operator located between the *lac* promoter and the transcription start site of *lacZYA* genes to suppress their expression. When *E. coli* is placed in a medium containing only lactose, it must produce β-galactosidase to convert lactose to galactose and glucose and, therefore, must turn on the *lac* operon. *E. coli* does so by first converting lactose to allolactose which binds and inactivates the *lac* repressor, rendering it unable to bind the *lac* operator. Without the *lac* repressor on the operator, the *lac* promoter becomes accessible to RNA polymerase which then binds and starts to transcribe the *lacZYA* genes to produce β-galactosidase and other proteins.

If glucose is added to this lactose-containing medium, the *lac* operon is turned off and stops producing β-galactosidase. Therefore, it appears that glucose suppresses β-galactosidase production.

Adenosine Monophosphate (AMP) cyclic AMP (cAMP)

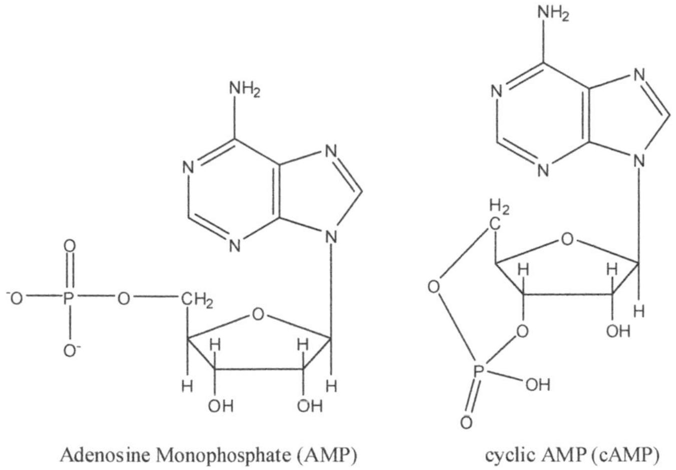

Fig. 5-9: AMP and cAMP.

Since glucose is a catabolite of lactose, this phenomenon is called catabolite repression. The mechanism by which glucose suppresses β-galactosidase production is related to cyclic AMP (cAMP) production (Fig. 5-9). The intracellular cAMP level is high when *E. coli* is grown in a medium containing no glucose but is low when it is grown in the presence of glucose. If cAMP is added to the glucose-containing medium, *E. coli* starts to produce β-galactosidase again, indicating that cAMP plays a major role in the regulation of β-galactosidase production. The enzyme responsible for making cAMP is adenylate cyclase. This enzyme is active only when it is associated with a phosphorylated regulatory protein, Rpr~P. When glucose is transported into the cell by glucose permease, glucose becomes phosphorylated with Rpr-P serving as the phosphate donor. The dephosphorylated Rpr is no longer able to bind to adenylate cyclase. Therefore, adenylate cyclase becomes inactive and unable to produce cAMP (Fig. 5-10). When lactose is transported into the cell, it does not dephophorylate Rpr~P; therefore, adenylate cyclase remains active to produce cAMP, and the cAMP level stays high.

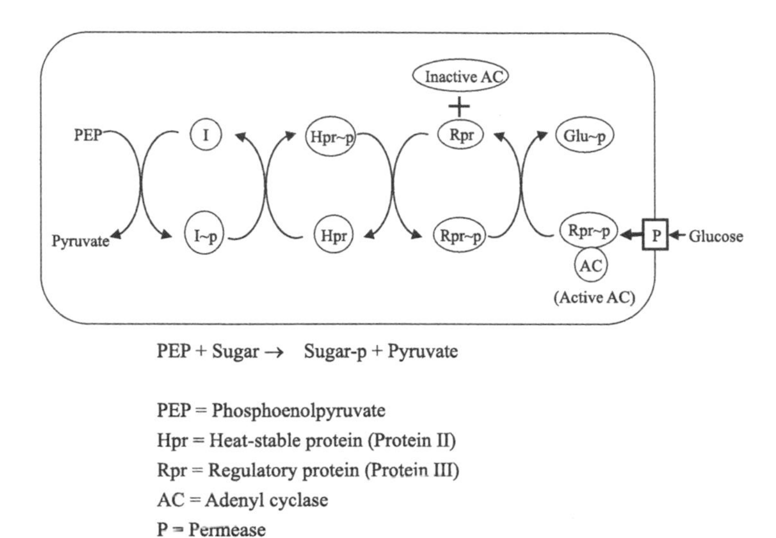

Fig. 5-10: Effect of Glucose on cAMP Production.

5.8: Activation of Lac Operon

The *lac* operon is repressed by the *lac* repressor under normal conditions. When lactose is converted to allolactose, the allolactose binds to the *lac* repressor and changes its conformation so that it cannot bind to the *lac* operator, allowing the *lac* promoter to become functional. This process is called de-repression or induction, and the inducer is allolactose or lactose. However, de-repression alone is not sufficient to activate *lac* operon. The *lac* operon also needs to be activated to become functional, and its activator is the cAMP and cAMP receptor protein (CRP) complex. CRP is also called catabolite activator protein (CAP). The cAMP-CRP (or cAMP-CAP) complex binds to a region upstream from the *lac* promoter and attracts RNA polymerase to bind to the promoter to initiate transcription and produce β-galactosidase. When glucose is present, adenylate cyclase is inactivated and no cAMP is produced to form cAMP-CRP complex. Without cAMP-CRP complex, the *lac* promoter is not activated and no β-galactosidase is produced.

Summary

Promoter is an essential element for transcription. A typical prokaryotic gene promoter is composed of two important sequences. The first, referred to as the -10 box, has a consensus sequence of TATAAT and is located approximately 10 bp upstream from the transcription start site. Since it was first discovered by Pribnow, it is also called the Pribnow box. The other sequence is TTGACA, referred to as the -35 box, because it is located approximately 35 bp upstream from the transcription start site. The -35 box is where RNA polymerase first binds to the template, and the -10 box is where the template starts to open for transcription. *E. coli* RNA polymerase is present in two forms: holoenzyme and core enzyme. The holoenzyme is composed of two α subunits and one each of β, β', ω, and σ subunit. The core enzyme lacks the σ subunit which is responsible for directing RNA polymerase to bind to the promoter. Once transcription is initiated, the σ subunit dissociates from the holoenzyme changing it to core enzyme which carries out the remaining task of the RNA synthesis.

Prokaryotic gene transcription terminates at terminators which usually have inverted repeat sequences. When the terminator region is transcribed, a stem-and-loop structure is formed. This structure slows down the movement of RNA polymerase on the template, allowing the Rho protein, which binds to and moves along the newly synthesized RNA, to catch up, bind, and inactivate RNA polymerase. Thus, the transcription is terminated. Since this type of terminator requires the Rho protein to help terminate transcription, it is called Rho-dependent terminator. The other type of terminator has a run of A's, and therefore, the newly synthesized RNA has a run of U's. Since the hydrogen bonds between these AU pairs are relatively weak, the RNA automatically falls off the template, and transcription terminates. This type of terminator does not require the Rho protein and is therefore called Rho-independent terminator.

There are two major types of eukaryotic gene promoters. One has the consensus sequence TATA(A/T)A(T/A), referred to as the TATA box, located at -25 to -30 and GC(C/T)CAATCT, referred to as the CAT box, located at -70 to -80 positions. The TATA box is functionally equivalent to the -10 box, and the CAT box is equivalent to the -35 box of prokaryotic promoters. The other type has several GC-rich sequences such as GGGCGGG. Some prokaryotic promoters also have enhancers. They are short DNA sequences of approximately 15 bp and are the locations where transcription factors bind to attract RNA polymerase to bind to the promoter. Some genes have silencers which function oppositely of enhancers. Eukaryotic genes do not have transcription terminators. Termination of eukaryotic transcription is coupled to polyadenylation during which a newly synthesized mRNA is cleaved 10 - 30 bases downstream from the polyadenylation signal AAUAAA, thus terminating transcription.

Genes that make proteins essential for the function of a cell are called structural genes, and those that regulate the expression of structural genes are regulatory genes. A structural gene and its regulatory gene together form an operation unit in transcription. This unit is called operon. The most well known operon is the *lac* operon which has three structural genes including *lacZ*, *lacY*, and *lacA* encoding β-galactosidase, β-galactoside permease, β-thiogalactoside acetyl transferase, respectively. These proteins are made only when *E. coli* is grown in a medium containing only lactose. Therefore, the *lac* operon is normally inactive due to suppression by the *lac* repressor which binds to the operator of *lac* operon thus preventing RNA polymerase from binding to the promoter to initiate transcription. When *E. coli* must use lactose as the carbon source, it first converts lactose to allolactose which then binds to the *lac* repressor making it unable to bind to the operator. RNA polymerase then binds to the promoter to transcribe *lacZYA* genes. This process is called induction or de-repression. The β-galactosidase thus produced converts lactose to glucose and galactose that are then used to generate energy and other essential substances in the cell. If the culture medium contains glucose, *E. coli* will not make β-galactosidase. Since glucose is a catabolite of lactose, this phenomenon is called catabolite repression. This is due to inactivation of the adenylate cyclase when glucose is transported into the cell and becomes phosphorylated. The phosphate group which phosphorylates glucose is from the regulatory protein which normally binds to adenylate cyclase and maintains it in an active form. Dephosphorylation of this regulatory protein renders adenylate cyclase inactive, and thus no cAMP is produced to form the cAMP-CRP complex which is required to activate the *lac* promoter.

Sample Questions

1. The -10 region of a prokaryotic promoter is: a) the first place where RNA polymerase makes contact with the template, b) the first place where template starts to open to become single stranded, c) the first place where transcription starts, d) the first place where Rho factor binds.
2. The -35 region of a prokaryotic promoter is: a) the first place where RNA polymerase makes contact with the template, b) the first place where template starts to open to become single stranded, c) the first place where transcription starts, d) the first place where Rho factor binds.
3. Which of the following is the consensus sequence of the -10 (Pribnow) box of prokaryotic promoters: a) GC[C/T]CAATCT, b) TATAAT, c) TATA[A/T]A[T/A], d) TTGACA
4. Which of the following is the consensus sequence of the -35 box of prokaryotic promoters: a) GC[C/T]CAATCT, b) TATAAT, c) TATA[A/T]A[T/A], d) TTGACA
5. Which of the following is the consensus sequence of the TATA box of eukaryotic promoters: a) GC[C/T]CAATCT, b) TATAAT, c) TATA[A/T]A[T/A], d) TTGACA
6. Which of the following is the consensus sequence of the CAT box of eukaryotic promoters: a) GC[C/T]CAATCT, b) TATAAT, c) TATA[A/T]A[T/A], d) TTGACA
7. Which of the following is required for initiation of transcription in prokaryotic cells: a) RNA polymerase core enzyme, b) RNA polymerase holoenzyme, c) Rho factor, d) F factor
8. Which of the following is the holoenzyme of *E. coli* RNA polymerase: a) $\alpha_2\beta\beta'\omega\sigma$, b) $\alpha_2\beta\beta'\omega$, c) $\alpha_3\beta\beta'\sigma$, d) $\alpha_3\beta\beta'$

9. Which of the following is the core enzyme of *E. coli* RNA polymerase: a) $\alpha_2\beta\beta'\omega\sigma$, b) $\alpha_2\beta\beta'\omega$, c) $\alpha_3\beta\beta'\sigma$, d) $\alpha_3\beta\beta'$
10. Which of the following sequence could be a transcription terminator in prokaryotic cells: a) ATTAAAGGCTCCTTTTGGAGCCTTTTTTT, a) TGAGTACCATCTGCT, b) AACGTCAAGGGTGC, d) GTAGCGTAGACTGGCTTAGGC
11. Which of the following sequence could be a transcription terminator in eukaryotic cells: a) ATTAAAGGCTCCTTTTGGAGCCTTTTTTT, a) TGAGTACCATCTGCT, b) AACGTCAAGGGTGC, d) none of above
12. Which of the following are true for enhancers: a) cis acting, b) trans acting, c) position independent, d) orientation independent
13. Polycistronic mRNA means a mRNA with: a) multiple introns, b) multiple exons, c) the potential to produce multiple peptides, d) multiple enhancers
14. For the Lac operon to become functional, it must be: a) de-repressed and activated, b) induced by glucose, c) induced by galactose, d) induced by sucrose
15. Which of the following can induce the Lac operon: a) glucose, b) galactose, c) lactose, d) IPTG
16. Which of the following is the activator of the Lac operon: a) glucose, b) galactose, c) cAMP, d) cAMP-CRP complex
17. Glucose inhibits the expression of Lac operon by: a) inactivating glucose permease, b) inactivating adenyl cyclase, c) inhibiting β-galactosidase, d) competing with lactose for lactose permease

Suggested Readings

1. Adhya, S. and Gottesman, M. (1978). Control of transcription termination. Annu. Rev. Biochem. 47: 967-996.
2. Blackwood, E. M. and Kadonaga, J. T. (1998). Going the distance: a current view of enhancer action. Science 281: 60-63.
3. Botsford, J. L. and Hartman, J. G. (1992). Cyclic AMP in prokaryotes. Microbiol. Rev. 56: 100-122.
4. Brennan, C. A., Dombroski, A. J., and Platt, T. (1987). Transcription termination factor Rho is an RNA-DNA helicase. Cell 48: 945-952.
5. Butler, J. E. and Kadonaga, J. T. (2002). The RNA polymerase II core promoter: a key component in the regulation of gene expression. Genes Dev. 16: 2583-2592.
6. Das, A. (1993). Control of transcription termination by RNA-binding proteins. Annu. Rev. Biochem. 62: 893-930.
7. Gilbert, W. and Muller-Hill, B. (1967). The lac operator is DNA. Proc. Natl. Acad. Sci. USA 58: 2415-2421.
8. Gourse, R. L., Ross, W., and Rutherford, S. T. (2006). General pathway for turning on promoters transcribed by RNA polymerases containing alternative σ factors. J. Bacteriol. 188: 4589-4591.
9. Helman, J. D. and Chamberlin, M. (1988). Structure and function of bacterial sigma factors. Annu. Rev. Biochem. 57: 839-872.
10. Ishihama, A. (2000). Functional Modulation of *Escherichia coli* RNA Polymerase. Annu. Rev. Microbiol. 54:499-518.

11. Jacob, F. and Monod, J. (1961). Genetic regulatory mechanisms in the synthesis of proteins. J. Mol. Biol. 3: 318-389.

12. Kolb, A. (1993). Transcriptional regulation by cAMP and its receptor protein. Annu. Rev. Biochem. 62: 749-795.

13. Matzke, M., Matzke, A. J., and Kooter, J. M. (2001). RNA: guiding gene silencing. Science 293: 1080-1083.

14. Meadow, N. D., Fox, D. K., and Roseman, S. (1990). The bacterial phosphoenolpyruvate: glucose phosphotransferase system. Annu. Rev. Biochem. 59: 497-542.

15. Mekler, V., Kortkhonjia, E., Mukhopadhyay, J., Knight, J., Revyakin, A., Kapanidis, A. N., Niu, W., Ebright, Y. W., Levy, R., and Ebright, R. H. (2002). Structural organization of bacterial RNA polymerase holoenzyme and the RNA polymerase-promoter open complex. Cell 108: 599-614.

16. Muller, M. M., Gerster, T., and Schaffner, W. (1988). Enhancer sequences and the regulation of gene transcription. Eur. J. Biochem. 176: 485-495.

17. Orphanides, G., Lagrange, T., and Reinberg, D. (1996). The general transcription factors of RNA polymerase II. Genes Dev. 10: 2657-2683.

18. Pribnow, D. (1975). Nucleotide sequence of an RNA polymerase binding site at an early T7 promoter. Proc. Natl. Acad. Sci. USA 72: 784-788.

19. Pribnow, D. (1975). Bacteriophage T7 early promoters: nucleotide sequences of two RNA polymerase binding sites. J. Mol. Biol. 99: 419-443.

20. Reinhart, B. J., Weinstein, E. G., Rhoades, M. W., Bartel, B., and Bartel, D. P. (2002). MicroRNAs in plants. Genes Dev. 16: 1616-1626.

21. Richardson, J. P. (1996). Structural organization of transcription termination factor Rho. J. Biol. Chem. 271: 1251-1254.

22. Sharp, P. A. (2001). RNA interference - 2001. Genes Dev. 15: 485-490.

23. Tijsterman, M., Ketting, R. F., and Plasterk, R. H. (2002). The genetics of RNA silencing. Annu. Rev. Genet. 36: 489-519.

24. Woychik, N. A. and Hampsey, M. (2002). The RNA polymerase II machinery: structure illuminates function. Cell 108: 453-463.

Chapter 6

Electrophoresis

Outline

Types of electrophoresis
Electrophoresis of RNA
Resolution limits of electrophoresis gels
Pulsed-field gel electrophoresis
Preparation of samples for PFGE
Factors affecting the efficiency of PFGE

6.1: Types of Electrophoresis

Electrophoresis is used to isolate DNA fragments, determine the size of a DNA or RNA molecule, sequence DNA, and assess the efficiency of enzymatic digestions. Two different substances, agarose and polyacrylamide, are used to make gels for electrophoresis. Polyacrylamide for electrophoresis of DNA is usually made of acrylamide and bis-acrylamide at a ratio of 29 to 1. Agarose gel electrophoresis is normally done with a horizontal gel apparatus, while polyacrylamide gel electrophoresis is performed with a vertical gel apparatus. Since DNA and RNA molecules are negatively charged, they migrate from cathode (- electrode) to anode (+ electrode) during electrophoresis. For agarose gel DNA electrophoresis, TAE (40 mM Tris-base, 20 mM NaOAc,1 mM EDTA, pH 7.2) or TBE (89 mM Tris-base, 89 mM boric acid, 2 mM EDTA, pH 8.3) buffer is used. Sharper DNA bands are produced if TAE buffer is used for electrophoresis; however, TAE buffer can only be used once, whereas TBE buffer can be reused multiple times. Although TBE has a better buffer capacity than TAE, the boric acid in TBE may cross link DNA to agarose, making it difficult to recover from the gel after electrophoresis. For polyacrylamide gel electrophoresis, TBE buffer is used.

6.2: Electrophoresis of RNA

Almost all RNA molecules have complicated coil-like secondary structures. In order to accurately determine the size of an RNA molecule, these secondary structures must be resolved to make RNA a linear molecule. This process is called denaturation. Chemical agents that can be used to denature RNA include methylmercury, glyoxyal, and combination of formamide and formaldehyde. Methylmecury is very toxic, and glyoxyal is relatively expensive. Therefore, the most commonly used agents are formamide and formaldehyde. Formamide weakens hydrogen bonds that hold RNA in its secondary structure, while formaldehyde cross links adjacent adenine bases so that the denatured RNA does not reform its secondary structure (Fig. 6-1). To

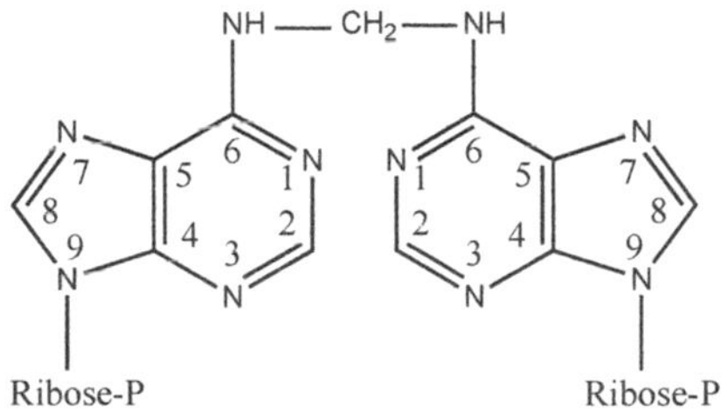

Fig. 6-1: Cross linking of two adjacent adenines by formaldehyde.

denature RNA prior to electrophoresis, formamide and formaldehyde are added to an RNA sample to final concentrations of 20% and 10%, respectively. The sample is then heated at 70°C for 10 minutes. Since RNA is readily degraded in an alkaline solution, electrophoresis of RNA samples is done in a buffer with a neutral pH (~7.0). Therefore, MOPS [3-(N-morpholino) propanesulfonic acid, sodium salt] instead of Tris buffer is used because Tris has very poor buffer capacity at a neutral pH. To maintain the RNA in a denatured state during electrophoresis, formaldehyde is added to electrophoresis buffer to a final concentration of 10% (2.2M); where the concentrated 37% formaldehyde is considered as 100%.

6.3: Resolution limits of Electrophoresis gels

Various concentrations of both agarose and polyacrylamide gels are used. Concentrations of agarose gels usually range from 0.5 to 2% and can separate DNA fragments from approximately 200 bp to 50 kb. Lower agarose concentrations are used to separate larger DNA fragments. Polyacrylamide gels are usually used at concentrations between 4 and 8% and can separate DNA fragments from 50 to 1000 bp. Polyacrylamide gels (6-15%) containing high concentration (8M) of urea are used for DNA sequencing.

6.4: Pulsed-field Gel Electrophoresis

To separate DNA fragments greater than 50 kb, pulsed-field gel electrophoresis (PFGE) is used. PFGE separates large DNA fragments by frequently changing the direction of migration during electrophoresis (Fig. 6-2). The original PFGE, orthogonal field alternation gel electrophoresis (OFAGE), has one cathode located at one end and two anodes located at approximately 45° angles at both sides of the other end of the gel apparatus. DNA sample is first allowed to run toward one anode for a short period of time (45 seconds to 1.5 minutes) by turning on the cathode and one of the two anodes. To change the direction of migration, this anode is turned off and the other anode located at opposite side of the gel apparatus is turned on. This on and off switching of these two anodes continues until the electrophoresis is done. Although this method can separate DNA fragments as big as 800 kb, only the samples loaded in the wells in the central portion of the gel run straight from top to bottom of the gel. Those that are loaded in the wells in the two sides of the gel run in a curved path at approximately a 45° angle, making it difficult to discern DNA bands. The contour clamped homogenous electric field (CHEF) apparatus is an improved PFGE apparatus. It has multiple cathodes and anodes, and a group of cathodes and anodes are turned on for each direction of migration. Therefore, distribution of the electric field is much more even throughout the gel than that of OFAGE, and DNA samples run straight from top to bottom throughout the gel. Field inversion gel electrophoresis (FIGE) is another type of PFGE which allows DNA samples to run back and forth in a gel. This method can separate DNA fragments up to approximately 500 kb. Rotating field gel electrophoresis places the gel on a turn table located between a fixed cathode and anode. Changes in migration direction are achieved by rotating the turn table left to right and right to left. CHEF is the most commonly used method at present although the FIGE and RFGE apparatuses are less expensive.

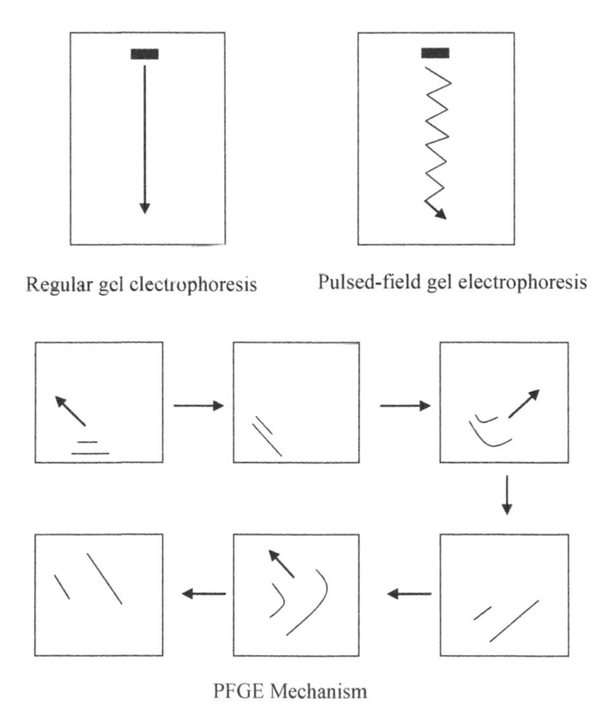

Fig. 6-2: Mechanism of PFGE.

6.5: Preparation of DNA Samples for PFGE

The most critical step in PFGE is the sample preparation. Since the purpose of PFGE is to separate big DNA fragments or even individual chromosomes, it is critical that the DNA stays intact during sample preparation. To minimize the mechanical force applied to DNA during sample preparation, cells are embedded in low melting point agarose in a cubic mold. The small agarose cubes thus formed are incubated in ESP solution (0.5 M EDTA, 1% sarcosyl, 1 mg/ml proteinase K) in which sarcosyl dissolves cell membranes and proteinase K digests proteins. The DNA released stays in the agarose cube and therefore is not subjected to any mechanical force. Restriction enzyme digestion can also be performed within the agarose cube by incubating it in a solution containing a certain restriction enzyme. After digestion, the agarose cubes are inserted into wells of a gel and electrophoresed.

6.6: Factors affecting the efficiency of PFGE

Many factors may affect the efficiency of PFGE including types and concentrations of agarose. Temperature, voltage, switch interval, and the buffer used for electrophoresis may also affect PFGE. PFGE generates an enormous amount of heat; therefore, all PFGE systems have cooling devices to control the temperature during electrophoresis. Some PFGE systems are computer controlled with preset conditions for optimal results.

Summary

Electrophoresis can be used to determine the length of a DNA or RNA molecule or the concentration of a DNA or RNA sample. It can also be used to determine the number of chromosomes of a microorganism or to verify completion of a restriction enzyme digestion. Electrophoresis can also be used to isolate DNA fragments. Agarose and polyacrylamide are used to make gels for electrophoresis of DNA or RNA. Polyacrylamide gels are used for separation of DNA fragments smaller than 500 bp, and agarose gels are used to separate DNA fragments approximately 500 bp to 50 kb. Vertical gel apparatus is used for polyacrylamide gel electrophoresis, and horizontal apparatus is used for agarose gels. The buffer used for electrophoresis of DNA is usually slightly alkaline such as TBE (Tris-borate-EDTA, pH 8.3) and TAE (Tris-acetic acid-EDTA, pH 7.2). Electrophoresis of RNA must be done under a neutral pH; therefore, MOPS [3-(N-morpholino) propanesulfonic acid, sodium salt] buffer is used. RNA needs to be denatured to become a linear molecule before electrophoresis. The most commonly used denaturant for RNA is combination of formamide and formaldehyde.

Different concentrations of agarose or acrylamide are used for electrophoresis of different lengths of DNA or RNA molecules. Commonly used agarose concentrations range from 0.5 to 2%, and polyacrylamide concentrations range from 4 to 8%. The polyacrylamide gel used for DNA sequencing is usually 6% containing 8M urea. For separation of DNA fragments greater than 50 kb, pulsed-field gel electrophoresis (PFGE) is performed. In PFGE, DNA is forced to constantly change its direction of migration in order to achieve separation. Since the purpose of PFGE is to separate big DNA fragments, DNA is kept intact by embedding cells in agarose during sample preparation. The agarose cubes formed are then incubated in a solution containing detergent to dissolve cell membranes and proteinase K to digest proteins. If cell walls are present, the cells are first digested with lysozyme (for bacteria) or zymolase and lyticase (for yeasts). Since many different factors including type of gel apparatus, gel concentration, temperature, voltage, switch interval, and the buffer may affect the efficiency of PFGE, many trials may be necessary to obtain an optimal result.

Sample Questions

1. Gel electrophoresis can be used to a) determine the length of DNA, b) determine the

length of RNA, c) determine the concentration of DNA, d) isolate DNA fragments.

2. Pulsed-field gel electrophoresis separates large DNA fragments by a) using very high voltage for electrophoresis, b) running a very long gel, c) running electrophoresis for very long time, d) frequent change of migration direction.
3. Which of the following can be used to isolate a 5-kb DNA fragment: a) 10% polyacrylamide gel, b) 1% agarose gel, c) 1% polyacrylamide gel, d) 10% agarose gel.
4. Which of the following can be used to denature RNA: a) methylmercury, b) glyoxal, c) combination of formamide and formaldehyde, d) phenol.
5. Which of the following is most commonly used to denature RNA: a) methylmercury, b) glyoxal, c) combination of formamide and formaldehyde, d) phenol.
6. The efficiency of pulsed-field gel electrophoresis may be affected by: a) switch interval, b) voltage gradient, c) temperature, d) agarose type and concentration.
7. Preparing DNA for pulsed-field gel electrophoresis is usually done by a) conventional phenol-chloroform extraction method, b) salt-chloroform extraction method, c) embedding cells in agarose gel and lyse cells and digest proteins in situ, d) alcohol precipitation.

Suggested Readings

1. Schwartz, D. C. and Cantor, C. R. (1984). Separation of yeast chromosome-sized DNAs by pulsed field gradient gel electrophoresis. Cell 37: 67-75.

Chapter 7

Cloning

Outline

Required elements of cloning vectors
pBR322 and pUC19 vectors
Preparation of DNA fragments for cloning
Determination of quality and quantity of DNA fragments
Determination of insert to vector ratio
Spin dialysis
Transformation
Screening of transformants
Ligation of insert with vector
Prevention of vector self ligation
Use of tailing, adaptors and linkers
E. coli genotypes

7.1: Required elements of cloning vectors

The purpose of cloning is to make a specific DNA, RNA, or protein. The first step in achieving this goal is to replicate the DNA fragment containing the gene of interest. A DNA must have a replication origin in order to replicate. However, most DNA fragments containing genes of interests do not have replication origins and cannot replicate by themselves. The way to replicate these DNA fragments is to link them to another piece of DNA that has a replication origin, and then introduce the recombinant DNA into *E. coli* to replicate. This kind of DNA is called cloning vector, and the process of linking two pieces of DNA fragments or two ends of a DNA fragment together is called ligation. Two types of cloning vectors, plasmid and bacteriophage, exist.

Plasmids are extrachromosomal genetic elements. They can replicate autonomously in bacteria because they all have replication origins. In addition to a replication origin, a plasmid must have unique restriction sites and at least two selection markers in order to become a useful cloning vector. Having a unique restriction site means that a certain restriction enzyme cuts the vector only once. An ideal vector has several unique restriction sites so that it can accept DNA fragments isolated by digestion with various restriction enzymes. To clone a DNA fragment, the vector is first cut with a certain restriction enzyme to become a linear DNA molecule. The DNA fragment containing the gene of interest is then ligated to the linearized vector, and the recombinant plasmid DNA is introduced into *E. coli* to replicate. Since *E. coli* cannot replicate linear DNA, only the plasmid DNAs that become circular during ligation can replicate in *E. coli*. When a plasmid is introduced into an *E. coli*, the phenotype of the *E. coli* may be changed; therefore, the process in which a DNA is introduced into *E. coli* is referred to as transformation. Since no transformation method is 100% efficient at getting plasmid DNA into *E. coli*, two types of *E. coli* transformants are present after transformation. One contains the plasmid; the other does not have the plasmid. The former is the desired one. Therefore, a method to distinguish between these two types of transformants is required. If the plasmid contains an antibiotic resistance gene, such as the *bla* gene which encodes the β-lactamase, the *E. coli* that has acquired this plasmid will become resistant to ampicillin or its derivatives. Therefore, an antibiotic resistance gene can be a selection marker for a vector.

When a DNA fragment is linked to a plasmid vector, two types of plasmids are generated. One contains the insert, and the other contains no insert. The plasmid containing no insert is the vector that has its two ends religated back together. To successfully isolate the plasimid with the insert, another selection marker is required. If a plasmid contains two selection markers such as ampicillin and tetracycline resistance genes and the DNA fragment is inserted into the tetracycline resistance gene, the *E. coli* cells that have taken up the vector will be resistant to both ampicillin and tetracycline, and those that have taken up the recombinant plasmid will be resistant to ampicillin but sensitive to tetracycline. In this manner, cells containing the desired recombinant plasmid are selected.

7.2: pBR322 and pUC19 vectors

Plasmid pBR322 is the first vector developed for cloning (Fig. 7-1). Its replication origin is derived from a plasmid encoding colicin E1 and is therefore called the colE1 origin. The two selection markers of pBR322 are ampicillin and tetracycline resistance genes. If a DNA fragment is inserted into its PstI site, *E. coli* cells containing the recombinant plasmid are sensitive to ampicillin but resistant to tetracycline. On the other hand, *E. coli* cells will be sensitive to tetracycline but resistant to ampicillin if they contain the recombinant plasmid with the DNA fragment inserted into its ClaI, HindIII, EcoRV, BamHI, SalI, HincII, or AccI site.

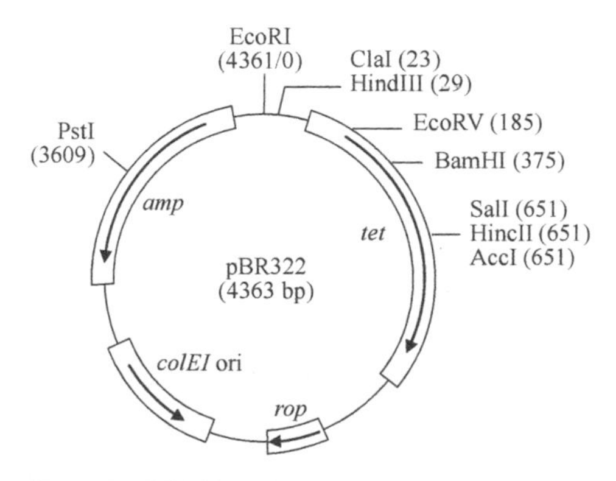

Fig. 7-1: pBR322.

Selection markers do not have to be antibiotic resistance genes; they can also be genes of certain enzymes. The most commonly used one is the gene encoding the β-galactosidase, which is the *lacZ* gene. pUC19 is one of the pUC series of plasmid vectors containing the *lacZ* gene as a selection marker; the second selection marker of pUC plasmids is the ampicillin resistance gene with its original PstI site mutated. *E. coli* cells containing pUC19 will produce β-galactosidase which not only can convert lactose to galactose and glucose but also 5-Br-4-Cl-3-Indolyl-β-D-galactopyranoside (X-gal) to 5-Br-4-Cl-3-Indol. Since 5-Br-4-Cl-3-Indol is blue, it will turn colonies of *E. coli* containing pUC plasmids blue when they are grown on agar plates containing X-gal. If a DNA fragment is inserted into the *lacZ* gene, *E. coli* cells containing the recombinant plasmid will no longer able to make β-galactosidase to convert X-gal to 5-Br-4-Cl-3-Indol; therefore, their colonies are colorless.

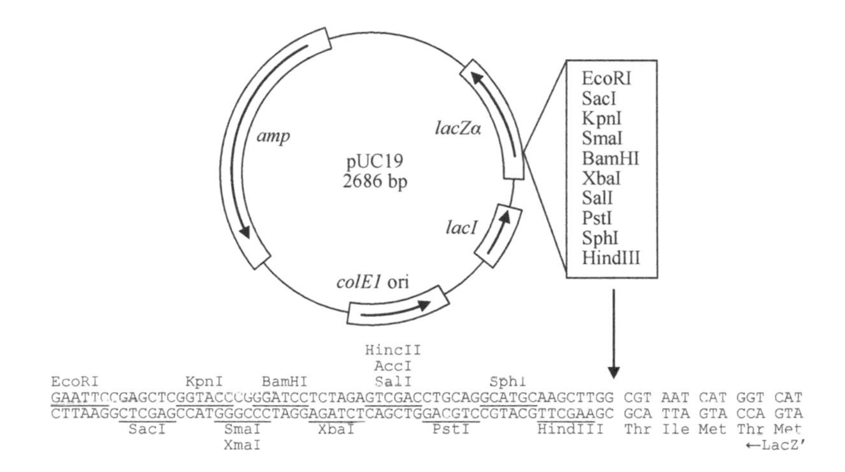

Fig. 7-2: pUC19. The sequence of multiple cloning sites is shown below the diagram. Underlined nucleotide sequences in the multiple cloning sites are recognition sequences of certain restriction enzymes.

The *lacZ* gene is regulated by *lacI*, and its expression is normally suppressed by the *lac* repressor. To enable cells containing pUC plasmids or their derivatives to produce β-galactosidase, an inducer such as lactose is needed to inactivate the *lac* repressor. However, lactose can be metabolized by β-galactosidase to become galactose and glucose. Glucose in turn inhibits cAMP production and thus no cAMP-CRP is available to maintain the *lac* operon in an active state to produce sufficient amounts of β-galactosidase to convert X-gal to 5-Br-4-Cl-3-Indol. Therefore, a non-metabolizable inducer is used to induce the *lac* operon in this cloning system. The most commonly used one is isopropyl-thio-β-galactoside (IPTG) which can bind to and inactivate the *lac* repressor but can not be metabolized by β-galactosidase. Therefore, the *lac* operon will stay on after induction and produce β-galactosidase.

Since selection by color is much easier than that by antibiotic sensitivity, pUC vectors are much more popular than pBR322. Another reason for their popularity is the presence of more unique cloning sites (Fig. 7-2). They contain an oligonucleotide with many restriction sites, referred to as multiple cloning sites (MCS). In addition, the copy number of pUC plasmids is at least 10 times higher than that of pBR322 which has a copy number of approximately 20. This is because pUC vectors do not have the *rop* (repressor of primer) gene which is also called *rom* (RNA I modulator). RNA I is one of the two small RNA molecules involved in pBR322 replication. The other RNA molecule is RNA II (555 nucleotides) which is processed to become the primer for replication of pBR322. This processing is carried out by RNase H which removes a portion of the 5' region of RNA II. Since RNA I (108 nucleotides) and RNA II are transcribed from the same region in opposite orientations and overlap at their 5' ends, RNA I will bind to RNA II when it is produced. This binding inhibits RNA II from being processed, making it unable to serve as the primer for replication; therefore, pBR322 stops replicating. The Rop protein enhances the binding between RNA I and RNA II. Without Rop, RNA I does not bind to RNA II efficiently, and therefore, pUC vectors have a much higher copy number than pBR322.

When ligating a DNA fragment with a vector, many different products are generated, including those with multiple vectors and a vector with multiple inserts in different orientations. However, only one product is present in *E. coli* after transformation. This is due to a phenomenon called plasmid incompatibility, meaning that two different plasmids with the same replication origin cannot coexist in the same *E. coli* cell. Plasmid incompatibility is related to control of plasmid replication. During transformation, an *E. coli* cell may take up two or more different plasmids. If these plasmids have the *colE1* replication origin, the RNA I produced from one plasmid will inhibit the replication of the other plasmid. Therefore, the copy numbers of these two plasmids are different. After several generations of cell division, the one which replicates slower is lost. Thus, only one type of plasmid exists in an *E. coli* cell. Cloning is possible mainly because of plasmid incompatibility,

7.3: Isolation of DNA Fragments for Cloning

If the gene of interest is located in a DNA fragment between two EcoRI sites, the gene can be isolated by digesting the DNA with EcoRI. The digested DNA is then electrophoresed in an agarose gel, and the DNA band containing the gene of interest is isolated. Many methods can be used to elute a DNA fragment out of agarose gel. Electro elution uses electrophoresis to do so. An agarose gel slice containing the DNA fragment is placed in a dialysis bag filled with buffer. The dialysis bag is then placed in an electrophoresis apparatus, and the DNA is electrophoresed out of the gel and onto the wall of the dialysis bag. The DNA is then dislodged from the wall of the dialysis bag by electrophoresing the bag in an opposite direction. The buffer containing the eluted DNA is transferred to a tube, and the DNA is precipitated with salt and alcohol. Another method is to cut a slit in front of the DNA band in the agarose gel. A DEAE-cellulose paper (NA-45) the same length as the DNA band is inserted into the slit, and the DNA is electrophoresed onto the NA-45 paper. The paper is then placed in a test tube containing 1.5 M NaCl. After heating at 65°C for 30 minutes, the NA-45 paper is removed, and ethanol is added to precipitate the eluted DNA.

If the agarose gel used to electrophorese DNA is made of low melting point agarose, the gel slice containing the DNA band is placed in a tube containing water (2-3 volumes of the gel slice) and then heated at 60°C for 5 - 10 minutes to dissolve. The DNA is then precipitated with salt and alcohol. Although this method is very efficient, it is not commonly used because low melting agarose is quite expensive. A very simple method is to wrap the gel slice in a paraffin paper and allow it to freeze in a -70°C freezer. The DNA is then squeezed out of the gel. This method recovers only approximately 50% of the DNA, but is commonly used if high concentration of the DNA is present in the gel slice and very little DNA is needed. The most popular method at present is to dissolve the gel slice containing the DNA fragment with a certain chemical such as NaI. The DNA is then bound to a certain type of resin and eluted from the resin with water after washing. Most commercial DNA elution kits use this method.

7.4: Determination of Quantity and Quality of DNA Fragments

For cloning, the ratio of the number of insert molecules to that of vector molecules is critical. Therefore, it is necessary to quantify the DNA eluted from the gel slice. A very popular method for DNA quantification is to use a spectrophotometer to determine the OD_{260} value of the DNA. If the OD_{260} value is 1, the concentration is 50 μg/ml for a double-stranded DNA sample. DNA concentration can also be determined by fluorometry. Bisbenzimidazole (Hoechst 33258) is added to a DNA sample to a final concentration of 0.1 μg/ml. The DNA sample is then excited with 365 nm light, and the amount of 456 nm light emitted is measured. The measured light value is then compared with those of a standard curve of a DNA sample whose concentration is known to determine the concentration of the unknown sample.

Both spectrophotometry and fluorometry only determine the quantity of DNA samples; they do not accurately determine the quality of DNA. The most practical method is electrophoresis. The DNA sample is electrophoresed in the same gel with a DNA sample of known concentration such as a HindIII digested lambda DNA (λ/HindIII). λ/HindIII contains 0.5, 2.0, 2.2, 4.0, 6.0, 9.6, and 23 kb fragments. If the intensity of the unknown DNA band is the same as that of the 4.0 kb λ/HindIII band, the concentrations of these two DNA bands are the same. If 100 ng of λ/HindIII is loaded into the gel for the electrophoresis, the concentration of the 4.0 kb band is 100 ng x (4/50) = 8 ng since the entire λ genome is approximately 50 kb. Therefore, the unknown DNA sample is also 8 ng. If the DNA sample is degraded, a smear instead of a distinct DNA band is observed. This DNA sample would not be suitable for cloning because it contains many DNA fragments of varying sizes.

7.5: Determination of Insert to Vector Ratio

The optimal insert to vector ratio for cloning is 2:1, and the easiest way to determine the insert to vector ratio is electrophoresis. If a 3.0-kb vector band and a 1.5-kb insert band have the same intensity on the gel after electrophoresis, the insert to vector ratio is 2:1. The most commonly used enzyme to ligate DNA fragments together is the T4 DNA ligase. The 10x ligation buffer contains 5 mM ATP, 100 mM dithiothreitol, 300 mM Tris (pH 7.5), and 100 mM $MgCl_2$. The optimal reaction temperature for T4 DNA ligase is 16°C. Although it also works at 37°C, it loses its activity rather quickly. ATP is required for the ligation reaction. If the ligation reaction contains dATP or phosphate, the reaction may not work. High salt concentration also inhibits the ligation reaction. Therefore, DNA samples used for cloning are usually subjected to purification before ligation. The easiest way to purify a small amount of a DNA sample is through spin dialysis.

7.6: Spin Dialysis

Spin dialysis is a small-scale column chromatography. A column can be prepared using an Eppendorf tube. The attached lid is first removed, and a hole is made at the bottom of the tube with a 20-gauge needle. A drop of siliconized fine glass beads of approximately 500 μm in diameter is placed in the tube to plug the hole. 0.5 – 1 ml of a 60% suspension of an appropriate resin is then added. The tube is placed on another Eppendorf tube, and the whole column apparatus is centrifuged in a table top centrifuge with a swing bucket at approximately 200 x g for 1 minute to remove the excess liquid. The DNA sample to be purified is then applied to the column, which is placed on another tube, and centrifuged again at 200 x g for 3 minutes. The solution thus collected in the bottom tube should contain purified DNA. Substances such as salt, protein, polysaccharide, and lipid are retained in the column inside the resin. Since DNA is too big to enter the resin, it flows around the resin and through the column during the centrifugation.

To ensure that the spin dialysis works, an indicator dye is mixed with DNA sample. A commonly used DNA loading solution called 5x stop mix can be used. This solution contains 2% SDS, 25% glycerol, 0.05% bromophenol blue, and 0.05% xylene cyanol, of which bromophenol blue is blue and xylene cyanol is dark green. A 40 μl DNA solution is mixed with 10 μl of the 5x stop mix and then heated at 70°C for 15 minutes prior to spin dialysis. If the solution collected after spin dialysis is blue, bromophenol blue and xylene cyanol are not removed. Since salt has a similar molecular weight as these dyes, salt and other small molecules are also not removed. This spin dialysis, therefore, did not work.

The volume of the solution collected after spin dialysis can also be used to judge whether a spin dialysis has worked. The volume of the solution recovered should be the same as the original input volume. If larger, excess water from the resin was not removed during the first centrifugation. If smaller, the first centrifugation may have been too long or too high in speed; and the liquid inside the resin is removed, making the resin too dry. This resin will absorb water from the DNA sample and trap the DNA in the resin. Some proteins may bind to DNA. Since the 5x stop mix solution contains SDS, addition of it to the DNA sample and heating the sample at 70°C for 15 minutes usually dissociates proteins from DNA. These proteins can then be removed from the DNA sample by spin dialysis.

Many different types of resins can be used for spin dialysis including Sephadex G50, Sepharose CL-6B, Sephacryl S-100, and Sephacryl S-400. These resins are used to purify DNA fragments greater than 100 bp. For DNA molecules smaller than 100 bp, Sephadex G25 is used. Sephadex is sold in powder form and needs to be suspended in water or buffer overnight before use. Sepharose CL-6B, Sephacryl S-100, and Sephacryl S-400 are in suspension form and usually contain sodium azide to prevent microorganisms from growing. It is washed with sterile water to remove sodium azide before use and is usually made as a 60% suspension.

7.7: Transformation

After ligating the insert with the vector, the recombinant plasmid is introduced into *E. coli* to replicate by transformation. *E. coli* that has the ability to take up foreign DNA is called competent cell. Therefore, the first step in transformation is making competent cells. A very common method is that of Stanley Cohen developed in early 70's. *E. coli* is grown to early phase with an OD_{600} value of 0.3 - 0.5. Cells are then collected by centrifugation and incubated in a solution containing 10 mM MOPS (pH 6.5), 100 mM $CaCl_2$, 30 mM dextrose, and 15% glycerol for one hour at 0°C. With this method, 1 µg of pBR322 DNA usually results in approximately 1 x 10^6 transformants. In the early 80's, Doug Hanahan developed another method which has a better transformation efficiency (J. Mol. Biol. 166:577-580, 1983). This method uses a buffer containing 10 mM KMES (pH 6.2), 100 mM RbCl, 45 mM $MnCl_2$,10 mM $CaCl_2 \cdot 2H_2O$, and 3 mM $HACoCl_3$ to make *E. coli* competent, in which MES is 2-N-morpholinoethane sulfonic acid and HA is hexamine. Most commercial transformation kits are made with this method. Since these methods use chemicals to make *E. coli* competent, they are referred to as chemical transformation.

Transformation can also be achieved by electroporation. *E. coli* cells are grown to early log phase. After washing thoroughly with ice-cold water, the cells are suspended in 10% glycerol to approximately 3 x 10^{10}/ml. A dried salt-free DNA sample is mixed with 20 μl of the cells. The sample is then placed in an electroporation cuvette and electrically shocked, making *E. coli* cell wall permeable to DNA. This method kills approximately 80% of the cells, but the majority of the survived cells are transformed. Electroporation is mainly used for plasmid DNA greater than 20 kb. Transformation efficiency of electroporation is at least 10 times better than that of chemical transformation, but additional costs are required to purchase electroporation cuvettes that are mostly disposable.

7.8: Screening of Transformants

Cells that have taken up the DNA during transformation are called transformants. Since the vector may self ligate back to a circular plasmid, the first step in screening the transformants is to distinguish *E. coli* cells containing recombinant plasmid from those containing the original vector. If the DNA fragment is inserted into the *lacZ* gene of the vector, the transformants are grown on agar plates containing X-gal and IPTG. Cells that contain the original vector will produce β-galactosidase and thus form blue colonies, and those containing the recombinant plasmid are colorless. If the DNA fragment is inserted into an antibiotic resistance gene such as the tetracycline gene on pBR322, cells containing the recombinant will be resistant to ampicillin but sensitive to tetracycline, and those containing the original pBR322 will be resistant to both ampicillin and tetracycline. Some vectors allow positive selection of cells containing recombinant plasmid. The vector pZErO-1 (Fig. 7-3), developed by Invitrogen, is one such vector which uses live and death selection. Only the cells containing recombinant plasmid will survive because pZErO-1 contains the *ccdB* gene; where *ccd* stand for control of cell death. The *ccdB* gene encodes a cytotoxin which is lethal to *E. coli*. This *ccdB* gene is fused to the *lacZ* gene and insertion of a DNA fragment into this *ccdB-lacZ* fusion destroys the *ccdB* gene; therefore, cells containing the recombinant plasmid do not produce the cytotoxin and will survive. Those that contain the original vector with the intact *ccdB* gene will produce the cytotoxin and kill themselves. In order to produce the vector pZErO-1, cells that contain the *ccdA* gene are used. The *ccdA* gene encodes an anti-toxin which neutralizes the cytotoxin; therefore, $ccdA^+$ cells containing pZErO-1 are not killed and are capable of replicating the plasmid.

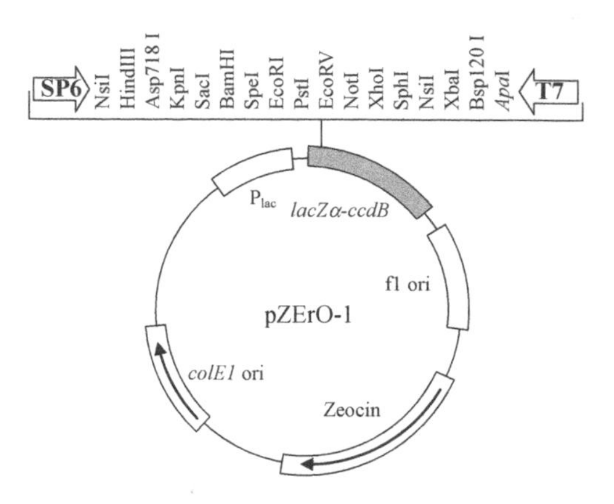

Fig. 7-3: pZErO-1

If the DNA fragment is inserted into a region outside any of the selection marker genes, none of the methods described above will work. In this case, colony hybridization is usually used to screen the transfromants. Colonies of transformants are duplicated on two agar plates. The bacterial colonies on one of the plates are transferred to a circular filter paper the same size as the agar plate. The filter is then placed colony side up on a filter paper pad saturated with 0.5 N NaOH/1.5 M NaCl for 15 minutes. This process lyses bacterial cells and denatures DNA to its single-stranded state. The filter paper is then transferred to a second filter paper pad saturated with 1.5 M NaCl/0.5 M Tris (pH 8.0) to neutralize the negative charges. After 15 minutes of incubation, the filter with colonies is transferred to the third filter paper pad saturated with 2x SSC (300 mM NaCl, 30 mM sodium citrate) or 2x SSPE (300 mM NaCl, 20 mM NaH2PO4, 1 mM EDTA) for 15 minutes to reduce the salt concentration on the filter. The filter is then dried and hybridized with a labeled insert DNA fragment as described in Chapter 8.

7.9: Ligation of DNA Fragments with Vector

To join two DNA fragments together, the ends of the fragments must be compatible, meaning that nucleotide sequence of the overhang ends of the two fragments must be identical. If a DNA fragment is isolated by digestion of the DNA containing the fragment with BamHI, the two ends of the fragment will have GATC overhangs at 5' ends. The linearized vector must also have the same overhangs to be able to accept the DNA fragment. This can be generated by also digesting the vector with BamHI. Ligation of DNA fragments with the vector that has the same overhangs at both ends will result in recombinants containing the insert in two different orientations, and the place where the two DNA fragments join will regenerate the original restriction site, in this case BamHI (Fig. 7-4).

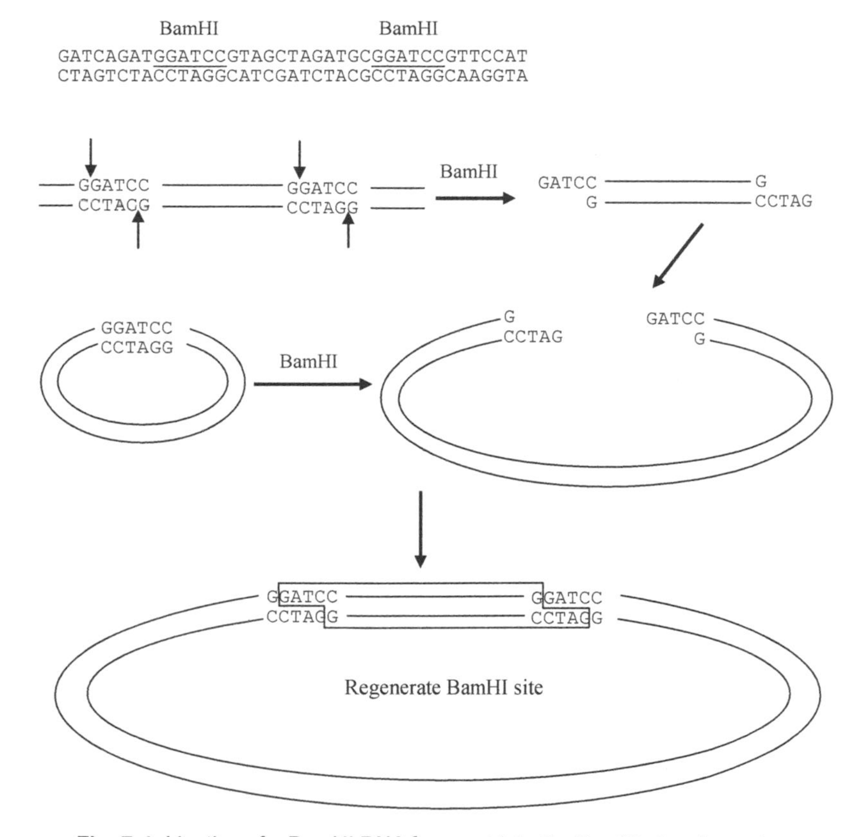

Fig. 7-4: Ligation of a BamHI DNA fragment into the BamHI site of a vector.

Some restriction enzymes generate the same overhangs although they recognize different sequences. For example, BamHI recognizes GGATCC, and BglII recognizes AGATCT. The overhangs generated by these two enzymes are the same, GATC. Therefore, a BglII DNA fragment can be ligated into the BamHI site of a vector. This method of ligation also produces recombinants with insert in opposite orientations (Fig. 7-5). Table 1 lists several restriction enzymes that generate compatible ends. A vector digested with AccI can accept DNA fragments that are isolated by digestion with HinPI, HpaII, or MsPI, and that digested with BamHI can accept DNA fragments isolated by digestion with MboI, Sau3A, BclI, BglII, or XhoII.

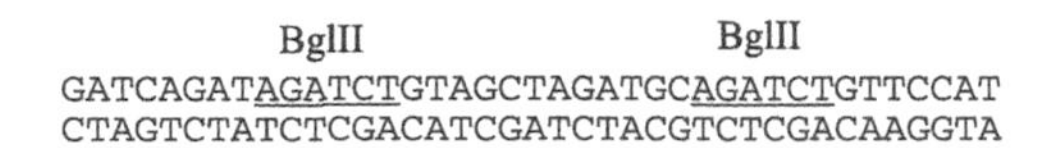

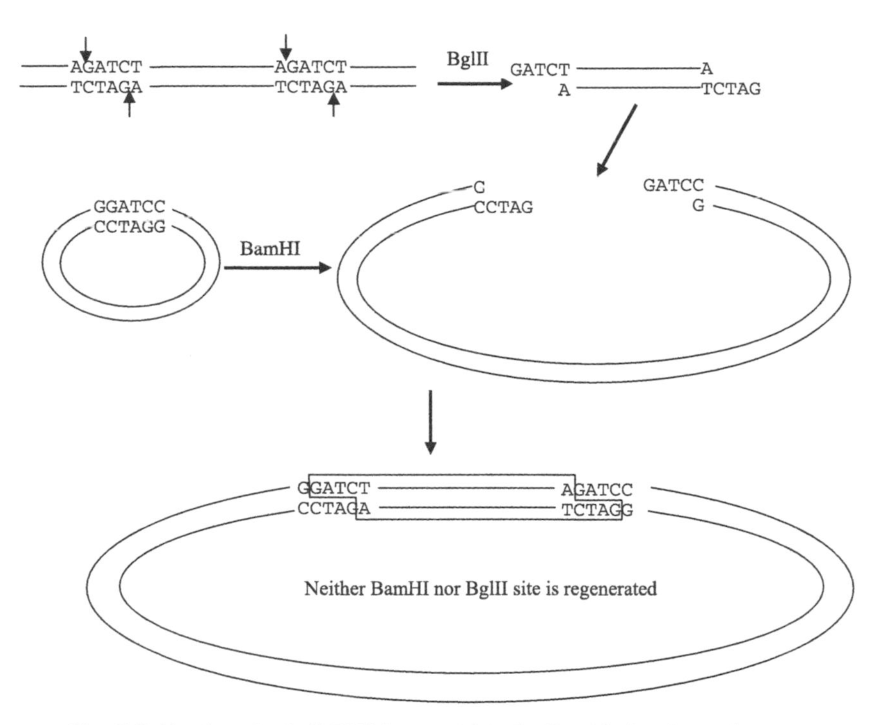

Fig. 7-5: Ligation of a BglII DNA fragment into the BamHI site of a vector.

Table 7-1: Compatible restriction sites.

Cloning site		Compatible site	
AccI	GT↓CGAC	HinPI	G↓CGC
		HpaII	C↓CGG
		MspI	C↓CGG
		MaeII	A↓CGT
		TaqI	T↓CGA
		AhaII	GPu↓CGPyC
		AsuII	TT↓CGAA
		ClaI	AT↓CGAT
		NarI	GG↓CGCC
BamHI	G↓GATCC	MboI	↓GATC
		Sau3A	↓GATC
		BclI	T↓GATCA
		BglII	A↓GATCT
		XhoII	Pu↓GATCPy
EcoRI	G↓AATTC	EcoRI*	↓AATT
HincII	GTC↓GAC	Any blunt-ended fragment	
PstI	CTGCA↓G	NsiI	ATGCA↓T
SalI	G↓TCGAC	XhoI	C↓TCGAG
SmaI	CCC↓GGG	Any blunt-ended fragment	
SphI	GCATG↓C	NlaIII	CATG↓
		Nsp7524I	PuCATG↓Py
XbaI	T↓CTAGA	AvrII	C↓CTAGG
		NheI	G↓CTAGC
		SpeI	A↓CTAGT

If the ends of a DNA fragment and the vector are not compatible, they can be made blunt ended and then ligated together. An example of such is to insert an EcoRI fragment into the BamHI site of a vector. Both EcoRI and BamHI generate 5' overhang ends. Two methods can be used to convert 5' overhang ends to blunt ends. The first method is to fill in the missing bases with the Klenow fragment of *E. coli* DNA polymerase I or T4 DNA polymerase in the presence of dNTP. The other method is to remove the overhangs with the single-strand specific nuclease S1 or mung bean nuclease. Both the vector and insert DNA molecules can be filled in or chewed up to become blunt ended. If the fill-in DNA fragment is ligated to the chew-up vector, EcoRI sites are regenerated at the two joints. On the other hand, BamHI sites are regenerated if the chew-up DNA fragment is ligated to the fill-in vector (Fig. 7-6).

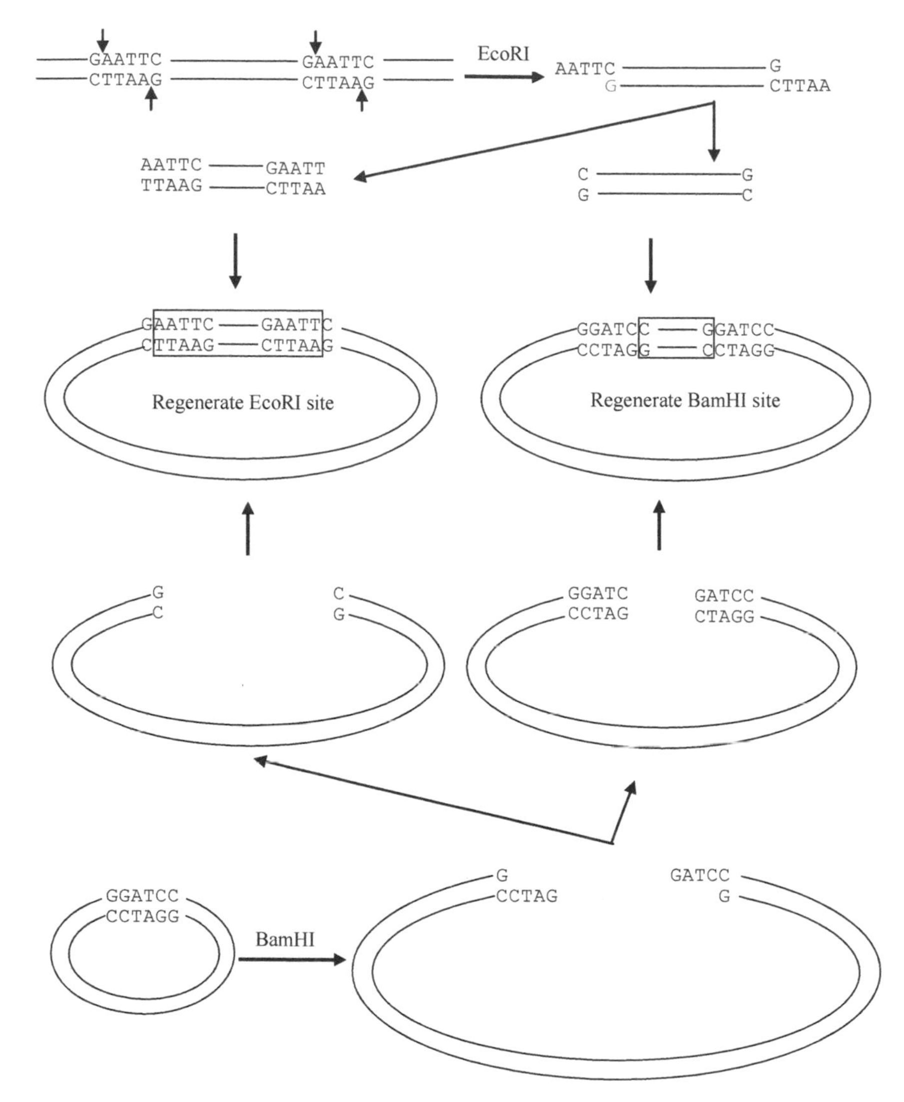

Fig. 7-6: Cloning of an EcoRI DNA fragment into the BamHI site of a vector by blunt end ligation.

7.10: Prevention of Vector Self Ligation

To construct a recombinant plasmid, a DNA fragment is ligated with a vector. During ligation, the two ends of a linear DNA can be ligated together to become a circular molecule if the DNA is digested with one restriction enzyme. Since the insert DNA does not have a replication origin, this DNA molecule will not replicate when introduced into *E. coli* and will be lost. However, if the two ends of a linearized vector are ligated together, a plasmid capable of replicating in *E. coli* is generated. Since the efficiency of intra-molecular ligation is much better than that of inter-molecular ligation, the number of self ligated plasmids is much greater than those of recombinant plasmids in which the insert is successfully ligated with the vector. To increase the chance of getting recombinant plasmids, it is usually necessary to prevent vector self ligation.

When ligase joins two pieces of DNA together, it links the 5' P group of a DNA fragment with the 3' OH group of another DNA fragment together to form a phosphodiester bond. If the 5' P groups located at both ends of the linearized vector are removed, this vector will not self ligate, but their 3' OH groups at both ends can still be ligated with the 5' P groups at both ends of the DNA fragment. Therefore, the chance of vector self ligation is minimized. The enzyme used to dephosphorylate linearized DNA is alkaline phosphatase. Calf intestinal and shrimp alkaline phosphatases are the two most commonly used ones. However, if the alkaline phosphatase is not completely removed after the reaction, it will also remove the 5' P groups from the insert DNA, rendering them unable to be joined with the vector. A better approach to prevent vector self ligation is to digest both the vector and the DNA fragment with two different restriction enzymes, e.g., EcoRI and BamHI. Since the ends generated by these two enzymes are not compatible, a linearized vector with EcoRI site at one end and BamHI site at the other end can not self ligate. With this approach, it is not necessary to use alkaline phosphatase to remove the 5' P groups from the linearized vector.

7.11: Use of Tailing, Adaptors and Linkers

Although blunt end ligation is possible, it is not very efficient. One way to improve the efficiency of this type of ligation is to add several C's to the 3' ends of the insert and G's to those of the vector using terminal deoxynucleotide transferase. This process is called GC-tailing. Since poly-C readily pairs with poly-G, vector and insert DNA molecules can be linked together rather easily. Occasionally, it is necessary to ligate a DNA fragment to a linearized vector that is digested with a different restriction enzyme. For example, the insert DNA is isolated by BamHI digestion and the vector is digested with EcoRI. To ligate such a BamHI fragment to an EcoRI digested vector, adaptors can be used to convert the BamHI ends of the DNA fragment to EcoRI ends. An adaptor is normally composed of two oligonucleotides. In the example shown in Fig. 7-7, the sequence of the first oligonucleotide is 5'-AATTCGTAGTG-3', in which AATT can pair with the EcoRI ends of the vector, and CGTAGTG can pair with the second oligonucleotide with the sequence 5'-GATCCACTACG-3'. Therefore, the ends of the insert DNA are converted from AATT to GATA overhangs that can be ligated to the BamHI ends of the vector.

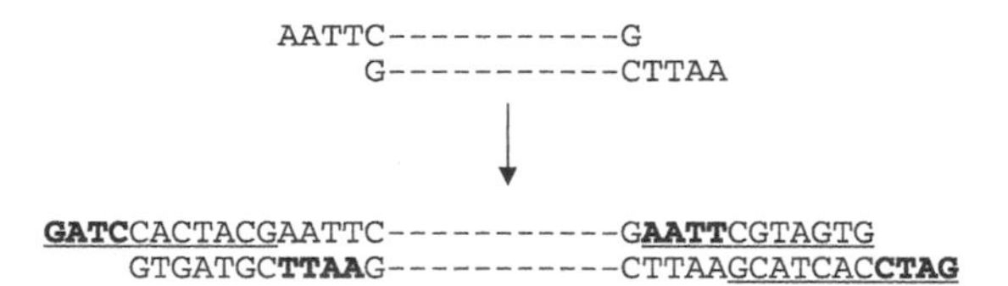

Fig. 7-7: Conversion of EcoRI ends to BamHI ends by adaptors.

There are situations where a certain restriction site must be generated. This can be achieved by the use of linkers. A linker is a very short piece of oligonucleotide with a certain restriction site. For example, the sequence of an EcoRI linker is GAATTC. Since this sequence is a palindrome, two molecules of the oligonucleotide can anneal together to form a double-stranded linker DNA. The EcoRI linkers are then ligated to blunt ended DNA fragments. Since more than one linker may be ligated to the ends of a DNA fragment, the excess linkers are digested away with EcoRI. This digestion also makes the DNA fragment contain the AATT 5' overhangs which can be easily ligated to the EcoRI ends of the vector.

7.12: *E. coli* Genotypes

Many different *E. coli* strains exist, and every strain has a certain character, known as genotype. In *E. coli* genotype, genes that are defective or unusual are listed. An *E. coli* gene symbol usually consists of 3 lower case letters followed by one capital letter or number, all written in italics (Table 7-2). For example, *recA* represents a gene involved in DNA recombination. DH5α is a very commonly used *E. coli* strain. Its genotype is [F' *endA1 hsdR17* (r_k^- m_k^+) *supE44 (glnV44) thi1 recA1 gyrA96* (Nal^r) *relA1 Δ(lacZYA-argF)U169 deoR* (Φ80dlacΔ(lacZ) M15)]. Among these symbols, F' represents a derivative of the F factor (or F plasmid) that contains genes responsible for making the sex pilus. An *E. coli* with a sex pilus is male. Usually, there is only one sex pilus per *E. coli*. The F factor may integrate into *E. coli* genome. When the integrated F factor is excised, it may carry a portion of the *E. coli* genome. Therefore, this F factor has a piece of foreign DNA and is called F'. The symbol *endA1* indicates that DH5α is defective in endonuclease A. "*hsdR17* (r_k^-, m_k^+)" denotes that DH5α has a defect in the type I restriction-modification system which is encoded by the *hsd* genes. The real meaning of a number in genotype such as 17 in this example is usually known only to the person who created the mutation. It may mean number 17 mutation, or mutation at the 17th amino acid or nucleotide of the gene. The information enclosed in parentheses explains the mutation. Therefore, (r_k^-, m_k^+) means that DH5α is K type and is defective in the restriction (R) subunit of the type I restriction-modification system, but its modification (M) subunit is normal.

Table 7-2: Genotypes of commonly used *E. coli* strains.

Strain	Genotype
BHB2688	F^- *recA* λ^r (λ *Eam4 b2 red3 imm434 cIts Sam7*)
BHB2690	F^- *recA* λ^r (λ *Dam15 b2 red3 imm434 cIts Sam7*)
BL21(DE3)	F^- *ompT gal dcm lon hsdS* (r_B^- m_B^-) with DE3, a λ prophage carrying the T7 RNA polymerase gene
DH1	F^- *glnV44 recA1 endA1 gyrA96* (Nal^r) *thi*-1 *hsdR*17 (r_k^- m_k^+) *relA1*
DH5αF'	F' *endA1 hsdR17* (r_k^- m_k^+) *supE44 (glnV44) thi1 recA1 gyrA96* (Nal^r) *relA1 Δ(lacZYA-argF)U169 deoR* (Φ80dlacΔ(lacZ)M15)
HB101	F^- Δ(*gpt-pro*)62 *leuB6 glnV44 ara14 galk2 lacY1* Δ(mcrC-mrr) *rpsL20* (Str^r) *xyl-5 mtl-1 recA13*
JM101	F' *traD36 lacI*^{q} Δ(*lacZ*)M15 *proA*$^+$*B*$^+$/*supE thi* Δ(*lac-proAB*)
JM103	F' *traD36 lacI*^{q} Δ(*lacZ*)M15 *proA*$^+$*B*$^+$/*endA1 glnV44 sbcBC thi1 rpsL* (Str^r) Δ(*lac-pro*) (P1) (r_k^+ m_k^+, r_{P1}^+ m_{P1}^+)
JM109	F' *traD36 lacI*^{q} Δ(*lacZ*)M15 *proA*$^+$*B*$^+$/*recA1 endA1 gyrA96* (Nal^r) *thi hsdR17* (r_k^- m_k^+) *supE44* e14$^-$ (*mcrA*$^-$) *relA1* Δ(*lac-proAB*)
LE392	F^- e14$^-$ (*mcrA*$^-$) *hsdR514* (r_k^- m_k^+) *glnV44 supF58 lacY1* or Δ(*lacIZY*) *galK2 galT22 metB1 trpR55*
Q358	F^- *hsdR* (r_k^- m_k^+) *glnV44 fhuA* ($Φ80^r$)
Q359	Q358 (P2)
WA803	F^- e14$^-$ (*mcrA*$^-$) *lacY1* or Δ(*lac*)6 *glnV44 galK2 galT22 rfbD1 metB1 mcrB1 hsdS3* (r_k^- m_k^-)
XL1-Blue	F'::Tn10 *proA*$^+$*B*$^+$ *lacI*^{q} Δ(*lacZ*)M15/*recA1 endA1 gyrA96* (Nal^r) *thi hsdR17* (r_k^- m_k^+) *supE44 relA1 lac*
Y1088	F^- Δ(*lac*)U169 *glnV supF hsdR* (r_k^- m_k^+) *metB trpR fhuA21 proC*::Tn5 (pMC9, Tet^r Amp^r) Note: pMC9 is pBR322 containing $lacI^q$
Y1089	F^- Δ(*lac*)U169 *lon-100 araD139 rpsL* (Str^r) *hsdR*17 (r_k^- m_k^+) *hflA150*::Tn10 (pMC9)
Y1090	F^- Δ(*lac*)U169 *lon-100 araD139 rpsL* (Str^r) *hsdR17* (r_k^- m_k^+) *supF mcrA trpC22*::Tn10 (pMC9)

The genotype *supE44* indicates that DH5α has abnormal tRNA. The abbreviation "*sup*" stands for suppressor, and the *supE* gene encodes one of the several different suppressor tRNAs which recognize stop codons as sense codons. For example, *supC* tRNA recognizes UAG or UGA as tyrosine codon. Several suppressor tRNAs exist as shown in Table 7-3.

Table 7-3: Suppressor tRNAs.

Suppressor tRNA	Stop Codon Recognized	Amino Acid Carried
supC	UAG and UGA	Tyrosine
supD	UAG	Serine
supE	UAG	Glutamine
supF	UAG	Tyrosine
supG	UAG and UAA	Lysine
supU	UGA	Tyrosine

The symbol "*thi*" is the abbreviation of thiamin which is vitamin B1. It indicates that DH5α cannot synthesize its own vitamin B1. "*recA*" denotes that the recombination A gene is defective. "*gyr*" stand for gyrase which is a type II topoisomerase. Gyrase causes DNA to become supercoiled. "*gyrA96*" indicates that the gyrase A gene is defective. *E. coli* cells with this mutation are resistant to nalidixic acid; therefore, the genotype is followed by (Nal^r). The *relA* mutation renders *E. coli* relaxed in the regulation of mRNA synthesis. Normally, mRNAs of certain proteins are made only when they are needed. *E. coli* with the *relA* mutation makes mRNAs even if the proteins are not needed. In genotype designation, Δ represents deletion, and Δ(*lacZYA-argF*) indicates that the region between *lacZYA* located at approximately 5 minutes and *argF* located at approximately 3 minutes in *E. coli* genome is missing (Fig. 7-8). *E. coli* genome is divided into 100 minutes because it takes 100 minutes to transfer the entire genome from one *E. coli* cell to another by the F factor. This (*lacZYA-argF*) deletion is called U169 deletion. "*deoR*" indicates that the gene encoding the repressor (DeoR) of the deoxyribose biosynthesis operon is defective; therefore, the *E. coli* constitutively makes deoxyribose. "(Φ80dlacΔ(lacZ)M15)" indicates that DH5α has the Φ80 bacteriophage, and this bacteriophage carries a defective *lac* operon (*dlac*). This *lac* operon is defective because of the M15 deletion which is a deletion in the α portion of the *lacZ* gene. The *lacZ* gene is composed of α and ω portions. A functional β-galactosidase is made of two α and two ω subunits. For DH5α to produce a functional β-galactosidase, a plasmid containing the α portion of the *lacZ* gene must be introduced. This process is called α complementation.

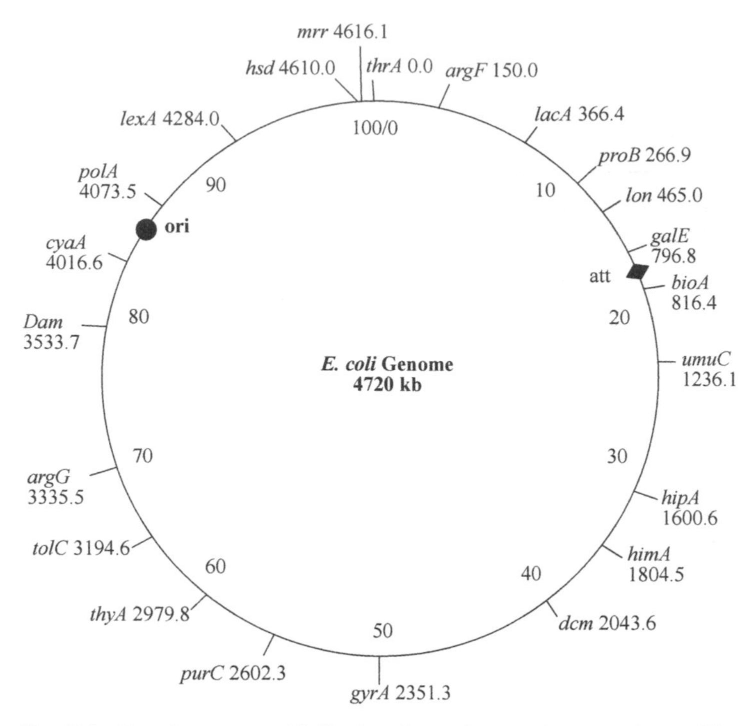

Fig. 7-8: *E. coli* genome with the locations of several genes shown. The unit of numbers shown outside the circle is kb, and that inside the circle is minutes.

HB101 is another commonly used *E. coli* strain. It has the Δ*(mcrC-mrr)* genotype, indicating that the region of the genome between *mcrC* and *mrr* genes is deleted. Therefore, it is defective in the methylation-dependent restriction system (MDRS). It also has the *rpsL20* mutation in which the S12 protein of the 30S ribosome subunit is mutated. This mutation renders *E. coli* resistant to streptomycin, thus the indication (Str^r). *E. coli* strain JM103 has the genotype (r_{p1}^+, m_{p1}^+) indicating that it harbors the P1 bacteriophage and thus has the P1 restriction-modification system. JM109 has the el4$^-$(*mcrA*$^-$) genotype; el4 is a bacteriophage-like element and carries the *mcrA* gene. This genotype indicates that JM109 does not have the el4 element and therefore is *mcrA*$^-$. WA803 has *hsdS3* (r_k^-, m_k^-); this is a mutation in the S subunit of the type I restriction-modification system. Therefore, WA803 is both restriction and modification defective because the enzyme cannot recognize the sequence of the type I restriction and modification system in order to cut or modify the DNA. *E. coli* strain XL1-Blue has the genotype F'::Tn10. The double colon sign indicates insertion. F'::Tn10 denotes that the transposon Tn10 is inserted into the F factor in XL1-Blue. Insertion of a transposable element into a DNA is called transposition. Y1089 has *hflA150*::Tn10 indicating that its *hfl* gene is mutated due to Tn10 transposition. In Y1088, its proline C gene has a Tn5 insertion and thus *proC*::Tn5. A complete genotype information of *E. coli* can be found in Bachman P. J., Linkage Map of *E. coli* k12. Edition 8. Microbiol. Rev. 54: 130-197, 1990 or Han, M. J. and S. Y. Lee. The *E. coli* proteome: past, present and future prospects. Microbiol. Mol. Biol. Review. 70:362-439, 2006.

Summary

In cloning, a DNA fragment containing the gene of interest is inserted into a vector in order to make more of the same DNA or to produce RNA or protein. Both Plasmid and bacteriophage can be used as cloning vectors. Plasmid usually resides in bacteria. All plasmids have replication origins and therefore can replicate autonomously in bacteria such as *E. coli*. The size of plasmid may range from 0.5 kb to greater than 100 kb. A plasmid must contain at least two selection markers to become useful for cloning. After inserting a DNA fragment into a vector, the recombinant DNA must be introduced into *E. coli* to replicate. One of the markers allows selection of *E. coli* cells that have taken up the plasmid. The other marker makes it possible to distinguish recombinant plasmid from the wild type vector with the two ends ligated back together, if the DNA fragment is inserted into the second marker. When ligating a DNA fragment with a vector, many different recombinant plasmids are generated, including a vector with one or more inserts in different orientations. However, when these DNA molecules are introduced into *E. coli*, only one type of plasmid DNA exists in one *E. coli* cell. This is due to plasmid incompatibility, meaning that two different plasmids with the same replication origin can not coexist in the same cell. Because of this, cloning of a certain DNA fragment is possible.

The first step in cloning is to prepare insert DNA, the DNA fragment containing the gene of interest. A bigger piece of DNA containing the fragment is digested with a certain restriction enzyme, and the fragment containing the gene of interest is separated from other DNA fragments by electrophoresis. The DNA fragment is then eluted off the gel slice containing the DNA fragment. The most popular method for this is to dissolve the gel slice with a certain chemical such as NaI. The DNA is allowed to bind to a certain resin and then eluted off the resin. The optimal insert to vector ratio is 2:1, and the easiest way to determine this ratio is electrophoresis. If the intensity of a 1.5-kb band insert is the same as the 3.0 kb vector band, the insert to vector ratio is 2:1. The most commonly used enzyme to ligate DNA is the T4 DNA ligase. Since many substances may interfere with ligation, both insert and vector DNA are purified before ligation. The easiest way to purify small amount of DNA is spin dialysis which is a small-scale column chromatography. A DNA sample is passed through a small column packed with Sephadex G-50, Sepharose CL-6, Sephacryl S-100, or Sephacryl S-400 by centrifugation. Small molecules such as salt, protein, polysaccharide, and lipid are retained in the resin, while DNA flows through the column without getting into the resin. After ligation, recombinant plasmids thus formed are introduced into *E. coli* to replicate. Since introduction of a foreign DNA into *E. coli* may change its phenotype, the process is called transformation. *E. coli* cells are first made competent to become able to take up DNA by incubating them in a solution containing $CaCl_2$. DNA can also be introduced into *E. coli* by electroporation.

After transformation, transformants are screened to select those containing recombinants. If a pBR322 derived plasmid is used as the vector and the DNA fragment is inserted into the tetracycline-resistance gene, transformants containing the recombinant plasmid are resistant to ampicillin but sensitive to tetracycline. If a pUC derived plasmid is used, and the DNA fragment is inserted into the *lacZ* gene. Colonies of *E. coli* cells containing the wild type vector will be blue when they are grown on agar plates containing X-gal and IPTG, and those containing the recombinant plasmid will be colorless. If the DNA fragment is not inserted into any of the selection markers, transformants are usually screened by colony hybridization in which the DNA fragment is labeled and used as the probe to determine which bacterial colony contains the recombinant plasmid.

The ends of insert and vector DNA must be compatible to be ligated together. If both DNA molecules have the same 5' or 3' overhangs, the DNA molecules can be ligated together rather easily. Therefore, these ends are called sticky ends. One example of this is inserting a BamHI DNA fragment into a BamHI digested vector. In this case, the ends of both types of DNA have the same GATC overhangs and can be linked to each other by base pairing. Some restriction enzymes may generate compatible ends. For example, the 5' overhang ends generated by BglII digestion is also GATA and can be inserted into the BamHI site of a vector. When a vector is digested with a certain restriction enzyme, the two ends thus generated have the same overhangs and therefore can be ligated back very easily and become unable to accept a DNA fragment. To prevent this from happening, the 5' P groups at both ends of the linearized vector are removed so that the vector can only be ligated with the insert. An alternative approach is to digest both the DNA fragment and the vector with two different restriction enzymes such as EcoRI and BamHI. Since EcoRI ends are not compatible with BamHI ends, the vector cannot self ligate.

If the ends of both insert and vector are not compatible, they can be converted to blunt ends because any two blunt ended DNA fragments can be linked together. To make a 5' overhang blunt, the overhangs can be removed by using single-strand specific nuclease such as S1 or mung bean nuclease or filled in with nucleotides by using Klenow fragment or T4 DNA polymerase. A DNA fragment with 3' overhangs is made blunt ended by removing the single-strand region with the single-strand specific nuclease. DNA fragments with blunt ends are much harder to ligate than compatible overhang ends. One way to improve the efficiency of blunt end ligation is to tail several G's to the 3' ends of the vector and C's to those of the insert. The insert and vector are then joined together by GC pairing. If a certain DNA fragment such as that isolated by BamHI digestion must be inserted into a certain site such as EcoRI on the vector, adaptors can be used to convert the BamHI ends of the insert to EcoRI ends. Another way to insert a DNA fragment into a specific site of the vector is to use linkers. The DNA fragment is first converted to blunt ends and ligated with linkers which are very small oligonucleotides with a certain restriction site.

After ligation, the recombinant plasmids are introduced into an appropriate *E. coli* host to replicate. To determine which *E. coli* strain can be used for a certain plasmid, *E. coli* genotypes are inspected. Every *E. coli* gene has a symbol which is usually composed of 3 lower case letters and a capital letter or number. For example, *lacZ* is the gene encoding β-galactosidase which metabolizes lactose to galactose and glucose. Genes that are listed in the genotype are mutated or have unusual properties. For example, the genotype of *E. coli* DH1 is [F' *gln44 recA1 endA1 gyrA96* (nar^r) *thi1 hsdR17* (r_k^- m_k^+) *relA1*], indicating that *gln, recA, endA, gyra, thi1, hsdR, and relA* genes of this *E. coli* are defective. F' indicates that the F factor in this *E. coli* carries foreign DNA. Information which is enclosed in parentheses explains the preceding genotype. For instance, *gyrA96* (nar^r) indicates that the *E. coli* has a mutation in the *gyrA* gene rendering it resistant to nalidixic acid. "*hsdR17* (r_k^- m_k^+)" indicates that the restriction (R) subunit of the type I restriction-modification system is defective but the modification (M) subunit is normal and that this *E. coli* is type K. In genotype, the Δ sign denotes deletion and the double colon (::) sign represents insertion. JM109 has Δ(*lac-proAB*) indicating that the region between the *lac* operon and the *proAB* genes is missing in the genome of this *E. coli.* "*hflA150*::Tn10" indicates that the *hflA* gene is defective due to Tn10 insertion.

Sample Questions

1. A plasmid must have the following to become a cloning vector: a) replication origin and at least two selection markers, b) Lac Z gene alone, c) ampicillin gene alone, d) Rop gene.
2. In agarose gel electrophoresis, a circular plasmid usually runs as a) one band, b) two bands, c) no bands, d) ten bands.
3. In bacterial genotype, Δ sign represents: a) transposition, b) insertion, c) deletion, d) restriction.
4. In bacterial genotype, the double colon sign (::) represents: a) transposition, b) insertion, c) deletion, d) restriction.
5. In bacterial genotype, the information enclosed in parentheses usually explains: a) transposition, b) insertion, c) deletion, d) the preceding genotype.
6. Suppressor tRNA is: a) super tRNA that can carry any amino acid, b) mutated tRNA that has no function, c) abnormal tRNA that recognizes stop codons as sense codons, d) abnormal tRNA that recognizes a sense codon as stop codon.
7. The Rop protein a) enhances pBR322 replication, b) enhances transcription of the ampicillin resistance gene, c) enhances binding of RNA I to RNA II, d) increases the copy number of pBR322.
8. Two different plasmids with the same replication origin cannot coexist in the same *E. coli* cell; this phenomenon is referred to as a) intolerance, b) incompatibility, c) immunity, d) discrimination.
9. To ligate two DNA fragments together, the ends must be a) compatible, b) blunt ended, c) 5' overhang, d) 3' overhang.
10. A naked plasmid DNA can be introduced into *E. coli* by a) conjugation, b) transduction, c) transformation, d) electroporation.
11. *E. coli* cells that have the ability to take up plasmid DNA by transformation is called: a) able cells, b) super cells, c) competent cells, d) hungry cells.
12. F' means an F factor containing a) protein, b) polysaccharide, c) exogenous DNA, d) fatty acid.

Suggested Readings

1. Bachmann, B. (1990). Linkage map of *Escherichia coli* K-12, Edition 8. Microbiol. Rev. 54: 130-197.
2. Bernard, P. and Couturier, M. (1992). Cell killing by the F plasmid CcdB protein involves poisoning of DNA-topoisomerase II complexes. J. Mol. Biol. 226: 735-745.
3. Bolivar, F., Rodriguez, R. L., Greene, P. J., Betlach, M. C., Heyneker, H. L., Boyer, H. W., Crosa, J. H., and Falkow, S. (1977). Construction and characterization of new cloning vehicles. II. A multipurpose cloning system. Gene 2: 95-113.
4. Calvin, N. M. and Hanawalt, P. C. (1988). High-efficiency transformation of bacterial cells by electroporation. J. Bacteriol. 170: 2796-2801.
5. Eggertsson, G. and Soll, D. (1988). Transfer RNA-mediated suppression of termination codons in *E. coli*. Microbiol. Rev. 52: 354-374.
6. Han, M. J. and Lee, S. Y. (2006). The *E. coli* proteome: past, present and future

prospects. Microbiol. Mol. Biol. Review. 70:362-439.

7. Hanahan, D. (1983). Studies on transformation of *Escherichia coli* with plasmids. J. Mol. Biol. 166: 557-580.
8. Ihler, G. and Rupp, W. D. (1969). Strand-specific transfer of donor DNA during conjugation in *E. coli*. Proc. Natl. Acad. Sci. USA 63: 138-143.
9. Jacoby, G. A. (2006). β-lactamase nomenclature. Antimicrob. Agents Chemother. 50: 1123-1129.
10. Lppen-Ihler, K. A. and Minkley, E. G. (1986). The conjugation system of F, the fertility factor of *E. coli*. Annu. Rev. Genet. 20: 593-624.
11. Masukata, H. and Tomizawa, J. (1990). A mechanism of formation of a persistent hybrid between elongating RNA and template DNA. Cell 62: 331-338.
12. Murgola, E. J. (1985). tRNA, suppression, and the code. Annu. Rev. Genet. 19: 57-80.
13. Nordstrom, K. and Austin, S. J. (1989). Mechanisms that contribute to the stable segregation of plasmids. Annu. Rev. Genet. 23: 37-69.
14. Novick, R. P. (1987). Plasmid incompatibility. Microbiol. Rev. 51: 381-395.
15. Pheiffer, B. H. and Zimmerman, S. B. (1983). Polymer-stimulated ligation: enhanced blunt- or cohesive-end ligation of DNA or deoxyribo-oligonucleotides by T4 DNA ligase in polymer solutions. Nucleic Acids Res. 11: 7853-7870.
16. Scott, J. R. (1984). Regulation of plasmid replication. Microbiol. Rev. 48: 1-23.
17. Tomizawa, J. I. and Itoh, T. (1981). Plasmid ColE1 incompatibility determined by interaction of RNA with primer transcript. Proc. Natl. Acad. Sci. USA 78: 6096-6100.
18. Vieira, J. and Messing, J. (1982). The pUC plasmids, an M13mp7-derived system for insertion mutagenesis and sequencing with synthetic universal primers. Gene 19: 259-268.

Chapter 8

Nucleic Acid Hybridization

Outline

Definition and application of nucleic acid hybridization
Southern hybridization
Probe labeling
Preparation of RNA probes
Non-radioactive probes
Prehybridization
Stringency
Northern hybridization
Types of probes

8.1: Definition and Application of Nucleic Acid Hybridization

If two double-stranded DNA fragments are denatured to become single-stranded and then mixed together, the plus strand of one fragment will anneal to the minus strand of the other if the sequences of the two DNA fragments are identical or very similar. This process is called hybridization. DNA hybridization can be used to detect microorganisms, identify cells, determine the structure of a certain region of the genome, etc. One example of its application is to determine whether the hepatitis B virus (HBV) DNA molecule is integrated into the genome of human liver cells. HBV DNA integration is a cause of hepatocellular carcinoma. HBV DNA may be integrated into multiple sites on the same chromosome or different chromosomes in liver cells. These integrated HBV DNAs may undergo homologous recombination, leading to deletion or rearrangements of portions of the genome of infected cells. The HBV genome is 3.2 kb in length with one EcoRI site and no HindIII site. If HBV DNA is integrated into the chromosome, it will introduce one extra EcoRI site into the genome of the infected cell. Therefore, two EcoRI fragments containing portions of HBV DNA are present if the chromosomal DNA of HBV infected cells is digested with EcoRI. If this chromosomal DNA is digested with HindIII, an HBV-containing HindIII DNA fragment greater than the entire HBV genome is present (Fig. 8-1). If HBV DNA is not integrated into the chromosome, it will remain circular in infected cells and display both relaxed circular and supercoiled forms when run on a gel by electrophoresis. HindIII digestion will not change this pattern since HBV DNA has no HindIII site, but EcoRI digestion will result in one band the same size as the linearized HBV DNA (Fig. 8-1).

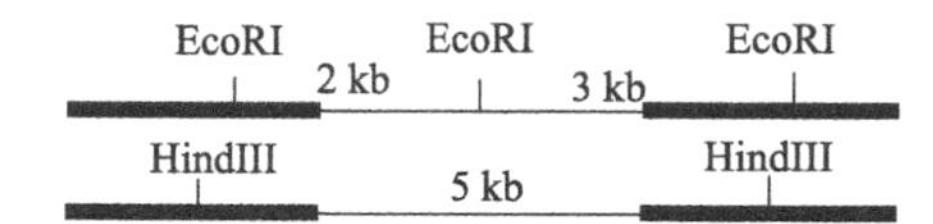

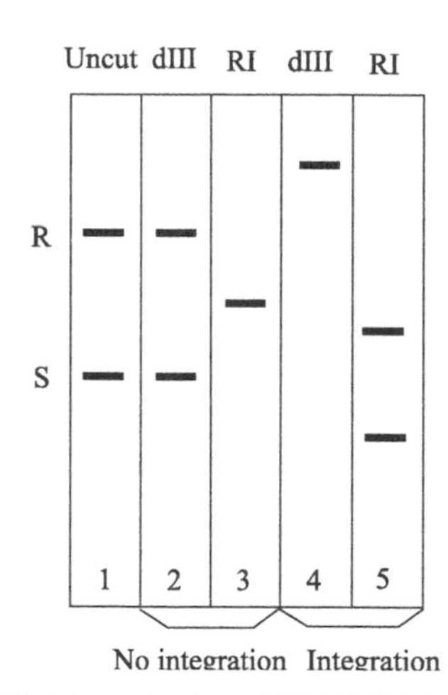

Fig. 8-1: Determination of HBV DNA integration. R, S, dIII, and RI represent relaxed circle, supercoiled, HindIII, and EcoRI, respectively. The thicker lines on top of the figure represent chromosomal DNA, and the thinner ones denote integrated HBV DNA.

It is very difficult to isolate only HBV DNA from infected cells for examination. Therefore, all DNAs, including chromosomal and HBV

DNA, in infected cells are isolated together. When this DNA mixture is digested with a certain restriction enzyme, numerous fragments are generated. These DNA fragments migrate one next to another in the gel and are seen as a smear instead of distinct bands after electrophoresis. HBV-containing DNAs are just one or two of these DNA fragments. Therefore, methods for detection of these DNA fragments are required. One such method is to use labeled HBV DNA to hybridize with the fragments containing HBV DNA in the gel in order to localize them. To be able to perform DNA hybridization, DNA in the gel is transferred to a membrane. Since Dr. Ed Southern is the first person to develop the method, this method is referred to as Southern transfer, and the process of performing hybridization on the membrane is called Southern hybridization.

8.2: Southern Hybridization

To perform a Southern hybridization, the DNA sample is first digested with a certain restriction enzyme and then electrophoresed in an agarose gel. The gel is then soaked in 0.25 N HCl for 15 minutes to remove some purines from DNA molecules and break them into smaller pieces so that they can be transferred from gel to membrane. After washing away excess HCl, the gel is soaked in 0.5 N NaOH/1.5 M NaCl for 15 minutes to denature DNA to single-stranded so that it will anneal with the probe. The gel is washed with water again and then soaked in 1.5 M NaCl/1 M Tris-HCl (pH 8.0) for 15 minutes to neutralize the negative charges of DNA. This gel is now ready for transfer.

A typical Southern transfer assembly is shown in Fig. 8-2. A glass plate is placed across a baking dish. A piece of filter paper is then placed across the glass plate and dipped into the buffer in the baking dish to serve as the wick. The gel is then placed on the filter paper upside down. A membrane, such as nitrocellulose membrane, the same size as the gel is placed on the gel, followed by a 3MM filter paper the same size as the membrane, a stack (~ 5 inches) of paper towels, and a weight of 300 – 500 g to ensure close contact of the paper towels with the gel (Fig. 8-2). The paper towels absorb water creating a capillary flow of buffer from the dish through the gel to the paper towel. When DNA is moved by the flow of buffer, it will adhere to the membrane. Nitrocellulose membrane is very fragile and is therefore quickly replaced by more sturdy membranes such as Nytran, Zeta Probe, Gene Screen, and Hybond. These membranes are mostly nylon backed nitrocellulose. The membrane is dried and baked at 80°C in a vacuum oven for two hours to crosslink DNA to the membrane so that it does not dissociate from the membrane during hybridization. DNA can also be cross-linked to the membrane by UV irradiation if positively-charged membranes, such as NytranPlus™, are used.

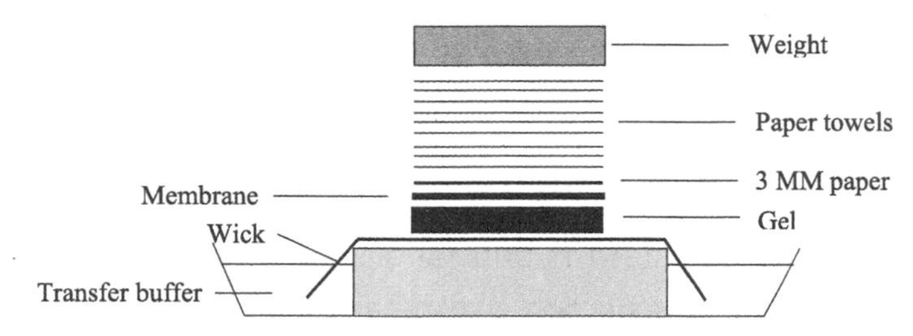

Fig. 8-2: Southern transfer.

DNA can also be transferred from gel to membrane by electrophoresis, commonly referred as electro blotting. It is done in a way very similar to Western blotting, where proteins are electrophoresed from gel to membrane. Vacuum blotting is another commonly used method. A device with two chambers separated by a porous support is used. A membrane is placed at the bottom of the upper chamber, and the gel is placed on the membrane. The upper chamber is then filled with transfer buffer such as 5x SSC (1x SSC is 15 mM sodium citrate, 150 mM sodium chloride). The DNA is trapped on the membrane when it is moved toward the lower chamber as the buffer is sucked by vacuum through the gel to the lower chamber. Although both electro and vacuum blotting are faster, the transfer efficiency may not be better than that of capillary transfer.

8.3: Probe Labeling

Labeled probes are required for DNA hybridization. Both DNA and RNA can be used as probes. Several methods are available to label DNA probes, including nick translation, primer extension, kinase reaction, terminal transferase reaction, fill in, and polymerase chain reaction. In nick translation, the DNA fragment to be used as the probe is first treated with very low concentration of DNase I (8 ng/ml) in the presence of Mg^{++} (0.01 M) to create nicks. The nicks are then expanded to gaps with *E. coli* DNA polymerase I by its 5' to 3' single-stranded exonuclease activity. Since *E. coli* DNA polymerase I also has 5' to 3' polymerase activity, it will fill in the gaps if dNTP's (dATP, dCTP, dGTP, and dTTP, 2 mM each) are present. If a radioactive dNTP such as α-^{32}P-dCTP is also present, the newly synthesized DNA is radioactively labeled. To label DNA fragments by primer extension, the DNA fragment is denatured by boiling in a water bath. Random primers are then added to anneal to denatured DNA. The annealed primers are extended by the Klenow fragment of the *E. coli* DNA polymerase I in the presence of a mixture of dNTP and radioactive dNTP. The random primers used are usually 6 nucleotides long (hexamers) with all possible sequence combinations of G, A, T, and C. Since each position has four possibilities, a total of 4^6 (4096) different hexamers are present. Some of them will anneal to the denatured DNA to serve as primers for DNA synthesis. This method is also called oligolabeling because it uses oligonucleotides as primers.

DNA fragments can also be labeled at their ends. One common method is to use kinase to transfer the radioactive phosphate group from γ-^{32}P-ATP to a 5' dephosphorylated DNA. Dephosphorylation of DNA is achieved by using alkaline phosphatase such as calf intestinal or shrimp alkaline phosphatase. Another method is to label DNA fragments at their 3' ends by tailing with terminal deoxynucleotide transferase in the presence of a radioactive dNTP such as α-^{32}P-dCTP. DNA fragments with 5' overhangs can be labeled by filling in the missing bases with radioactive nucleotides. Polymerase chain reaction can also be used to make DNA probes (described in Chapter 13). In DNA labeling, α-^{32}P-dNTP is used for any reaction that involves DNA synthesis. α-^{32}P-dCTP is most commonly used since it is more stable than other dNTP's. γ-^{32}P-ATP is used only for kinase reactions.

8.4: Preparation of RNA Probes

RNA probes are made by in vitro transcription because none of the methods used for DNA labeling can be used. The DNA fragment from which an RNA probe will be made is first cloned into a vector. The recombinant plasmid is then isolated from *E. coli* and used for in vitro transcription. The plasmid is placed in a test tube containing RNA polymerase and NTP (ATP, CTP, GTP, and UTP). If a radioactive NTP, such as α-^{32}P-UTP, is also present in the reaction, the RNA synthesized is radioactive and can be used as a probe.

Since RNA is single-stranded, either the plus or the minus strand RNA is made during in vitro transcription. DNA is mostly double stranded, and both plus and minus strands are present. Either plus or minus strand of the RNA probe will react with the DNA target. If the target is mRNA, a minus strand RNA probe must be used since mRNA is plus stranded. A promoter is required for transcription; therefore, the DNA fragment containing the gene of interest is cloned behind a promoter in order to perform in vitro transcription. If the orientation of the gene is the same as the transcription direction of the promoter, a plus strand RNA is made. The DNA fragment must be cloned behind the promoter in an orientation opposite to the transcription direction of the promoter in order to produce a minus strand RNA.

Most cloning vectors have more than one promoter; therefore, several different radioactive RNAs may be produced during in vitro transcription if *E. coli* RNA polymerase is used. These extra RNAs may cause false-positive hybridization reactions. Therefore, special vectors for in vitro transcription are developed (Fig. 8-3). These vectors allow insertion of the DNA fragment containing the gene of interest behind SP6, T3, or T7 promoter. SP6, T3, or T7 RNA polymerase is then used to perform in vitro transcription to synthesize RNA. Since these RNA polymerases only recognize their respective promoter, no non-specific RNAs are produced in the in vitro transcription reaction. SP6 is a bacteriophage of *Salmonella typhimurium*, and T3 and T7 are *E. coli* bacteriophages.

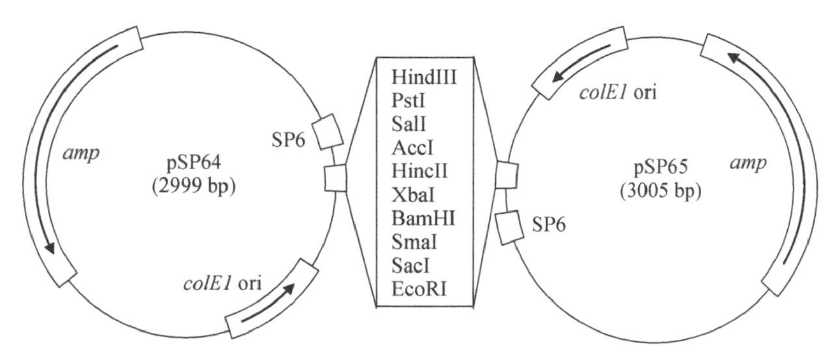

Fig. 8-3: pSP64 and pSP65

Since the orientation of a gene is not known until the gene is sequenced, it is necessary to re-clone the DNA fragment in an opposite orientation if the RNA synthesized is not the right polarity. Some vectors, such as pSP72, pSP73, pGEM-3Zf(+), pGEM-3Zf(-) (Fig. 8-4), and pBluescript contain SP6 and T7 or T3 promoters flanking the multiple cloning sites (MCS). The DNA fragment containing the gene of interest is inserted into MCS, and two separate in vitro transcription reactions are performed with the recombinant plasmid. One reaction uses SP6 RNA polymerase, and the other uses T7 or T3 RNA polymerase to synthesize RNA. In this manner, one reaction makes the plus strand, and the other produces the minus strand RNA.

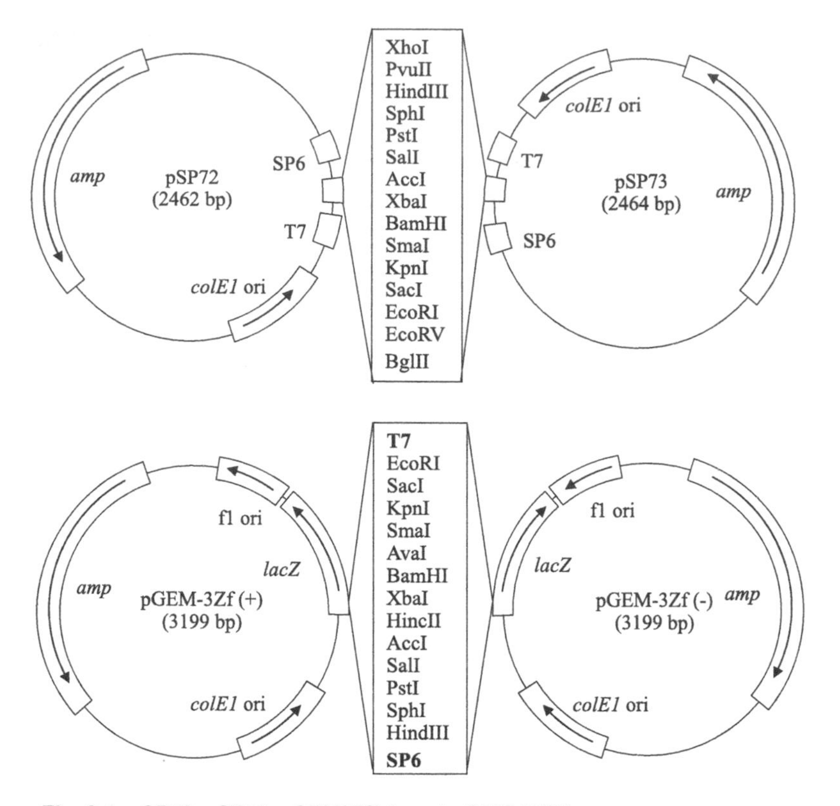

Fig. 8-4: pSP72, pSP73, pGEM3Zf(+), and pGEM-3Zf(-).

8.5: Non-radioactive Probes

Although probes are commonly labeled with radioactive nucleotides, the high cost and troublesome waste disposal limit their use. Therefore, non-radioactive probes are developed. Non-radioactive probes are labeled with modified nucleotides such as biotin-11-dUTP (Fig. 8-5) in which a biotin molecule is linked to dUTP through a linker arm of 11 carbon and nitrogen atoms. Biotin-11-dCTP is another modified nucleotide. These modified nucleotides can be incorporated into DNA by commonly used DNA polymerases such as the Klenow fragment and Taq DNA polymerase. Since avidin specifically binds to biotin, the targets which hybridize with the biotin labeled probes can be detected with labeled avidin. The most commonly used avidin is streptavidin from *Streptomyces avidinii*. If streptavidin is conjugated to horseradish peroxidase (HRP), hybridized target can be detected by adding H_2O_2 and diaminobenzidine (DAB) or 4-chloro-1-napthol. HRP reduces H_2O_2 to H_2O and oxidize DAB to a phenazine polymer which forms a brown precipitate (Fig. 8-6) or 4-chloro-1-napthol to benzo-4-chlorohexadienone which is blue in color (Fig. 8-7).

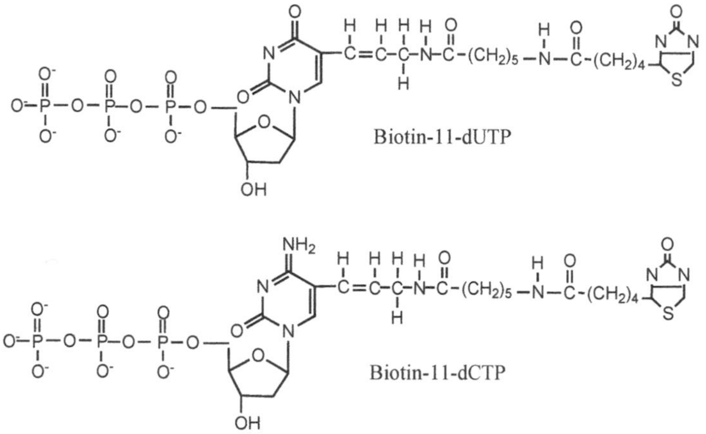

Fig. 8-5: Biotin-11-dUTP and biotin-11-dCTP.

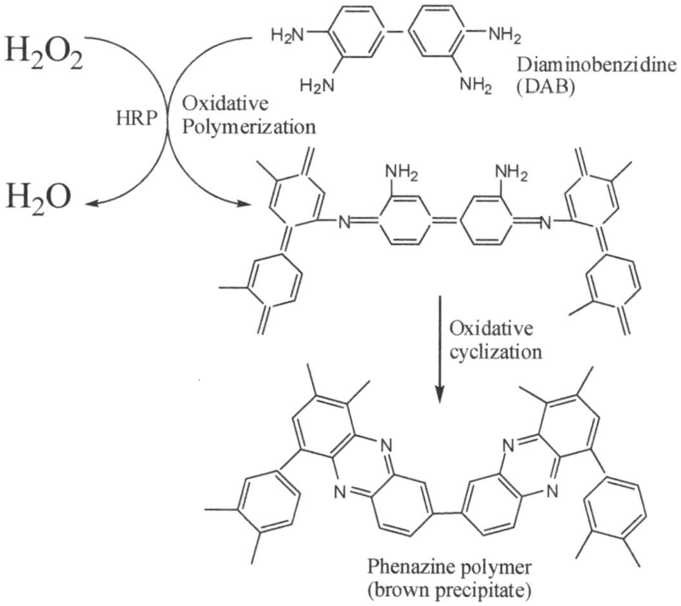

Fig. 8-6: Oxidation of diaminobenzidine (DAB) to brown precipitate. HRP, horseradish peroxidase.

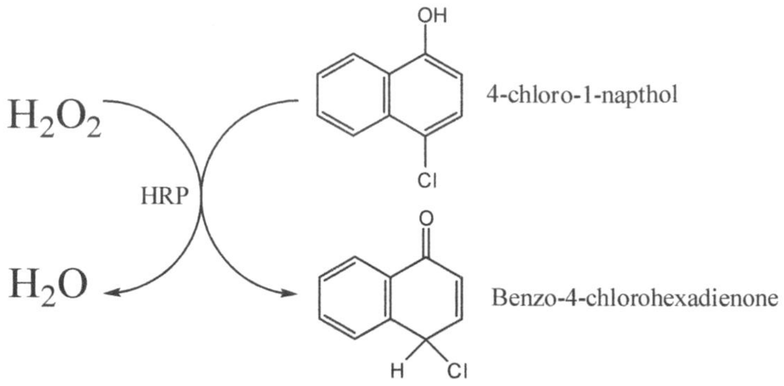

Fig. 8-7: Oxidation of 4-chloro-1-napthol to blue compound benzo-4-chlorohexadienone. HRP, horseradish peroxidase.

Alkaline phosphatase-conjugated HRP is also commonly used. Alkaline phosphatase can convert 5-Br-4-Cl-3-indolyl phosphate (BCIP) to 5,5'-dibromo-4,4'-dichloro indol which is colorless under an alkaline condition. Since alkaline phosphatase is only active under alkaline conditions, this reaction is usually coupled to reduction of nitroblue tetrazolium (NBT) (Fig. 8-8). Reduced NBT forms a blue precipitate, allowing the hybridized target to be visualized.

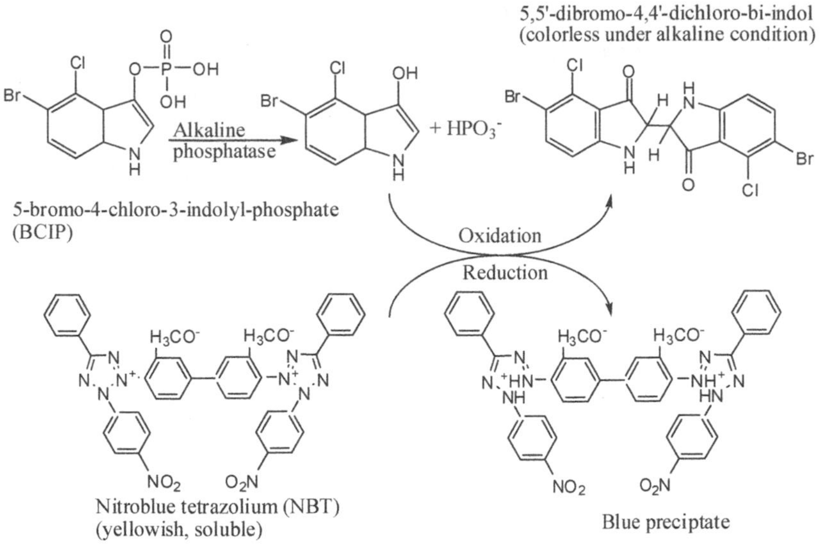

5,5'-dibromo-4,4'-dichloro-bi-indol
(colorless under alkaline condition)

Alkaline phosphatase

$+ HPO_3^-$

5-bromo-4-chloro-3-indolyl-phosphate
(BCIP)

Oxidation

Reduction

Nitroblue tetrazolium (NBT)
(yellowish, soluble)

Blue preciptate

Fig. 8-8: oxidation of 5-bromo-4-chloro-3-indolyl-phosphate and reduction of nitroblue tetrazolium.

Theoretically, uracil is present only in RNA. Since most DNA polymerases used in the lab cannot distinguish dUTP from dTTP, dUTP is often used in vitro to label DNA probes because it is easier to chemically synthesize than other nucleotides. In vivo, when dUTP is incorporated into DNA, uracil is removed by uracil-N-glycosylase creating an apurinic site which is then repaired by DNA repair systems. Therefore, uracil is not present in DNA in vivo.

Digoxigenin-11-dUTP is another commonly used nucleotide derivative for preparation of non-radioactive probes (Fig. 8-9).

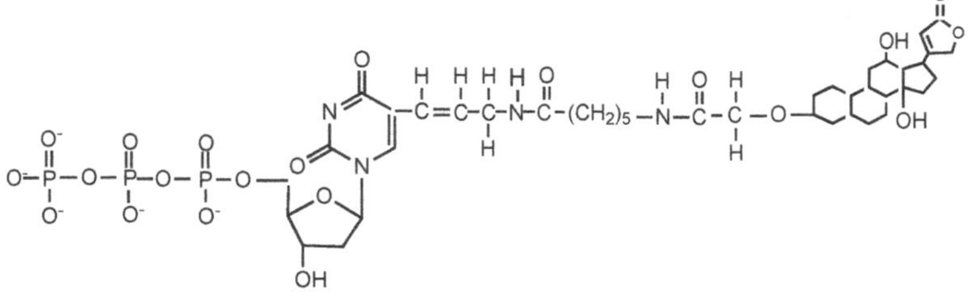

Fig. 8-9: Digoxigenin-11-dUTP.

Horseradish peroxidase (HRP) or alkaline phosphatase (AP)-conjugated antibody against digoxigenin is used to detect hybridized targets. Since digoxigenin is not present in animal cells, false-positive reaction in this system is minimal. DNP-11-dUTP is also commonly used to label DNA probes (Fig. 8-10). Similar to digoxigenin-labeled probes, HRP or AP-conjugated antibody against DNP is used to detect hybridized targets.

DNP-11-dUTP
(dinitrophenyl-11-dUTP)

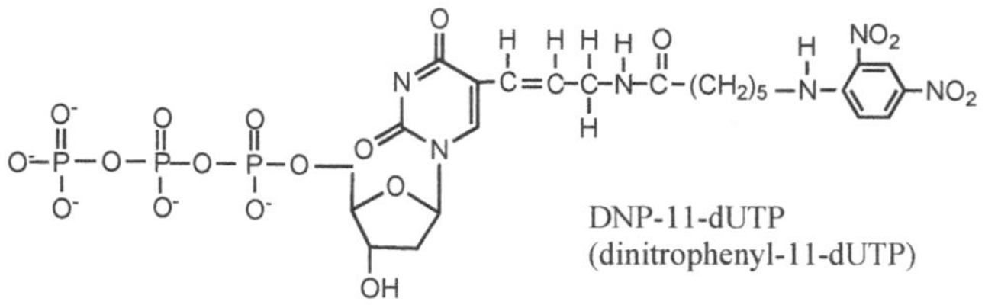

Fig. 8-10: DNP-11-dUTP.

After hybridization, excess probes must be washed away. If ^{32}P-labeled probes are used, the reduction in radioactivity can be used to monitor the washing process. However, no indicators can be used to monitor the washing process if non-radioactive probes are used. It cannot be known if excess probes are removed until the entire hybridization process is finished. If non-specific hybridization exists, it is impossible to undue. The only recourse is to start the entire process all over again. Non-radioactive probes in general are not as sensitive as radioactive probes; however, the use of chemiluminescence has improved their sensitivity. Lumi-Phos (4-methoxy-4-(3-phosphatephenyl) spiro[1,2-dioxetane-3,2'-adamantane], disodium salt) releases light when its phosphate group is removed by alkaline phosphatase (Fig. 8-11). An X-ray film is used to detect the released light. This method has a sensitivity comparable to that of radioactive probes.

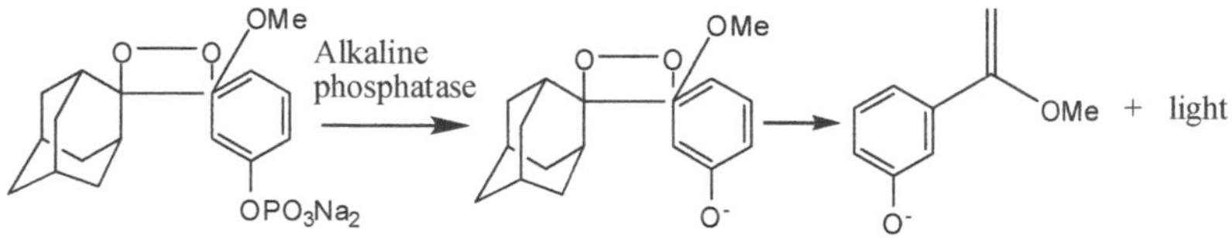

Fig. 8-11: Dephosphorylation of Lumi-Phos and release of light.

8.6: Prehybridization

When performing a Southern hybridization, DNA bands are transferred from agarose gel to a membrane. Probes are then added to the membrane to hybridize to the targets. Probes are DNA or RNA; they can also bind to the membrane. Therefore, it is necessary to block this non-specific binding before hybridization. This process is called prehybridization. The membrane is soaked in a solution containing substances that can bind to the membrane so that the only way by which the probes can bind to the membrane is to hybridize with their specific targets. The most commonly used blocking agent is 10x Denhardt's solution (0.2% Ficoll, 0.2% polyvinylpyrrolidone, 0.2% bovine serum albumin). Another commonly used prehybridization solution contains 7% SDS, 1 mM EDTA, 0.25 M phosphate (pH 7.0). Under such a high concentration of SDS, probes cannot bind to the membrane but can still hybridize with the targets.

8.7: Stringency of hybridization

In Southern hybridization, a probe is intended to bind to its target. If the nucleotide sequence of another DNA fragment on the membrane is similar to that of the target, the probe may also bind to it, resulting in a false-positive result. Therefore, it is necessary to perform the hybridization under a condition that does not allow non-specific hybridization to occur. The way to achieve the goal is to make it difficult for non-specific hybridization to occur. The degree of difficulty for hybridization is referred to as stringency.

Stringency is mainly determined by salt (e.g., NaCl) and temperature. Salt is used to neutralize the negative charges of DNA so that the probe and target can hybridize. The higher the salt concentration in a hybridization reaction the more negative charges are neutralized, making it easier for the probe to bind to its target. Therefore, high salt concentration in hybridization is a low stringency condition. Under a low salt condition, the negative charges of both probe and target DNA are not completely neutralized. Unless the nucleotide sequences of the probe and the target are identical or very similar, the probe cannot bind to the target. Therefore, low salt is a high stringency condition.

When the probe binds to its target, the two strands of DNA are held together by hydrogen bonds. The strength of hydrogen bonds is affected by temperature; the higher the temperature the weaker the hydrogen bonds are. Therefore, high temperature is a high stringency condition, and low temperature is a low stringency condition for hybridization. Formamide also affects the strength of hydrogen bonds and is commonly used to adjust the stringency condition for hybridization. The strength of hydrogen bonds between the two DNA strands is reflected in the Tm of DNA. Tm is the melting temperature of DNA and is the temperature where 50% of double-stranded DNA molecules in a solution become single-stranded. A double-stranded DNA fragment greater than 500 bp has a Tm of approximately 95°C, and 1% of formamide decreases the Tm of a DNA by 0.6°C. If a hybridization reaction is to be done at 80°C, it can be done at 50°C if the hybridization solution contains 50% formamide.

If the nucleotide sequences of the probe and target are identical or very similar (> 80% homology), a high stringency condition is used for hybridization. An example of such is performing the hybridization in a solution containing 2x SSC (30 mM sodium citrate, 300 mM sodium chloride), 1% SDS, 10x Denhardt's at 65°C. If the sequences of probe and target are not very similar, e.g., 70% homologous, a low stringency condition is used. Using the mouse alcohol dehydrogenase (*ADH*) gene to hybridize to human *ADH* gene is one such example. A hybridization performed in 5x SSC, 1% SDS, and 10x Denhardt's at 42°C is a low stringency condition.

In addition to the hybridization step, the step in which excess probes are washed away also needs to be done under an appropriate stringency condition. If the sequences of probe and target are identical, washing is done under high stringency condition such as 0.1x SSC, 0.5% SDS at 70°C. A low stringency washing condition can be done with 2x SSC, 0.5% SDS at 42°C.

8.8: Northern Hybridization

Northern hybridization is used to determine the length and concentration of a certain mRNA molecule. The major difference between Southern and Northern hybridizations is the target. The target for Southern hybridization is DNA, whereas the target for Northern hybridization is RNA. To perform Northern hybridization, an RNA sample is electrophoresed in an agarose gel, and the RNAs in the gel are then transferred to a membrane. The membrane is prehybridized and hybridized with the probe exactly the same way as Southern hybridization. Since RNA usually has extensive secondary structures, it is denatured with formamide and formaldehyde before electrophoresis as described in Chapter 6. Another difference is that the gel does not need to be treated with acid and alkaline. Acid treatment is used to depurinate nucleic acids thus breaking them into smaller fragments. Most mRNAs are relatively short and do not need to be made smaller before transferring from gel to membrane. They are single stranded in nature and therefore do not need to be treated with alkaline to become single-stranded. Neutralization of negative charges of RNA is done during transfer by using a buffer with higher concentrations of salt such as 10x SSC to transfer RNA from gel to membrane.

8.9: Types of Probes

Probes can be classified as identical, homologous, and degenerate probes. Identical probes have the same sequences as those of the targets. Using HBV probe to detect HBV DNA in human liver cells is one application of identical probes. The use of mouse *ADH* gene probe to detect human *ADH* gene is one example of hybridization with homologous probes. If the gene to be isolated or detected is novel, degenerate probes may be used. Degenerate probes are a mixture of labeled oligonucleotides that are synthesized based on the nucleotide sequence deduced from the amino acid sequence of a certain protein. Since most amino acids have more than one codon, it is necessary to make oligonucleotides with all possible sequence combinations. For example, glycine is encoded by GGG, GGA, GGT or GGC, and alanine is encoded by GCG, GCA, GCT, or GCC. Therefore, a total of 16 different nucleotide sequences can code for the amino acid sequence glycine-alanine, but only one of them is real. Since which one is real is not known until the DNA is sequenced, oligonucleotide probes with all 16 sequences are made. This kind of probe mixture is called a degenerate probe.

The minimum length of an oligonucleotide probe is approximately 18 nucleotides. Therefore, the sequence of a stretch of six amino acids of the protein is required to deduce the nucleotide sequences of the degenerate probe. If the sequence Gly-Ala-Gly-Ala-Thr-Gly of a peptide is chosen, a total of 4096 (4^6) different oligonucleotides need to be made, and this probe mixture would have a degeneracy of 4096 (Fig. 8-12). In contrast, if the sequence Tyr-His-Met-Asp-Trp-Pro of another stretch of 6 amino acids of the peptide is chosen, only 32 (2 x 2 x 1 x 2 x 1 x 2) different probes need to be made because the numbers of codons encoding these amino acids are 2, 2, 2, 1, 1, and 4, respectively. The mRNA sequence of this 6 amino acid sequence is UA(U/C)-CA(U/C)-AUG-GA(U/C)-UGG-CCN. If only the first 2 bases of the last codon CCN are used, the sequence becomes UA(U/C)-CA(U/C)-AUG-GA(U/C)-UGG-CC and only 8 (2 x 2 x 1 x 2 x 1 x 1) different probes need to be made. If a probe mixture has 3200 oligonucleotides and its degeneracy is 32, only 100 oligonucleotide probes can hybridize with the target. In contrast, 400 oligonucleotides probes can hybridize with the target if the degeneracy of the probe is 8. Therefore, an important point in designing degenerate probes is to reduce the degeneracy as much as possible. Since it is much easier to synthesize DNA than RNA,

DNA oligonucleotides are most commonly used. Therefore, the mRNA sequence of the selected amino acid sequence is converted to DNA sequence. Because a minus-strand probe can hybridize to both DNA and mRNA targets, it is more commonly used. Therefore, the sequence is converted to the anti-sense sequence, and it is the anti-sense (minus strand) probe which is synthesized and used.

```
Met  Gly  Ala  Gly  Ala  Thr  Gly  Tyr   His   Met Asp   Trp Pro Arg
 1    4    4    4    4    4    4    2     2     1   2     1   4   6

                         mRNA: 5'- UAU/C CAU/C AUG GAU/C UGG CCN - 3'

                          DNA: 5'- TAT/C CAT/C ATG GAT/C TGG CCN - 3'

                   Anti-sense: 3'- ATA/G GTA/G TAC CTA/G ACC GGN - 5'
                                     2     2     1   2     1   4 (32)

                   Anti-sense: 3'- ATA/G GTA/G TAC CTA/G ACC GG - 5'
                                     2     2     1   2     1  1 (8)
```

Fig. 8-12: Design of degenerate probes.

Summary

Single-stranded DNA derived from two DNA fragments with identical or very similar nucleotide sequences can anneal to its complementary strand. This phenomenon is called hybridization. If a single-stranded DNA is labeled with isotopes, the hybridized target will become radioactive. The location and amount of the hybridized target can then be determined by the radioactivity associated with the target. A labeled DNA used to detect targets is called a probe. Both DNA and RNA can be used as probes. DNA probes can be labeled by several different methods. In nick translation, the DNA is treated with DNase I in the presence of Mg^{++} to generate nicks. The nicks are expanded to gaps with *E. coli* DNA polymerase I by its 5' to 3' single-stranded exonuclease activity. The gaps are then filled in with radioactive nucleotides by the 5' to 3' polymerase activity of the same enzyme. DNA can also be labeled by primer extension. The DNA fragment to be used as a probe is denatured by heating in a boiling water bath. Random hexamers, which contain all possible sequence combinations of the 6 nucleotides, are added. The hexamers that anneal to denatured DNA are then extended in the presence of radioactive dNTP and the Klenow fragment of *E. coli* DNA polymerase I. The newly synthesized DNAs are radioactive and can be used as probes.

DNA fragments can also be labeled at their 5' or 3' ends. Those with 5' overhangs can be labeled by filling in the missing bases with radioactive nucleotides or by replacing the phosphate group with the radioactive γ phosphate of γ-^{32}P-ATP. To label a DNA fragment at its 3' ends, terminal deoxynucleotide transferase is used to add radioactive nucleotides. This is the most commonly used method for labeling synthetic oligonucleotides.

RNA probes are made by in vitro transcription. The DNA fragment containing the gene of interest is cloned between SP6 and T3 or T7 promoter of a vector. The recombinant plasmid DNA is then placed in a tube containing SP6, T3, or T7 RNA polymerase and radioactive NTP. The RNAs thus synthesized are labeled probes.

Probes can also be labeled with biotin-, digoxigenin-, or dinitrophenol (DNP)-conjugated nucleotides. If it is biotin-labeled, horseradish peroxidase or alkaline phosphatase-conjugated streptavidin is used to detect the hybridized targets. If digoxigenin or dinitrophenol-labeled probes are used, horseradish peroxidase or alkaline phosphatase-conjugated antibody against digoxigenin or DNP is used to detect the hybridized targets.

To detect targets in an agarose gel after electrophoresis, the DNA fragments are first transferred to a membrane, such as NytranPlus™. Prior to transfer, the gel is soaked in 0.25 N HCl for approximately 15 minutes to break DNA into smaller pieces. After washing with water, the gel is treated with 0.5 N NaOH/1.5 M NaCl for 15 minutes to denature DNA to single-stranded. The gel is washed with water again and then treated with 1.5 M NaCl/1 M Tris (pH 8.0) to neutralize the negative charges of DNA. The gel is then placed upside down on a filter paper, which is placed on a glass plate across a baking dish, with its two sides dipped in 5x SSC buffer in the baking dish. A membrane, such as NytranPlus™, the same size as the gel is placed on the gel, followed by a piece of filter paper the same size as the membrane, a stack of paper towel, and a weight of approximately 300 g. When the paper towels absorb the water, the buffer will move from the baking dish through the gel. The DNAs in the gel are moved along by the flowing buffer and trapped on the membrane. Since this method was developed by Southern, it is referred to as Southern transfer. The membrane is then baked in an 80°C vacuum oven for two hours or irradiated with UV for 1 minute to cross link DNA to the membrane. The membrane is then soaked in a blocking solution such as 10x Denhardt's overnight, so that the probe can only bind to the targets and not to the membrane. This process is called prehybridization. Labeled probes are then added to hybridize to the targets.

Hybridization must be done under an appropriate stringency condition to minimize non-specific reactions. Stringency conditions are adjusted with salt, temperature, and formamide. The higher the salt concentration the more negative charges of DNA are neutralized, allowing probes to hybridize to their targets more easily. Therefore, high salt concentration is a low stringency condition for hybridization. Under a low salt condition, the negative charges on both the probes and the targets may not be completely neutralized, and hybridization does not occur unless the nucleotide sequences of the probe and the target are identical or very similar. Therefore, low salt condition is a high stringency condition. Since high temperature weakens the hydrogen bonds between the probe and target, high temperature is a high stringency condition, and low temperature is a low stringency condition. Formamide also weakens the hydrogen bonds between DNA strands; therefore, hybridization performed in high concentration of formamide is high stringency. High stringency conditions are used to detect targets with identical or very similar nucleotide sequences to those of the probes, and low stringency conditions are used to detect targets with sequences homologous to those of probes. In Southern hybridization,

DNA targets are detected. The same technique can also be used to detect RNA targets. To distinguish these two methods, detecting RNA targets on the membrane is referred to as Northern hybridization. After hybridization, excess probes are washed away. This washing step also needs to be done under appropriate stringency conditions.

A probe is usually a small piece of DNA with the same nucleotide sequence as the target. If such probe is not available, homologous probes are used. One example of this is to use the mouse *ADH* probe to detect human *ADH* gene. If the gene to be detected or isolated is novel, oligonucleotides of approximately 18 bases synthesized based on amino acid sequence of the protein are used. Therefore, the sequence of a stretch of 6 amino acids of the protein is used to deduce oligonucleotide sequences. The mRNA sequence of these 6 amino acids is first converted to DNA sequence. Since minus strand oligonucleotides are usually used as probes, the DNA sequence is converted to its anti-sense sequence which is then used to synthesize oligonucleotides. Since most amino acids have more than one codon, oligonucleotides with all possible sequence combinations are synthesized. If each amino acid of the 6 amino acid-sequence chosen has 4 codons, a total of 4096 different oligonucleotides need to be synthesized, but only one of the 4096 oligonucleotides can hybridize to the target. This type of probe mixture is called degenerate probes, and 4096 is the degeneracy of this probe mixture. For degenerate probes, the lower the degeneracy the better the probes are. Therefore, the amino acid sequence of the region that allows making the least degenerate probes is used.

Sample Questions

1. Labeling DNA by nick translation requires: a) Mg^{++}, b) Mn^{++}, c) DNase I, d) *E. coli* DNA polymerase I, e) dNTP
2. Labeling DNA by primer extension requires: a) random primers, b) Klenow enzyme, c) DNase I, d) *E. coli* DNA polymerase I, e) dNTP
3. Which of the following can be used to label DNA by primer extension: a) α-^{32}P-ATP, b) α-^{32}P-dATP, c) γ-^{32}P-ATP, d) α-^{32}P-dCTP

4. Which of the following can be used to label DNA by kinase reaction: a) α-^{32}P-ATP, b) α-^{32}P-dATP, c) γ-^{32}P-ATP, d) α-^{32}P-dCTP
5. RNA probes are usually made using: a) primer extension, b) terminal deoxynucleotide trnasferase, c) nick translation, d) in vitro transcription
6. In Southern hybridization, which of the following are required: a) limited depurination of DNA in the gel, b) denaturation, c) neutralization, d) transcription
7. In Northern hybridization, which of the following are required: a) limited depurination of RNA in the gel, b) denaturation, c) transcription, d) non of above
8. Which of the following are required to probe a target with 75% DNA sequence homology with the probe: a) high salt, b) low salt, c) high temperature, d) low temperature
9. Stringency is degree of difficulty for: a) transcription, b) replication, c) hybridization, d) translation.
10. Northern hybridization can be used to a) determine the length of a DNA fragment, b) determine the length of an mRNA, c) determine the concentration of an mRNA, d) determine the intron-exon junction.

Suggested Readings

1. al-Hakim, A. H. and Hull, R. (1988). Chemically synthesized non-radioactive biotinylated long-chain nucleic acid hybridization probes. Biochem. J. 251: 935-938.
2. Chenal, V. and Griffais, R. (1994). Chemiluminescent and colorimetric detection of a fluorescein-labelled probe and a digoxigenin-labelled probe after a single hybridization step. Mol. Cell. Probes. 8: 401-407.
3. Denhardt, D. T. (1966). A membrane-filter technique for the detection of complementary DNA. Biochem. Biophys. Res. Com. 23: 641-646.
4. Franci, C. and Vidal, J. (1988). Coupling redox and enzymatic reactions improves the sensitivity of the ELISA-spot assay. J. Immunol. Meth. 107: 239-244.
5. McGadey, J. (1970). A tetrazolium method for non-specific alkaline phosphatase. Histochemie 23: 180-184.
6. Southern, E. M. (1975). Detection of specific sequences among DNA fragments separated by gel electrophoresis. J. Mol. Biol. 98: 503-517.
7. Viale, G. and Dell'Orto, P. (1992). Non-radioactive nucleic acid probes: labeling and detection procedures. Liver. 12(4 Pt 2):243-251.
8. Melton, D. A., Krieg, P. A., Rebagliati, M. R., Maniatis, T., Zinn, K. and Green, M. R. (1984). Efficient in vitro synthesis of biologically active RNA and RNA hybridization probes from plasmids containing a bacteriophage SP6 promoter Nucleic Acids Res. 12: 7035-7056.

Chapter 9

Library

Outline

Definition and types of library
Life cycle of lambda bacteriophage
Lambda vectors
Transcription regulation of lambda bacteriophage
Induction of lambda lysogen
Factors affecting the life cycle of lambda bacteriophage
Clear mutants of lambda bacteriophage
Property and application of λgt10
Property and application of λgt11
Generalized and specialized transducers
cDNA synthesis
Packaging of lambda bacteriophage
In vitro lambda packaging
cDNA library screening
Unidirectional cDNA cloning
Subtractive cDNA library
Differential cDNA library screening

9.1: Definition and Types of Library

To study a gene, the gene must be isolated, usually from a gene library. Two types of gene libraries exist: genomic and cDNA libraries. If the chromosomal (genomic) DNA of a cell is isolated and then digested with a certain restriction enzyme, numerous DNA fragments are generated and thus numerous recombinant clones are produced when these fragments are cloned into a vector. The collection of these clones is called a genomic library. A cDNA library is the collection of recombinant clones of cDNAs derived from mRNAs of a certain type of cells. Since RNA cannot be inserted into a vector directly, it is converted to DNA and then cloned. This DNA is called cDNA because its nucleotide sequence is complementary to that of the mRNA used to make the cDNA. Since different types of cells, such as liver, muscle, and nerve cells, express different genes, many cDNA libraries exist. In contrast, only one genomic library of a certain organism exists because all cells in the organism have the same genome. To isolate the promoter or introns of a gene, a genomic library is used, whereas a cDNA library is used if the coding region, 5'UTR, or 3'UTR of a gene is to be isolated.

The basic technology for library construction is cloning in which a DNA fragment is inserted into a vector such as plasmid. The recombinant plasmid is then introduced into *E. coli* to replicate. The method used to introduce a plasmid into *E. coli* is transformation which has an efficiency of only approximately 0.1%. A mammalian cell has at least 1000 different mRNAs. Therefore, more than 1000 cDNA clones are generated when a cDNA library is constructed. If these recombinant clones are introduced into *E. coli* by transformation, only one per 1000 recombinant plasmids is taken up by *E. coli*. The great majority of them are lost, and it is very likely the clone that carries the gene of interest is among the ones that are lost during transformation. Therefore, a different kind of cloning vector that can be introduced into *E. coli* more efficiently is used. Bacteriophages enter *E. coli* by infection that has an efficiency of almost 100%; therefore, they are used as vector for library constructions.

9.2: Life Cycle of Lambda Bacteriophage

The most common bacteriophage used as a cloning vector is the lambda (λ) bacteriophage, usually referred to as lambda phage. When a lambda phage infects *E. coli*, its genomic DNA is injected into the cell. Lambda DNA then replicates in the infected cell and produce approximately 200 progenies. The infected *E. coli* is subsequently lysed and killed. This life pathway is therefore referred to as the lytic cycle. Alternatively, the injected lambda genome integrates into *E. coli* genome. Integration of the lambda genome does not affect the viability of infected *E. coli*; it actually behaves like a normal *E. coli*. However, if this *E. coli* is irradiated with ultra violet light (UV), the integrated lambda genome is excised. The excised lambda DNA then replicates to produce lambda progenies that are released when the infected cell is lysed. Since the *E. coli* containing the integrated lambda genome has the potential to allow the lambda phage to become lytic, it is called lysogen. The life pathway in which lambda genome integrates into *E. coli* genome is referred to as lysogenic cycle. A bacteriophage which has both lytic and lysogenic life cycles is a temperate bacteriophage, and the integrated bacteriophage genome is called prophage.

To grow lambda phage, host *E. coli* cells are placed in a test tube and then infected with lambda phage, followed by addition of molten 0.75% agar. The mixture is then poured on an agar plate and spread evenly. After an overnight incubation, the plate will exhibit a lawn of *E. coli* with some translucent small circular zones, referred to as plaques. Plaques are formed because infected *E. coli* cells are killed. However, lambda plaques are not completely clear because lambda phage has both lytic and lysogenic cycles. In a plaque, some lambda phages are in lysogenic state, and the lysogens continue to replicate forming a film of bacteria in the background of the plaque. This type of plaque is called a turbid plaque.

Lambda genome is a double-stranded DNA of approximately 50 kb with a 12-base overhang at the 5' end of each strand. It is linear but becomes circular when it is injected into *E. coli* due to annealing of the two complementary overhangs. Since the overhang at each end allows the two ends of the lambda genome to readily join together, it is called cohesive site (cos) (Fig. 9-1). Major lambda proteins include A – J that make lambda head and tail; Int and Xis that control the integration and excision of the lambda genome; CIII, N, CI, Cro, and CII that regulate

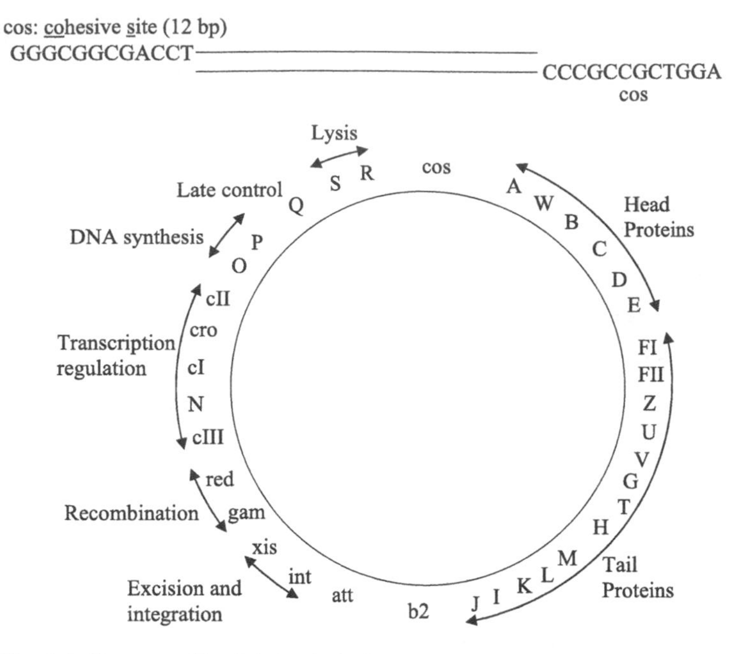

Fig. 9-1: Genome of lambda bacteriophage

transcription of lambda genes; O and P that initiate lambda DNA replication; and R (holin) and S (endolysin) that lyse infected *E. coli*. Lambda genome reproduces by both θ and rolling-circle replications (Fig. 9-2).

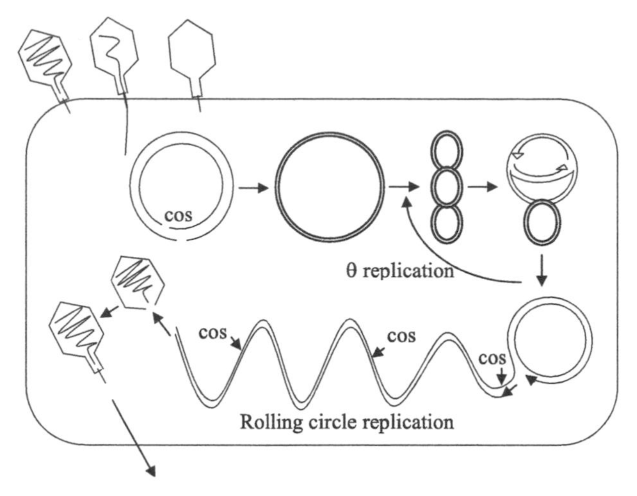

Fig. 9-2: Replication of the genome of lambda phage.

9.3: Lambda Vectors

Several lambda vectors have been developed; one example of such is the Charon 3A vector (Fig. 9-3). In the lambda genome, the region between J and *gam* genes is not essential for lambda replication and thus can be replaced with a foreign DNA fragment. A DNA fragment can also be inserted into its SalI, XbaI, XhoI, or EcoRI site. Most of the currently used lambda vectors are derived from λgt10, λgt11, EMBL3, and EMBL4 (Fig. 9-4) because they have better markers for selection of recombinant clones. λgt10 and λgt11 are used to construct cDNA libraries, and EMBL3 and EMBL4 are for construction of genomic libraries. λgt11 contains the *lacZ* gene, and insertion of a DNA fragment into the EcoRI site on the *lacZ* gene renders λgt11-infected *E. coli* unable to produce β-galactosidase. Therefore, plaques of recombinant λgt11 are colorless when infected cells are grown on agar plates containing X-gal and IPTG, while those of wild type λgt11 are blue. λgt10 has an EcoRI site in its *cI* gene. Insertion of a DNA fragment into this EcoRI site will inactivate the *cI* gene. A cI^- λgt10 produces clear plaques, whereas the wild type cI^+ λgt10 forms lysogens.

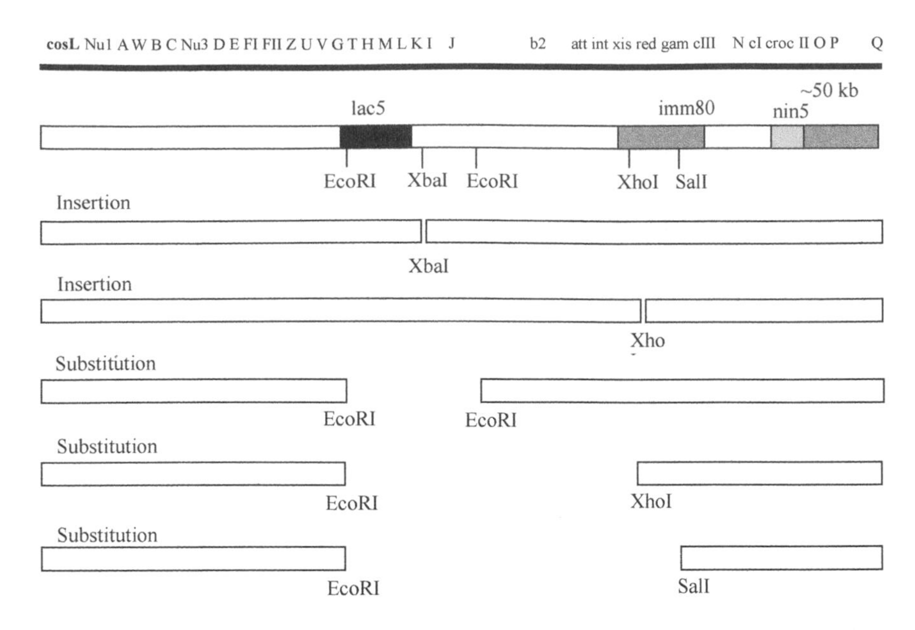

Fig. 9-3: Lambda phage vector Charon 3A.

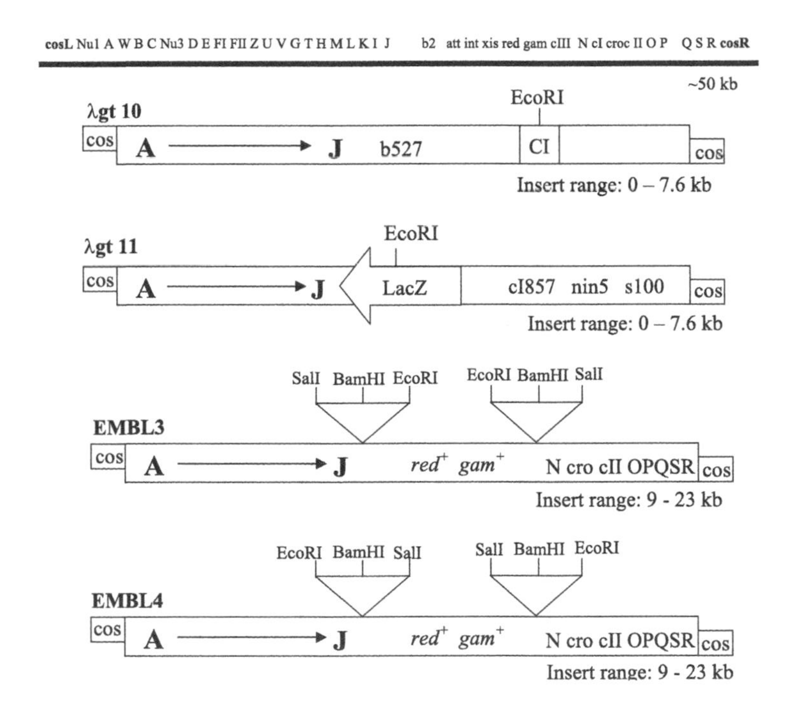

Fig. 9-4: λgt10, λgt11, EMBL3, and EMBL4 vectors.

9.4: Transcription Regulation of Lambda Bacteriophage

The genome of lambda phage has two major promoters: P_R and P_L. The transcription operators associated with these two promoters are O_R and O_L, respectively. Several transcription terminators including tR1, tR2, tR3, tR4, and tL1 are present (Fig. 9-5). When lambda DNA is injected into *E. coli*, transcription is initiated from P_R and terminated at tR1. Another transcription is initiated from P_L and terminated at tL1. The mRNA transcribed from the P_L promoter encodes the N protein which is an anti-terminator and anti-terminates tR1 so that the transcription initiated from P_R is extended to tR2. This anti-termination allows the production of Cro, CII, O, and P proteins. O and P proteins then initiate lambda DNA replication. When more N protein is produced, P_R transcription is further extended to tR3 to produce the Q protein, in addition to the Cro, CII, O, and P proteins.

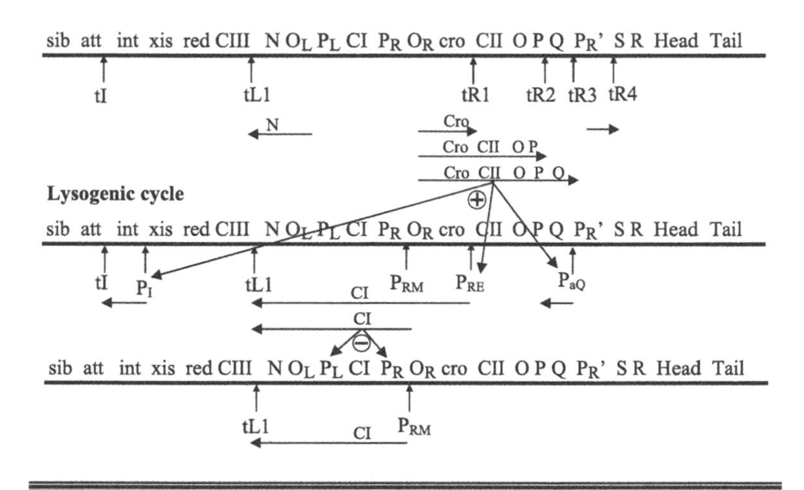

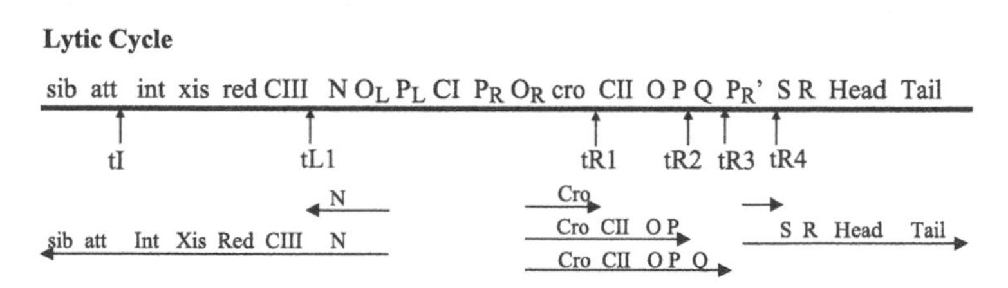

Fig. 9-5: Transcription regulation of lambda phage.

$P_{R'}$ is another promoter located on the right hand side of the lambda genome. It also becomes active as soon as the lambda DNA is injected into *E. coli*. This transcription, however, terminates immediately at tR4 and produces no proteins. When the Q protein is produced, it anti-terminates tR4 and extends the $P_{R'}$ transcription to the right-hand end of the lambda genome. The resulting mRNA encodes R, S, and all head and tail proteins of lambda phage. The head and tail proteins package the replicated lambda genome, thus producing infectious lambda particles. The S protein is holin which forms holes on the cell membrane allowing the R protein which is endolysin to digest the cell wall and lyse infected *E. coli*. This is the lytic life cycle of lambda phage.

The CII protein is a very important transcription regulator of lambda phage. It activates P_I, P_{RE}, and P_{aQ} promoters. The P_I promoter transcribes the *int* gene to produce integrase that inserts lambda genome into *E. coli* genome. The P_{aQ} promoter transcribes a small anti-sense RNA that anneals to the Q region of the longest P_R transcript, thus inhibiting the production of the Q protein. Without the Q protein, tR4 is not anti-terminated, and R, S, and all head and tail proteins are not made. Therefore, lambda particles are not produced. When the P_{RE} promoter is activated, the CI protein is produced. The CI protein in turn activates the P_{RM} promoter to produce more CI protein. CI protein is also called lambda repressor. It binds to O_R and O_L and turns off both P_R and P_L promoters, shutting down the production of almost all lambda proteins. This is the lysogenic stage of the lambda life cycle. During this stage, the only lambda protein produced is the CI protein since the P_{RM} promoter remains active once it is activated until lambda switches to lytic cycle. When the P_{RM} promoter becomes active, the P_{RE} promoter is turned off. Therefore, the P_{RE} promoter initiates production of the lambda repressor, and the P_{RM} promoter maintains continuous production of the repressor, thus RE and RM for repressor establishment and repressor maintenance, respectively.

9.5: Induction of Lambda Lysogen

When a lambda lysogen is irradiated with UV light, its RecA protein is activated. Activated RecA degrades the CI protein (lambda repressor). Without the CI protein to bind to O_R and O_L, the two major lambda promoters P_R and P_L regain their functions. The N protein is again produced to anti-terminate tL1, tR1, and tR2. When the function of tL1 is blocked, the P_L transcript is extended to the left-hand end of the lambda genome, and excisionase together with other lamda proteins are produced. The excisionase cuts the integrated lambda genome out of the *E. coli* genome. When tR1 and tR2 are anti-terminated, the mRNA encoding Cro, CII, O, P, and Q proteins are produced. The Q protein again anti-terminates tR4, allowing R, S, and all head and tail protein to be produced. O and P proteins allow the excised lambda DNA to replicate, and the replicated lambda DNAs are packaged by the lambda head and tail proteins to produce lambda progenies. The lambda phage thus enters the lytic cycle again.

Conversion of lambda from lysogenic to lytic cycle is called induction, and agents that can induce lambda prophage to become lytic are called inducers. In addition to UV, the following physical or chemical agents can also induce lambda lysogens: X-ray, α-ray, benzapyrene, 7, 12-dimethylbenzanthracene, aflatoxin, bleomycin, daunorubicin, mitomycin C, neocarzinostatin, aminopterin, trimethoprim, nalidixic acid, oxolinic acid, and 5-fluorouracil. Introduction of an F factor or P1 bacteriophage into a lysogen may also induce lambda prophage.

When tL1 is anti-terminated by the N protein, the P_L transcript is extended all the way to the end of the lambda genome. This mRNA encodes both integrase and excisionase. However, the lambda phage is entering lytic cycle and does not want integrase to be produced at this stage. Therefore, lambda phage uses a mechanism called retroregulation to prevent the production of integrase. The full-length P_L transcript can form a special conformation in the *sib* (site in the b) locus located near the 3' end of the transcript. This conformation is recognized by RNase III which degrades the *sib* region in which the integrase gene is located. Therefore, no integrase is produced during the lytic stage of the life cycle. In the lysogenic stage, the integrase gene is transcribed from the P_I promoter and terminated at the terminator tI (Fig. 9-5). This mRNA does not have the *sib* region and therefore is not degraded by RNase III.

9.6: Factors affecting the Life Cycle of Lambda Phage

Whether a lambda phage enters lytic or lysogenic cycle was believed to be controlled by Cro (control of repressor and other stuff) and CI as both proteins bind to O_R and O_L. It was thought that binding of CI and Cro to O_R and O_L has an opposite effect with Cro activating and CI inhibiting P_R and P_L. Therefore, the competition between Cro and CI for O_R and O_L determines whether a lambda phage becomes lytic or lysogenic. If Cro binds first, the lambda phage becomes lytic, but lysogenic if CI binds first. This theory persisted for at least 25 years until 2005. It is now believed that the CI protein alone controls whether a lambda phage goes lytic or lysogenic.

Both O_R and O_L are composed of three regions: O_R1, O_R2, and O_R3 and O_L1, O_L2, and O_L3 (Fig. 9-6). P_R is located between O_R1 and O_R2, and P_L is located between O_L1 and O_L2. P_{RM} is located between O_R2 and O_R3 and transcribes toward O_R3. Although both CI and Cro can bind to O_R and O_L, their binding affinity of these operator regions are different. The binding affinity for CI is $O_R1 > O_R2 > O_R3$ and $O_L1 > O_L2 > O_L3$, and that for Cro is $O_R3 > O_R2 > O_R1$ and $O_L3 > O_L2 > O_L1$. When P_{RE} is activated, the CI protein thus produced binds to O_R1 and O_L1 and turns off both P_R and P_L. When more CI protein is produced, O_R2 is also bound by CI thus activating P_{RM} to produce more CI. If excess CI is present, O_R3 is also bound by CI. This binding turns off P_{RM} and the production of CI. In the absence of CI, P_R and P_L become active, and Cro is produced. Cro has a very high affinity to O_R3 and O_L3; its binding to O_R3 prevents P_{RM} from being activated. Therefore, no CI protein is produced, and the lambda phage goes lytic.

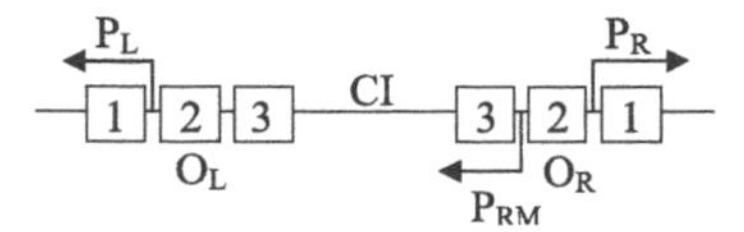

Fig. 9-6: O_R and O_L of lambda phage and locations of P_R, P_L, and P_{RM}. Arrows indicate transcription directions.

9.7: Clear Mutants of Lambda Bacteriophage

Approximately 1 million lambda phages are present in a single lambda plaque. Some of these phages are lytic and some are lysogenic. Since lysogens behave like normal *E. coli* and can proliferate, plaques of wild type lambda are not completely clear. However, some lambda mutants always make clear plaques. These mutants can be classified as CI, CII, and CIII mutants. The CI mutant is defective in the *cI* gene and thus produces no CI or non-functional CI. Therefore, both O_R and O_L are open, and P_R and P_L are active. The lambda phage, therefore, is always lytic and produces clear plaques. The CII mutant does not produce a function CII protein. Therefore, P_{RE}, P_I, and P_{aQ} promoters are not activated. If P_{RE} is not activated, no CI is produced; the lambda phage is therefore lytic. The CIII mutant is defective in the *cIII* gene. *E. coli* has a protein called Hfl (high frequency of lysogeny) protease which specifically degrades CII. Since CII protein is a very important regulatory protein, lambda produces the CIII protein that binds to CII and protects it from being degraded by Hfl. If CIII is deficient, CII will be degraded by Hfl. Without CII, no CI protein is produced, and therefore, the lambda phage goes lytic. If the *hfl* gene of *E. coli* is mutated, there will be no Hfl to degrade CII and the abundance of CII will be increased, leading to production of more CI protein and a high frequency of lambda phage becoming lysogenic. This is why the *hfl* gene is so named.

9.8: Property and Application of λgt10

λgt10 is a commonly used vector for cDNA library construction, and *E. coli* strain Y1089 is its host. The genotype of Y1089 is [F^-, Δ(*lac*) U169, *lon100, ara139, rpsL* (str^r), *hsdR17* (r_k^- m_k^+), *hflA150*::Tn10 (pMC9, *tet^r*, *amp^r*)]. "*lon*" is the abbreviation of long. The *Lon* protein is a protease involved in *E. coli* cell division. If this protein is defective, *E. coli* cells can elongate but do not divide, resulting in cells that are longer than those of wild type *E. coli*. "*hsdR17* (r_k^- m_k^+)" indicates that Y1089 is defective in the restriction function of its type I restriction-modification system. Therefore, lambda DNA is not degraded by this system when introduced. "*hflA150*::Tn10" denotes that Y1089 is defective in the *hfl* gene due to Tn10 insertion; therefore, wild type lambda phage enters lysogenic cycle at a very high frequency. λgt10 has the *b527* mutation which is a deletion in the *b2* region. The *b2* region is where the excisionase recognizes to cut the integrated lambda genome out of *E. coli* genome. A lambda phage with the *b527* mutation cannot be excised once it is integrated into *E. coli* genome. When λgt10 is used as the vector, the DNA fragment is inserted into the EcoRI site in its *cI* gene. Therefore, the recombinant λgt10 is *cI^-* and the wild type is *cI^+*. The *cI^+* λgt10 DNA will integrate into *E. coli* genome and become lysogenic in Y1089 because of the *hfl* mutation. Furthermore, the integrated λgt10 cannot be excised to become lytic because of the *b527* mutation. Therefore, only the recombinant λgt10 can form plaques. This is the mechanism by which the *cI* gene is used as a selection marker in this cloning system.

9.9: Property and Application of λgt11

E. coli strain Y1090 is a host of λgt11. Its genotype is [F^-, Δ(*lac*) *U169, lon100, ara139, rpsL* (str^r), *supF, mcrA, hsdR17* (r_k^- m_k^+), *trpC22*::Tn10 (pMC9, *tet^r*, *amp^r*)]. Y1090 has both *hsdR* and *mcrA* mutations; therefore, both the type I and MDRS restriction systems are defective. These are desired mutations so that the recombinant λgt11 introduced will not be degraded. The genotype of λgt11 is [*cI857*, *nin5* (ΔtR2), *sam100*]. Any lambda phage with the *cI857* mutation makes a temperature sensitive CI protein which is inactivated at 42°C. The ΔtR2 mutation renders transcription from the P_R promoter independent of anti-termination by the N protein. Its transcription always goes all the way to tR3, thus the genotype *nin* for N-independent. "*sam100*" indicates a mutation in the S gene due to conversion of a normal amino acid codon (sense codon) to the stop codon TAG. Since TAG is referred to as Amber codon, this mutation is designated *sam*. Because of this mutation, λgt11 does not form plaques unless its host cell has a suppressor tRNA to suppress the mutation. This is the major reason why Y1090 is used as the host for λgt11 because it has the *supF* suppressor tRNA. Y1090 also contains the plasmid pMC9 which has a *lacI*^{q} gene and over produces the LacI protein (*lac* repressor).

The three translation stop codons, TAG, TAA, and TGA are named Amber, Ochre, and Opal, respectively. The TAG codon was named by Seymour Benzer at the California Institute of Technology after his student Harris Bernstein who first discovered the TAG stop codon. Since Bernstein means Amber in German, Amber is used for the TAG codon. Amber is a golden-yellow fossil resin and is considered as a gemstone. To be consistent with the colored gemstone nomenclature, the TAA is named Ochre which is also a golden-yellow color, and TAG is named Opal which is a gemstone.

9.10: Generalized and Specialized Transducers

The abbreviation "gt" in λgt10 and λgt11 stands for generalized transducer. However, wild type lambda phage is not a generalized transducer. Transduction is one of the three ways genes can be transferred from one bacterium to another. The other two are transformation and conjugation. Transformation refers to the phenomenon in which a bacterium takes up foreign DNA, resulting in a change in its phenotype. In conjugation, DNA is transferred from one bacterium to another through the sex pilus during mating. In transduction, genetic materials of one bacterium are carried by bacteriophage and transferred to another when the bacteriophage infects other bacterium. During lysogenic cycle, the lambda genome always integrates into the attachment (*att*) site of *E. coli* genome. When the lambda genome is excised during induction, the excised lambda genome may carry a small portion of the *E. coli* genome. This lambda DNA is then packaged. When the resulting lambda progeny infects other *E. coli*, the infected *E. coli* would express the transferred gene. This process is called transduction. Since lambda phage always integrates into the same site, the genes that can be transduced by lambda phage are the ones located around the integration site, such as *bio* and *gal*. Therefore, lambda phage is considered a specialized transducer. In contrast, bacteriophage P1 can integrate into many different sites in *E. coli* genome, and different P1 progenies may carry different portions of the *E. coli* genome. When these progenies infect other *E. coli*, various genes are transduced. Therefore, P1 bacteriophage is a generalized transducer. Since both λgt10 and λgt11 can carry any foreign DNA inserted into their EcoRI site, these lambda vectors are therefore considered as generalized transducers, thus the name "gt."

9.11: cDNA Synthesis

To construct a cDNA library, mRNAs isolated from a certain type of cells are first converted to cDNA because it is not possible to directly clone RNA. DNA synthesis must have a primer. Since most mRNAs have a poly-A tail, an oligonucleotide with a run of approximately 20 T's, referred to as oligo-dT, is used to anneal to the poly-A tail of an mRNA to serve as the primer. In the presence of reverse transcriptase and dNTP, the mRNA is reversely transcribed to DNA. Since the sequence of this DNA is complementary to that of the mRNA, it is called cDNA. The original mRNA is then degraded with alkali, and the resulting single-stranded cDNA is converted to double-stranded. Since the 3' end of a single-stranded DNA may loop back, the open end of the loop is used as the primer, and the single-stranded region is used as the template to synthesize the second strand of the cDNA in the presence of the Klenow fragment of *E. coli* DNA polymerase I and dNTP (Fig. 9-7). After

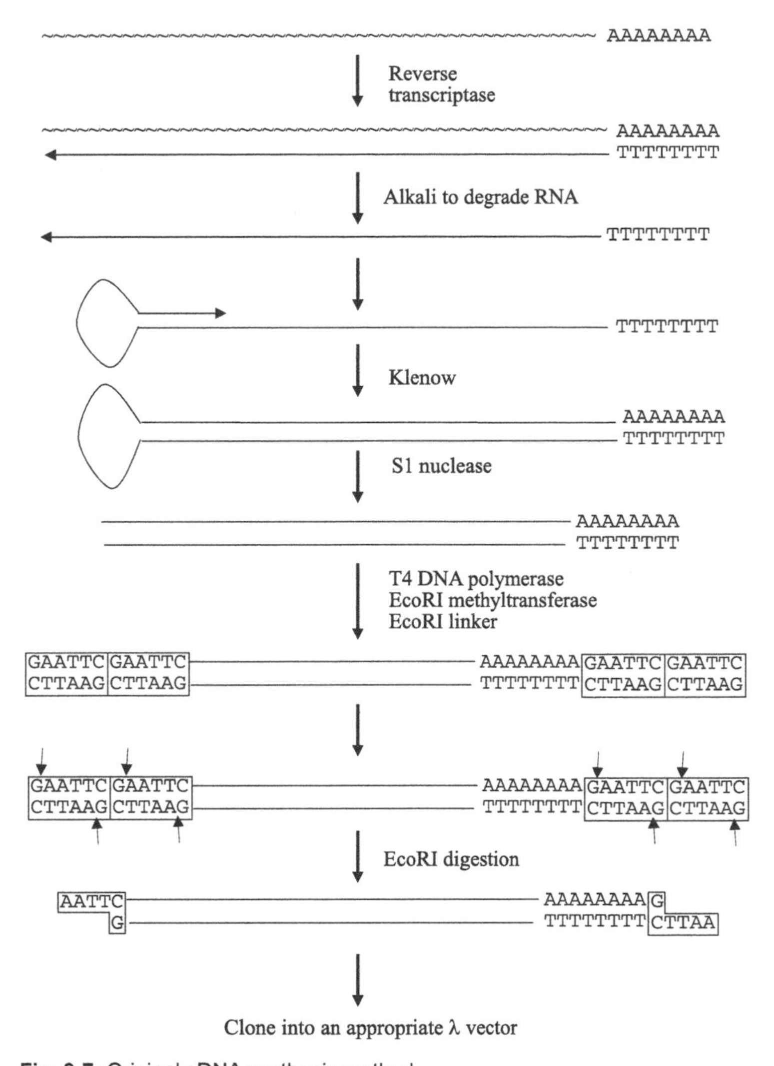

Fig. 9-7: Original cDNA synthesis method.

removal of the single-stranded loop by digestion with a single-strand specific nuclease such as S1, a double-stranded cDNA is formed. Since this DNA fragment is to be inserted into the EcoRI site of λgt10 or λgt11, EcoRI linkers are ligated to both ends of the fragment. Because

EcoRI linkers are blunt ended, the cDNA fragment is treated with T4 DNA polymerase in the presence of dNTP to ensure that it is also blunt ended. Since multiple EcoRI linkers are always ligated to each end, the excess linkers are removed by digestion with EcoRI. If EcoRI sites are also present in the cDNA fragment, this EcoRI digestion will cut the cDNA fragment into several pieces, making the cDNA too small to be cloned. To prevent this from happening, the cDNA fragment is treated with EcoRI methyltransferase to methylate these internal EcoRI sites so that the cDNA will not be cut to smaller pieces. After EcoRI digestion, the cDNA is ready to be cloned into λgt10 or λgt11.

Since the loop region, which is later removed, is the 5' end of the gene, this method of cDNA synthesis does not allow cloning of the entire gene. A modification of this method is not to degrade the mRNA by alkali treatment after the first strand cDNA is synthesized. Instead, RNase H is used to generate nicks on the mRNA (Fig. 9-8). The nicks are then extended to gaps by the 5' to 3' exonuclease activity of *E. coli* DNA polymerase I. These gaps are filled in by the 5' to 3' polymerase activity of the same enzyme. The newly synthesized DNA fragments are ligated with ligase, thus generating a double-stranded cDNA. After treatment with T4 DNA polymerase and EcoRI methyltransferase, EcoRI linkers are ligated to both ends. The excess linkers are digested away with EcoRI, and the digested cDNA fragment is cloned into λgt10 or λgt11.

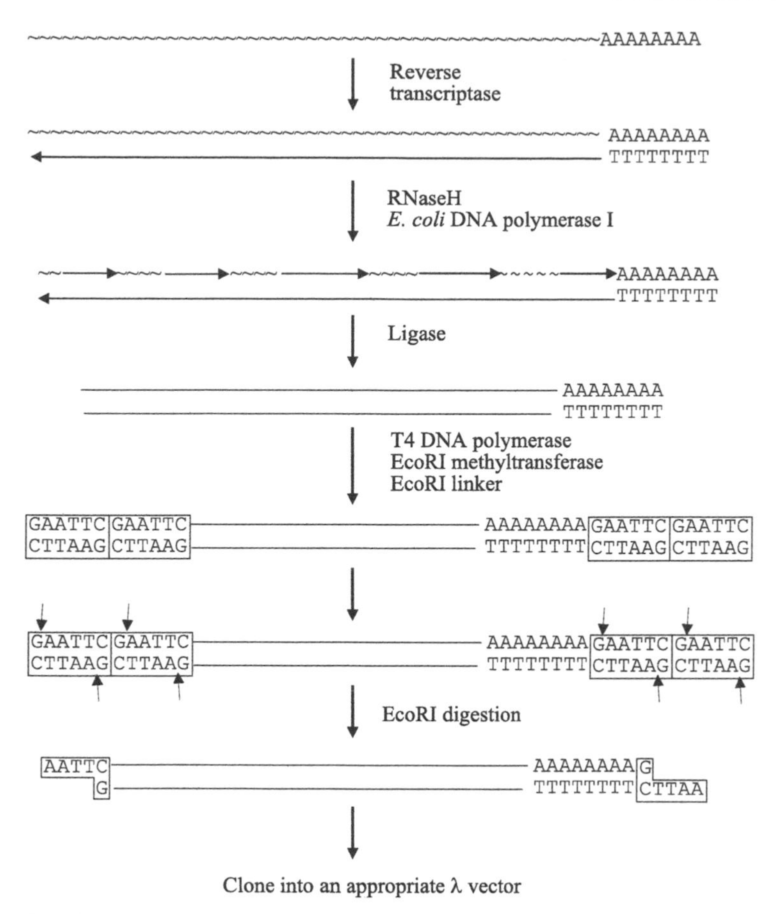

Fig. 9-8: cDNA synthesis using RNA fragments generated by RNase H digestion as primers.

If an mRNA is very long, reverse transcriptase may not copy the entire mRNA to cDNA; therefore, the cloned gene is missing the 5' end. To circumvent this problem, random primers, instead of oligo-dT, are used to prime the reverse transcription (Fig. 9-9). Since different primers may anneal to different regions of the mRNA, overlapping cDNA fragments corresponding different regions of the gene are generated. The primer that anneals close to the 5' end of the mRNA would likely lead reverse transcription all the way to the 5' end. If a gene contains a, b, c, and d regions, some cDNA fragments thus produced may contain a and b regions, others may have b and c or c and d regions. These cDNA fragments are then treated with T4 DNA polymerase and EcoRI methyltransferase, ligated with EcoRI linkers, digested with EcoRI, and separately cloned into λgt10 or λgt11. Recombinant clones containing different portions of the gene are then isolated, and overlapping fragments of the genes are joined to generate the complete gene.

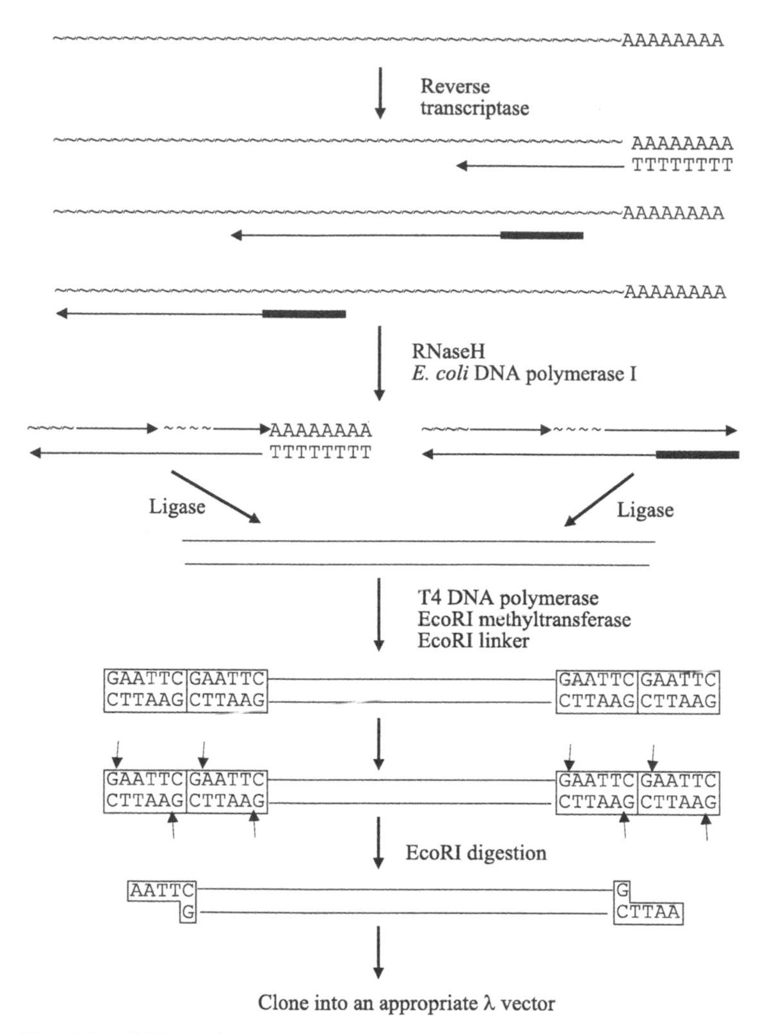

Fig. 9-9: cDNA synthesis using random primers.

9.12. Packaging of Lambda Bacteriophage

After a cDNA fragment has been inserted into a lambda vector, the recombinant lambda DNA is packaged by lambda head and tail proteins to become infectious phage particles. The packaged phage particles are then introduced into *E. coli* by infection to obtain plaques, and the plaques are screened to find the ones containing the gene of interest. To package lambda DNA, lambda head proteins E, B, Nu3, and C first form a scaffold (Fig. 9-10). The *E. coli* protein GroE is then added to the scaffold to form a prehead. Lambda DNA is then pulled into the prehead by NuI and A proteins. At the same time, FI and D proteins are incorporated into prehead to form a complete lambda head. A lambda head can package one unit of lambda genome of approximately 50 kb. When lambda DNA replicates in *E. coli*, a lambda concatemer is formed. Concatemer is a chain-like molecule containing many units of lambda genome. It is the lambda concatemer, not monomeric lambda genome, that is used for packaging. One unit of lambda genome is the lambda DNA located between two *cos* (cohesive site) sites. The A protein recognizes and cut at the *cos* site so that one unit of the lambda genome is packaged. After the lambda genome is packaged, the FII protein seals the lambda head. While the lambda head is being formed, lambda tail is also being assembled. Lambda tail proteins J, I, L, K, G, H, and M first form a stalk. The V protein is then stacked on the stalk, followed by addition of the U protein. Z and W proteins then link the head and tail together to form a complete lambda particle.

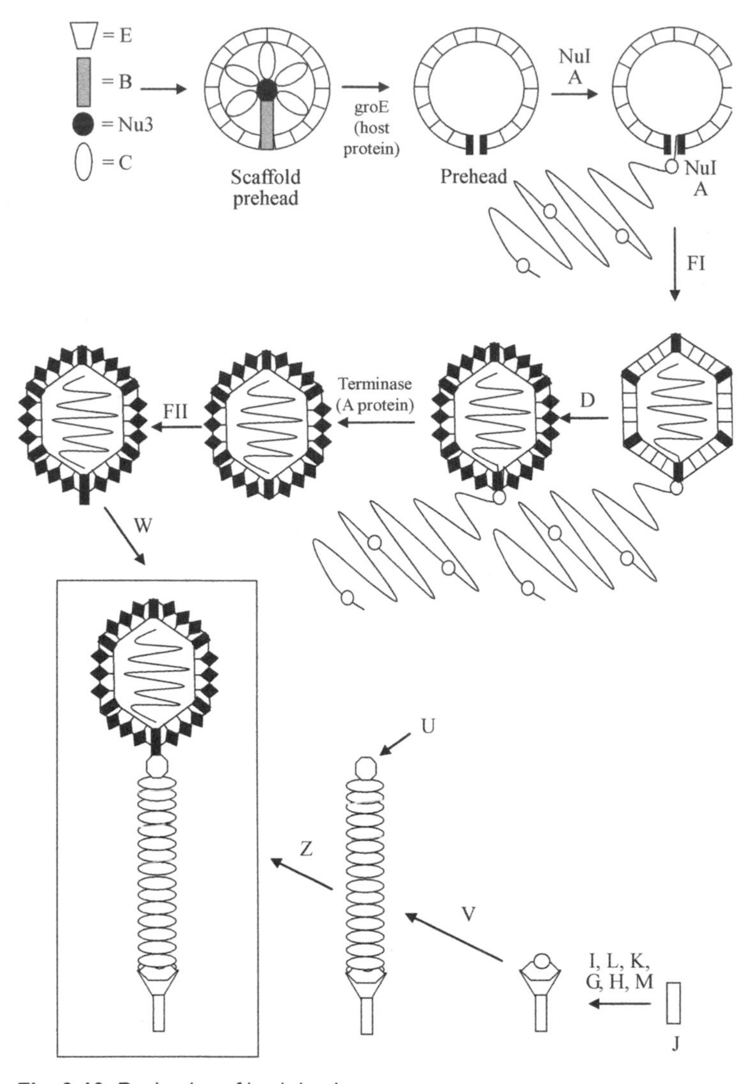

Fig. 9-10: Packaging of lambda phage.

9.13: In Vitro Lambda Packaging

Although lambda packaging is a very complicated process, it can take place in vitro if a lambda DNA concatemer and all lambda head and tail proteins are present. When a cDNA fragment is ligated with a lambda vector, a lambda concatemer may be formed, and can subsequently be packaged by the lambda packaging system. Lambda head and tail proteins are produced using lambda lysogens, such as *E. coli* strains BHB2688 and BHB2690. The genotype of BHB2688 is [F^-, *recA* (λ *imm434, cI857, b2, red3, Eam4, Sam7*)]. This genotype indicates that BHB2688 is defective in the *recA* gene, contains no F factor, and carries a lambda prophage with the genotype [*imm434, cI857, b2, red3, Eam4, Sam7*]. "*imm*" is the immunity gene of lambda phage. It makes a lambda lysogen immune to the same type of lambda phage. In other words, a lambda lysogen cannot be super infected by another lambda phage that has the same *imm* gene, e.g. the *imm434* gene. The *cI857* mutation renders the CI protein temperature sensitive. Since this lambda prophage has the *b2* mutation, it cannot be excised out of *E. coli* chromosome after induction. "*red3*" indicates that the prophage is defective in the *red* (recombination deficient) gene which is functionally equivalent to the *E. coli recA* gene. "*Eam4*" indicates that the E head protein of the lambda phage is defective, and this mutation is due to conversion of a sense codon to the Amber stop codon (TAG). The prophage also has the *Sam* mutation on the S gene and therefore does not make endolysin to lyse the lysogen; this mutation also results from a change of a sense codon to the Amber stop codon. The genotype of BHB2690 is [F^-, recA (λ *imm434, cI857, b2, red3, Dam4, Sam7*)]. The only difference between BHB2688 and BHB2690 is that BHB2690 has the *Dam4* mutation instead of *Eam4*. "*Dam4*" means that the D head protein of the prophage in BHB2690 is defective.

To make lambda proteins for in vitro packaging, BHB2688 and BHB2690 are separately grown in an appropriate medium such as LB broth at 30°C to early log phase. The cultures are then heated at 42°C for 15 minutes to inactivate the CI protein which is temperature sensitive because of the *cI857* mutation. Without the CI protein, both P_R and P_L become active, and the prophage makes all lambda proteins. BHB2688 makes all head proteins except the E protein, and BHB2690 makes all except the D protein. Since the S gene of the prophage in both lysogens is defective, the cells are not lysed. Furthermore, the prophage is not excised because of the *b2* mutation. Therefore, the two lysogens become a factory of lambda proteins once they are induced by heating. The cells are further grown at 37°C for one hour and harvested by centrifugation. The cells are then lysed by sonication, and the two cell lysates are mixed together to package lambda DNA. The major reason why two different lysogens with mutations in different head protein genes are used is to prevent self packaging of the prophage if it is accidentally excised.

9.14: Screening of cDNA Libraries

The packaged recombinant lambda phages are then used to infect an appropriate *E. coli* host. When the infected *E. coli* cells are plated on an agar plate as a lawn, plaques are formed. Each plaque may contain a recombinant lambda clone of a certain gene. If a cell has 3000 different mRNAs, there should be approximately 3000 different clones. To determine which plaque contains the desired clone, a labeled DNA probe of the gene of interest is used to hybridize with the plaques. The ones that reacted with the probe are isolated and identified. This process is called cDNA library screening. To perform cDNA library screening, lambda plaques on an agar plate are first transferred to a circular nitrocellulose membrane by placing it on the plate to bind the plaques. The membrane is then lifted and placed, plaque side up, on a filter paper saturated with 0.5 M NaOH/1.5 M NaCl for approximately 5 minutes. This process lyses all the bacteria and lambda phages on the membrane, exposes their DNA, and denature the DNA to single-stranded. The membrane is then transferred to a second filter paper saturated with 1.5 M NaCl/0.5 M Tris (pH 7-8) for 5 minutes to neutralize the negative charges of DNA. The salt concentration on the membrane is then reduced by placing the membrane on the third filter paper saturated with 2x SSPE (300 mM NaCl, 20 mM NaH_2PO_4, 2 mM EDTA, pH 7.4) for 5 minutes. The membrane is air dried, baked in an 80°C vacuum oven, prehybridized, and hybridized in the same manner as Southern hybridization.

The gene that is cloned into a lambda vector may be expressed in infected *E. coli*. Therefore, cDNA libraries can also be screened for desired clones with antibodies against the protein encoded by the gene of interest. For this purpose, λgt11 or its derivatives are used as the vector. The cDNA is cloned into the *lacZ* gene of the vector. If the gene is fused in frame with *lacZ*, a fusion protein is produced which may react with the antibody. Since a cDNA fragment with EcoRI at both ends may be inserted in opposite orientations into the EcoRI site of the vector, two different types of recombinant clones are produced. One has the DNA fragment inserted in the same orientation as the *lacZ* promoter, and the other has the fragment inserted in a wrong orientation. This clone will not produce a peptide that can be recognized by the antibody. To prevent the cDNA fragment from being inserted in the wrong orientation, a different cDNA cloning method is used in which the cDNA fragment is cloned in the same orientation as the transcriptional direction of the promoter used to drive the fusion gene.

9.15: Unidirectional cDNA Cloning

To perform unidirectional cloning, two different restriction sites such as EcoRI and XhoI are generated at the two ends of a cDNA fragment (Fig. 9-11). The primer with the sequence CTCGAGTTTTTTTTTT is used to anneal to the mRNA to synthesize the first strand cDNA by reverse transcriptase, in which CTCGAG is the recognition sequence of XhoI. RNase H, *E. coli* DNA polymerase I, and T4 DNA ligase are then used to synthesize the second strand cDNA. This double-stranded cDNA now has the XhoI site at its 3' end. An adaptor composed of two oligonucleotides with the sequences AATTCGGCAG and CTGCCG are ligated to both ends of the cDNA fragment, creating the AATT overhang at the 5' end. The adaptor that is ligated to the 3' end of the cDNA fragment is removed by digestion with XhoI, creating the TCGA overhang. Since AATT is compatible with EcoRI end and TCGA is compatible with XhoI end, the cDNA fragment can be inserted between EcoRI and XhoI sites of an appropriate lambda vector, thus ligating the fragment with the vector in only one orientation. This will increase the chance of the gene being expressed. Since internal XhoI sites may be present in the cDNA fragment, methylated dCTP is used to synthesize the second cDNA strand so that all the C's incorporated are methylated, and thus the cDNA fragment will not be cleaved by XhoI when it is used to remove the adaptor ligated to its 3' end.

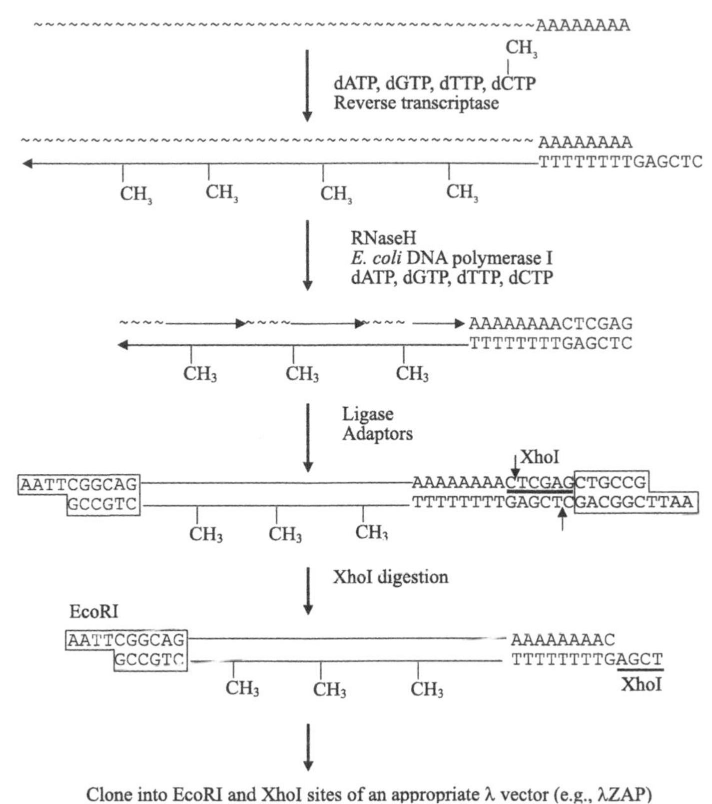

Fig. 9-11: Unidirectional cloning of cDNA.

9.16: Subtractive cDNA Library

A subtractive cDNA library is used to isolate genes that are expressed when the cell is stimulated with physical or chemical agents such as UV, hormones, or drugs. The basic principle is to remove the mRNAs that are not needed. Assuming that a cell normally expresses four different mRNAs designated A, B, C, and D and that treatment of the cell with a certain drug induces the expression of one extra mRNA designated E, a subtractive cDNA library can be constructed to specifically isolate the E gene. The A, B, C, and D mRNAs are first isolated from untreated cells and then converted to cDNAs. These mRNAs are then degraded with alkali, and the resulting single-stranded a, b, c, and d cDNAs are annealed with the mRNAs isolated from drug treated cells. Since there is no E mRNA in untreated cells, the E mRNA will not anneal to any of the cDNAs derived from the mRNAs of untreated cells. This E mRNA thus remains single-stranded, whereas the A, B, C, D mRNAs, which are not wanted, hybridized with the cDNA and are double-stranded. When this mixture is passed through a column packed with hydroxyapatite, the double-stranded cDNA-mRNA hybrids are retained in the column, and the single-stranded E mRNA flows through the column. The cDNA library constructed using the E mRNA-enriched preparation is a subtractive cDNA library from which the desired E gene can be easily isolated.

9.17. Differential cDNA Library Screening

Genes that are expressed after induction such as treatment with a certain drug are called differentially expressed genes. In addition to the subtractive cDNA library, these genes can also be isolated by differential cDNA library screening. A cDNA library is constructed using mRNA from drug treated cells and plated on an agar plate to form plaques. The plaques are transferred to two nitrocellulose membranes in an identical manner. One membrane is then hybridized with labeled cDNA probes from drug treated cells. In this hybridization, all plaques on the membrane will hybridize with the probe. The other membrane is hybridized with labeled cDNA probes from untreated cells. Plaques that do not hybridize with this probe are likely the ones containing the differentially expressed genes.

Summary

Two types of gene libraries, genomic and cDNA, exist. A genomic library is used to isolate promoter, regulatory regions, 5' and 3' untranslated regions, or introns of a gene, whereas a cDNA library is used to isolate the coding region of a gene. Genomic library is a collection of recombinant clones containing DNA fragments of genomic DNA of a certain organism or cell. A cDNA library is a collection of recombinant clones of cDNAs derived from mRNAs of a certain type of cells. A cloning vector is required to construct a library. If plasmid vectors are used, the recombinant plasmids are introduced into *E. coli* to replicate by transformation. Since the efficiency of transformation is only approximately 0.1%, the great majority of the recombinant plasmids are lost. Therefore, bacteriophages are used as the vector for library construction because they enter *E. coli* by infection which has an efficiency of approximately 100%. The most commonly used bacteriophage vectors are lambda phage vectors such as λgt10, λgt11, EMBL 3, and EMBL 4 or their derivatives. The lambda phage has two different life cycles: lysogenic and lytic. In the lysogenic cycle, the genome of lambda phage is integrated into *E. coli* genome. This integrated lambda genome is called prophage, and *E. coli* which carries a prophage is called lysogen which behaves like a normal *E. coli*. If a lysogen is irradiated with UV or treated with certain physical or chemical agents, the prophage is excised and replicated to produce approximately 200 lambda progenies. The *E. coli* host is then lysed, and the lambda progenies are released to infect other *E. coli* cells. This life pathway is called lytic cycle. When lambda-infected *E. coli* cells are plated as an overlay on an agar plate, plaques are seen. Plaques are formed because infected *E. coli* cells are killed. Therefore, the area that contains killed *E. coli* is translucent. However, lambda plaques are not completely clear. They are turbid because some lambda phages in the plaque become lysogenic, and the lysogen continues to proliferate, forming a thin film in the background of the plaque.

Some lambda phage mutants, such as CI, CII, and CIII mutants, always produce clear plaques. CI mutant has a defect in the *cI* gene and produces no lambda repressor to inhibit the function of the two major promoters P_R and P_L of lambda phage. Therefore, the lambda phage is always lytic and produces clear plaques. The CII protein activates P_{RE}, P_I, and P_{aQ} promoters of lambda phage. When the P_{RE} promoter is activated, lambda phage produces the CI protein which is the lambda repressor. CI binds to O_R and O_L, thus suppressing both P_R and P_L promoters so that proteins required for lambda DNA replication are not made. When the P_I promoter is activated, lambda integrase is produced which inserts the lambda genome into *E. coli* genome. When P_{aQ} promoter is activated, a small anti-Q RNA is produced. This RNA anneals to the Q region of the longest mRNA derived from the P_R promoter, thus inhibiting the production of the Q protein. Without the Q protein, lambda head and tail proteins and R and S proteins required to package lambda phages and to lyse host *E. coli* are not made, and no infectious phage particles are produced. The infected *E. coli* thus becomes a lysogen. Since CII is required for lambda phage to become lysogenic, its defect will render the lambda phage always in the lytic cycle. The CII protein is readily degraded by the Hfl protein of *E. coli*. Therefore, lambda phage makes the CIII protein to bind to CII and protect it from being degraded by Hfl. If the *cIII* gene is defective, CII will be degraded and lambda phage can only go lytic cycle. Therefore, any lambda phage with mutations on *cI*, *cII*, or *cIII* gene always produces clear plaques.

When a cDNA fragment is inserted into the EcoRI site of λgt10, its *cI* gene is destroyed. Therefore, the recombinant λgt10 will produce clear plaques when an appropriate host such as Y1089 is used. Wild type λgt10 will become lysogenic because it can produce the CI protein. λgt10 also has the *b527* mutation; its DNA cannot be excised once it is integrated into *E. coli* genome. In addition, Y1089 has the *hfl* mutation and does not make the Hfl protein to degrade CII. Therefore, the level of CI protein is very high and the wild type λgt10 becomes lysogenic at a very high frequency. Because of these reasons, all plaques formed with Y1089 contain recombinant λgt10. When λgt11 is used as the vector, the cDNA fragment is inserted into the EcoRI site of its *lacZ* gene. Therefore, the recombinant will not be able to produce β-galactosidase, and its plaques will be colorless when plated on an agar plate containing X-gal and IPTG. The wild type λgt11, on the other hand, will produce blue plaques on agar plates containing X-gal and IPTG.

To construct a cDNA library, mRNAs in a certain type of cells are isolated. Since most mRNAs have a poly-A tail, oligo-dT is used to anneal to the poly-A tail to copy mRNA to DNA by reverse transcription. The sequence of the DNA thus synthesized is complementary to that of the mRNA; therefore, it is called cDNA. This single-stranded cDNA is then converted to double-stranded. The most commonly used method employs RNase H to create nicks on the mRNA of the mRNA-cDNA hybrid. The nicks are extended to gaps and then filled with nucleotides by *E. coli* DNA polymerase I. The DNA fragments thus synthesized are ligated with ligase, and made blunt ended by treating with T4 DNA polymerase in the presence of dNTP. EcoRI linkers are then ligated to the ends of the cDNA fragment after it has been treated with the EcoRI methyltransferase. Excess EcoRI linkers ligated are removed by EcoRI digestion, and the digested cDNA fragment is inserted into the EcoRI site of λgt10 or λgt11. The recombinant lambda phages are then packaged to become infectious phage particles by in vitro packaging in which the lambda head and tail proteins are produced using appropriate lambda lysogens.

The packaged recombinant lambda phages are mixed with an appropriate host at 37°C for 30 minutes. Molten 0.75% agar is added, and the mixture is spread evenly on an agar plate. After an overnight incubation at 37°C, plaques will be seen. The plaques are transferred to a nitrocellulose membrane, and the membrane is reacted with labeled DNA probes derived from the gene of interest to determine which plaque contains the desired recombinant clone. The membrane can also be reacted with antibody against the protein encoded by the gene of interest to isolate the desired recombinant clones.

Sample Questions

1. Which of the following is used if one wishes to study a promoter: a) cDNA library, b) genomic library, c) cDNA microarray, d) none of above.
2. Which of the following is used if one wishes to study a coding region: a) cDNA library, b) genomic library, c) cDNA microarray, d) none of above.
3. Genomic or cDNA library is usually constructed using: a) pBR322, b) pBluescript, c) lambda bacteriophage vectors, d) pUC19.
4. A normal lambda bacteriophage produces a) clear plaques, b) turbid plaques, c) no plaques, d) triangular plaques.
5. The CI protein of lambda phage: a) activates P_R and P_L promoters, b) inactivates P_R and P_L promoters, c) cause lambda phage to go lytic cycle, d) cause lambda phage to go lysogenic cycle.
6. The CII protein of lambda phage activates: a) P_R and P_L promoters, b) $P_{R'}$ promoter, c) P_{RE} promoter, d) PI promoter.
7. The CIII protein of lambda phage protects which of the following protein from degradation by the Hfl protease: a) CI, b) CII, c) integrase, d) Cro.
8. Which of the following is responsible for the integration of lambda DNA into *E. coli* chromosome: a) excisionase, b) integrase, c) RecA protein, d) CI protein.
9. Which of the following is responsible for the excision of lambda DNA out of *E. coli* chromosome: a) excisionase, b) integrase, c) RecA protein, d) CI protein.
10. The O and P proteins of lambda phage are involved in: a) transcription initiation from P_R and P_L promoters, b) replication of lambda DNA, c) lysis of *E. coli* host, d) production of lambda head and tail proteins.
11. The R and S proteins of lambda phage are involved in: a) transcription initiation from P_R and P_L promoters, b) replication of lambda DNA, c) lysis of *E. coli* host cells, d) production of lambda head and tail proteins.
12. Lambda proteins required for making an infectious lambda particle are usually: a) isolated from lambda particles, b) made using certain type of lambda lysogens, c) in vitro translation, d) in vitro transcription.

Suggested Readings

1. Black, L. W. (1989). DNA packaging in dsDNA bacteriophages. Annu. Rev. Immunol. 43: 267-292.
2. Chauthaiwale, V. M., Therwath, A., and Deshpande, V. V. (1992). Bacteriophage lambda as a cloning vector. Microbiol. Rev. 56: 577-591.
3. Enquist, L. and Sternberg, N. (1979). In vitro packaging of λ Dam vectors and their use in cloning DNA fragments. Meth. Enzymol. 68: 281-309.
4. Johnson, A. D., Poteete, A. R., Lauer, G., Sauer, R. T., Ackers, G. K., and Ptashne, M. (1981). λ repressor and cro - component of an efficient molecular switch. Nature 294: 217-223.
5. Herskowitz, I. and Hagen, D. (1980). The lysis-lysogeny decision of phage λ: explicit programming and responsiveness. Ann. Rev. Genet. 14: 399-445.
6. Huynh, T. V., Young, R. A., and Davis, R. H. (1985). Construction and screening of

cDNA libraries in λgt10 and λgt11. DNA Cloning: A practical approach (D. M. Glover, ed) Vol. 1, pp. 49-78. IRL Press Limited, Oxford, England.

7. Martinson, H. G. (1973). The basis of fractionation of single-stranded nucleic acid on hydroxyapatite. Biochemistry 12: 2731-2736.
8. Pabo, C. O. and Lewis, M. (1982). The operator-binding domain of lambda repressor: structure and DNA recognition. Nature 298: 443-447.
9. Ptashne, M. (1992). Genetic switch: Phage lambda and higher organisms (Cell press and Blackwell Scientific, Cambridge).
10. Ptashne, M. (1967). Specific binding of the lambda phage repressor to DNA. Nature 214: 232-234.
11. Short, J. M., Fernandez, J. M., Sorge, J. A., and Huse, W. D. (1988). λ ZAP: a bacteriophage λ expression vector with in vivo excision properties. Nucleic Acids Res. 16: 7583-7600.
12. Svenningsen, S. L, Costantino, N., Court, D. L., and Adhya, S. (2005). On the role of Cro in λ prophage induction. Proc. Natl. Acad. Sci. USA 102: 4465-4469.
13. Young, R. A. and Davis, R. W. (1983). Efficient isolation of genes by using antibody probes. Proc. Natl. Acad. Sci USA 80: 1194-1198.

Chapter 10

Restriction Mapping

Outline

Verification of recombinant plasmid

Identification of restriction sites on a plasmid

10.1: Verification of recombinant plasmids

Restriction mapping is the technique used to determine the cutting site of a certain restriction enzyme on a plasmid or DNA fragment. It is most often used to verify recombinant plasmids. During cloning, a DNA fragment is ligated with a certain vector. Since many different recombinant plasmids with various vector and insert combinations may be generated, the recombinant plasmids are isolated from *E. coli* and screened for the correct one. The intact plasmid DNA is first electrophoresed along with the original intact vector in an agarose gel. If the recombinant plasmid is found to be bigger than the original vector, its insert is cut out and further examined. For example, if a 1.5-kb EcoRI fragment is inserted into the EcoRI site of a vector, a 1.5-kb fragment is generated when the recombinant plasmid is digested with EcoRI.

Since the insert may be ligated into the vector in different orientations, its orientation is then determined. If the vector is 3.0 kb in size and has a BamHI site located 0.2 kb from the EcoRI site (Fig. 10-1), a BamHI digestion of the plasmid will result in two fragments of 1.2 kb (1.0 kb + 0.2 kb = 1.2 kb) and 3.3 kb (2.8 kb + 0.5 kb = 3.3 kb) when the DNA fragment is inserted into the EcoRI site in a left-to-right orientation (Fig. 10-1A). In contrast, 0.7 kb (0.2 kb + 0.5 kb = 0.7 kb) and 3.8 kb (2.8 kb + 1.0 kb = 3.8 kb) fragments will be seen if the insert is ligated in a right-to-left orientation (Fig. 10-1B).

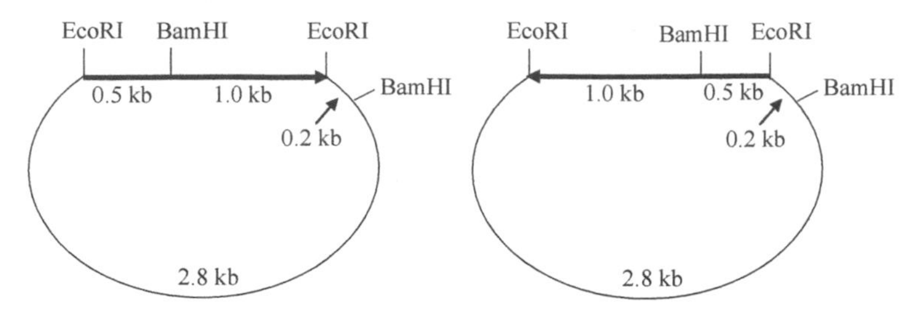

Fig. 10-1: DNA fragment inserted into the vector in two different orientations.

Some recombinant plasmids may contain multiple inserts. If two DNA fragments are ligated in tandem in a left-to-right orientation in the vector (Fig. 10-2a), digestion of the recombinant plasmid with BamHI will generate 3.3 kb, 1.5 kb, and 1.2 kb fragments. If these two fragments are ligated in tandem in a right-to-left orientation, 3.8 kb, 1.5 kb, and 0.7 kb fragments will result after BamHI digestion (Fig. 10-2c). On the other hand, 3.8 kb, 1.2 kb, and 1.0 kb fragments will be generated by BamHI digestion (Fig. 10-2b) if the two fragments are joined with the vector in a tail-to-tail orientation. In contrast, 3.3 kb, 2.0 kb, and 0.7 kb fragments will be seen after BamHI digestion if the two fragments are joined with the vector in a head-to-head orientation (Fig. 10-2d).

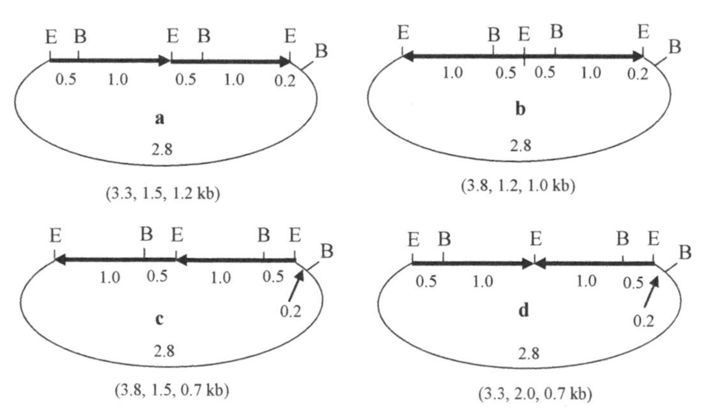

Fig. 10-2: Two DNA fragments are inserted into the vector in different orientations. E and B represent EcoRI and BamHI, respectively.

10.2: Identification of restriction sites on a plasmid

The following is an example of exercise to determine whether a plasmid (pIUMC123) has EcoRI, BamHI, PstI, HindIII, and HincII sites and where these sites are located. The first step is to digest the plasmid with EcoRI, and then electrophorese the digest in an agarose gel. As seen in Fig. 10-3, EcoRI digestion results in only one 4.3 kb band. This

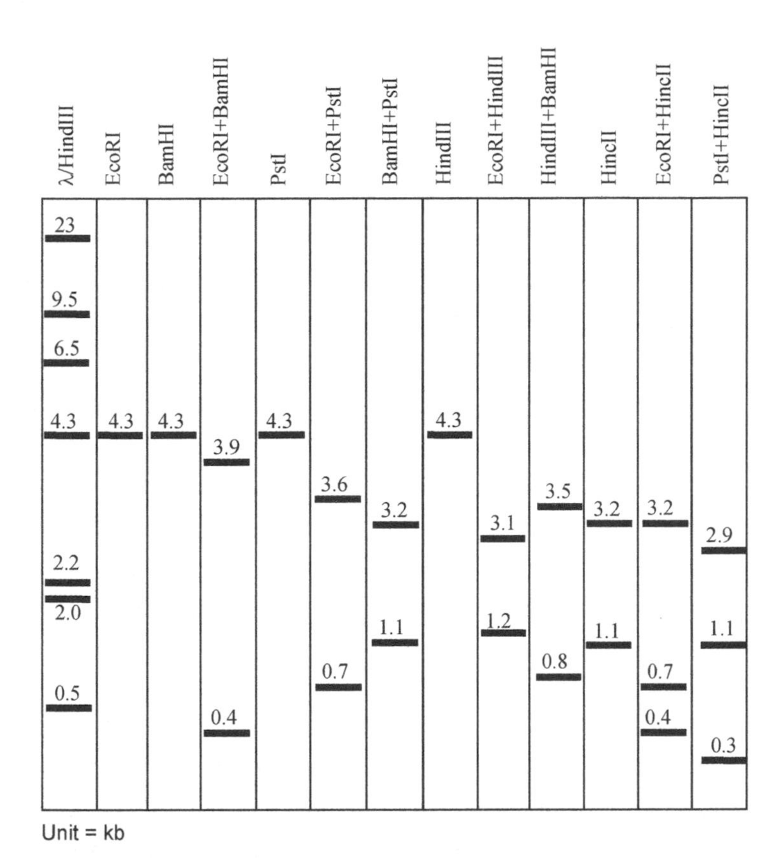

Fig. 10-3: Electrophoregram of pIUMC123 digested with various restriction enzymes.

information indicates that the plasmid is 4.3 kb in size and has only one EcoRI site. The plasmid is then digested with BamHI in a separate reaction. If this digestion also results in one 4.3-kb band, the plasmid has only one BamHI site. At this time, a circle representing the plasmid can be drawn, and the EcoRI site is placed at the 12 o'clock position to serve as a reference point for further restriction mapping (Fig. 10-4). The next step is to determine the location of BamHI site relative to that of EcoRI by digesting the plasmid with both EcoRI and BamHI. If two fragments of 3.9 kb and 0.4 kb are seen, the distance between EcoRI and BamHI would be 3.9 kb or 0.4 kb (Fig. 10-4), meaning that the BamHI site is located 0.4 kb to the right or 3.9 kb to the left of the EcoRI site. To serve as the second reference point, the BamHI site is tentatively placed at approximately 1:30 position on the circle, 0.4 kb away from the EcoRI site (Fig. 10-4).

To determine the PstI site, the plasmid is digested with PstI. If a 4.3-kb fragment is again seen, there is only one PstI site on the plasmid. To locate the PstI site, the plasmid is digested with both PstI and EcoRI. If two fragments of 3.6 kb and 0.7 kb are seen, the PstI site is located 0.7 kb away from the EcoRI site. However, it is not known at this point whether it is located 0.7 kb to the right or to the left of the EcoRI site. Hence, the plasmid is digested with both PstI and BamHI. As shown in Fig. 10-3, two fragments 3.2 kb and 1.1 kb are seen. This result indicates that the distance between PstI and BamHI is 1.1 kb. Therefore, the PstI site is located to the left of the EcoRI site at approximately 9:30 position on the circle (Fig. 10-4). If it is located to the right of the EcoRI site, the distance between PstI and BamHI sites would be 0.3 kb.

In the same manner, the location of HindIII site can be determined. The plasmid is digested with HindIII which also generates a 4.3 kb band in this example, indicating that the plasmid has one HindIII site. The plasmid is then digested with both EcoRI and HindIII to determine the distance between the cutting sites of these two enzymes. As shown in Fig. 10-3, this digestion generates two fragments of 3.1 kb and 1.2 Kb, indicating that the distance between EcoRI and HindIII is 1.2 kb and that the HindIII site is located either 1.2 kb to the left or to the right of the EcoRI site. The plasmid is then digested with HindIII and BamHI to determine the exact location of the HindIII site. If the result is as shown in Fig. 10-3 that two fragments of 3.5 kb and 0.8 kb are seen, the

HindIII site is located 0.8 kb away from the BamHI site on the right side of the EcoRI site at approximately 3:30 position on the circle (Fig. 10-4). Otherwise, the distance between HindIII and BamHI would be 1.6 kb.

When the plasmid is digested with HincII, two fragments of 3.2 and 1.1 kb are seen indicating that the plasmid has two HincII sites. The plasmid is then digested with both HincII and EcoRI to determine which of these two fragments contains the EcoRI site. The result shown in Fig. 10-3 indicates that double digestion with HincII and EcoRI generates 3.2 kb, 0.7 kb, and 0.4 kb bands, indicating that EcoRI cuts the 1.1 kb fragments into two fragments. Two possibilities exit: the 0.4 kb band may be located on the left or on the right of the EcoRI site. To determine this, the plasmid is cut with both HincII and PstI, and 2.9 kb, 1.1 kb, and 0.3 kb bands are seen. This result indicates that the distance between the PstI and the HincII site located on the left of the EcoRI site is 0.3 kb and that the 0.4 kb band is located on the left of the EcoRI site. Therefore, one HincII site is located at approximately 2:30 position and the other at approximately 10 o'clock position on the circle (Fig. 10-4).

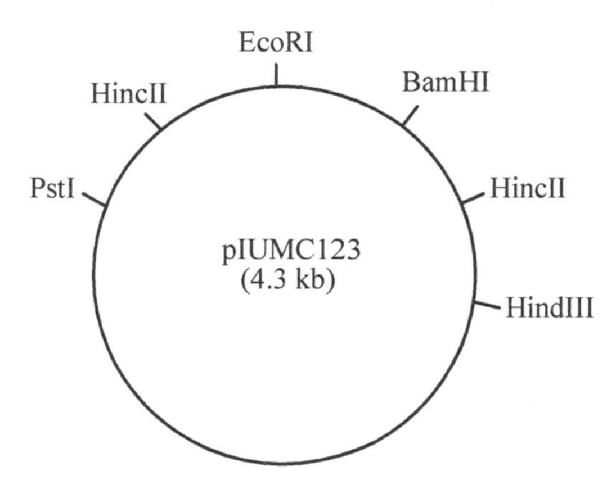

Fig. 10-4: Restriction map of pIUMC123

Summary

To verify a recombinant plasmid, restriction mapping is performed. The plasmid is first digested with a certain restriction enzyme. The digest is then electrophoresed in an agarose gel. The number of bands seen in the gel indicates the number of cutting sites of the restriction enzyme. To determine the location of a second restriction site, the plasmid is digested with the first and the second restriction enzymes simultaneously. If both restriction enzymes cut the plasmid once, two fragments are seen. If one of the fragments is 500 bp, the distance between these two cutting sites is 500 bp. The same principle is used to map the location of a third restriction site. By cutting the plasmid with the first and second, second and third, and first and third restriction enzymes, the location of their cutting sites can be determined.

Sample Questions

The following is a representation of an agarose gel on which digests of a plasmid DNA were run. Please map EcoRI, BglII, HindIII, BamHI, HincII, and AluI sites on the plasmid.

Lane	Bands
EcoRI	3.6
BglII	3.6
HindIII	3.6
BamHI	3.6
EcoRI + BamHI	2.4, 1.2
EcoRI + BglII	2.4, 1.2
EcoRI + HindIII	3.0, 0.6
BamHI + BglII	2.4, 1.2
BamHI + HindIII	1.8
BglII + HindIII	3.0, 0.6
HincII	2.3, 1.3
AluI	1.8, 1.1, 0.7
EcoRI + HincII	1.9, 1.3, 0.4
BamHI + HincII	2.3, 0.8, 0.5
BglII + HincII	1.6, 1.3, 0.7
HindIII + HincII	1.3, 1.0
EcoRI + AluI	1.1, 0.9, 0.7
BamHI + AluI	1.8, 0.8, 0.7, 0.3
BglII + AluI	1.8, 1.1, 0.4, 0.3
HindIII + AluI	1.5, 1.1, 0.7, 0.3
HincII + AluI	1.3, 0.8, 0.7, 0.5, 0.3

Chapter 11

DNA Sequencing

Outline

Principles of DNA sequencing
Maxam and Gilbert DNA sequencing method
Sanger's DNA sequencing method
DNA labeling in Sanger's sequencing method
Properties of M13 bacteriophage
Development of M13 vectors
Development of phagemids
Common sequencing problems
Cycle sequencing
Automated DNA sequencing
Pyrosequencing
In vivo excision of pBluescript from lambda ZAP

11.1: Principles of DNA Sequencing

A DNA molecule is composed of a set of G, A, T, and C nucleotides. The order of the nucleotides within a DNA molecule is called DNA sequence. Different genes have different lengths and sequences. Determination of the nucleotide sequence of a certain gene is called DNA sequencing. If a DNA molecule with the sequence 5'-GATCA-3' is cut progressively at every base from its 3' end, five different DNA fragments with a common 5' end are produced. If these fragments can be separated and the last base of each fragment is known, the sequence of this DNA can be determined. In this example, the last base of the DNA fragments with 1, 2, 3, 4, and 5 nucleotides long are G, A, T, C, and A, respectively. Therefore, the sequence of this DNA molecule is

GATCA.

From this example, it is clear that three problems need to be solved in order to sequence a DNA fragment. The first is to break the DNA fragment progressively at every base from its 3' end. The second is to separate the resulting smaller DNA fragments that differ in length from each other by only one base. The third is to identify the last base of each of the fragments. The second problem can be easily solved by denaturing polyacrylamide gel electrophoresis. The methods for solving the first and third problems were developed by Maxam, Gilbert, and Sanger in the early 70's. For their discovery, Gilbert and Sanger won the Nobel Prize in 1976.

11.2: Maxam and Gilbert DNA Sequencing Method

The Maxam and Gilbert sequencing method uses chemicals to modify DNA at specific bases. Several chemicals are used including dimethyl sulfate (DMS), formic acid, and hydrazine. DMS methylates guanine (G) residues in DNA. When the methylated DNA is treated with another chemical piperidine, the methylated bases are removed, thus breaking the DNA at specific bases. A critical point in DNA sequencing is that different molecules of the DNA fragment are methylated by the chemicals at different places. A DNA sample for sequencing may have millions of molecules. If the DNA fragment to be sequenced has 5 G's, some of the molecules are methylated at the first G, and others are methylated at the second, third, fourth, or fifth G. When the modified DNA are treated with piperidine, some DNA molecules are broken at the first G, and others are broken at the second, third, fourth, or fifth G. Therefore, five different smaller DNA fragments with a common 5' end are generated. These DNA fragments are then separated by electrophoresis in a polyacrylamide gel containing a high concentration (8 M) of urea. Since these DNA fragments are generated by removing the methylated G residues, the base next to the last base on these fragments is a G (Fig. 11-1). This is how the Maxam and Gilbert method

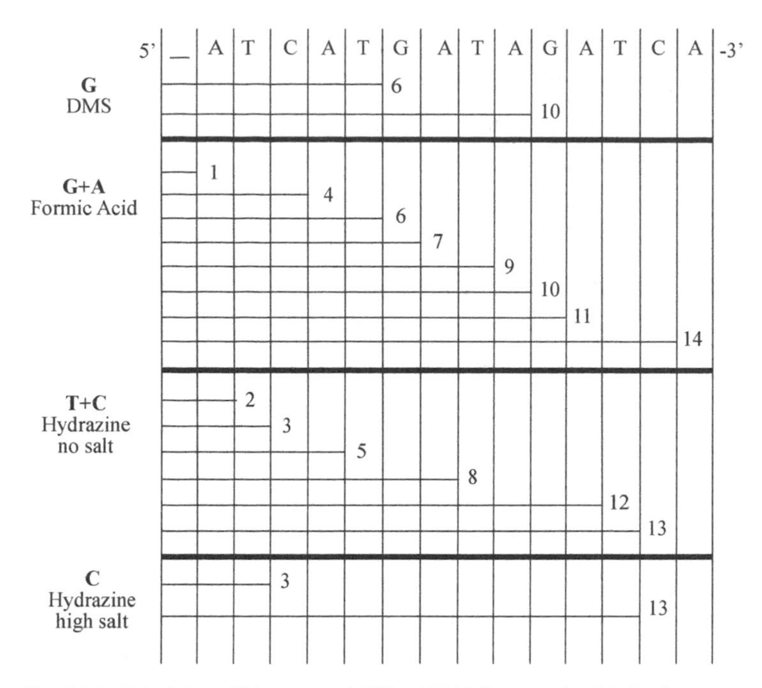

Fig. 11-1: Principles of Maxam and Gilbert DNA Sequencing Method.

breaks DNA at specific sites and identifies the last base of the resulting DNA fragments. If the DNA is treated with formic acid, both G and A residues are methylated, and all the DNA fragments generated after piperidine treatment will have G or A at their 3' ends. Under no salt conditions, hydrazine methylates T and C, but only the C residues are methylated by hydrazine under high salt (5 M NaCl) conditions.

To sequence a DNA fragment by the Maxam and Gilbert method, the DNA fragment is first labeled with ^{32}P at its 5' end by kinase. Since kinase will label the 5' end of both the upper and lower strands of the DNA fragment, the labeled DNA is cut with a certain restriction enzyme into two pieces so that each of the resulting two DNA fragments is labeled at only one end. These two smaller DNA fragments are then sequenced in two separate reactions. Each labeled DNA fragment is divided into 4 tubes marked G, G+A, T+C, and C. The DNA in the G tube is treated with DMS at a final concentration of 0.5%, and that in the G+A, T+C, and C tubes are treated with formic acid, hydrazine under no salt condition, and hydrazine under high salt conditions, respectively, all at a final concentration of 60%. The DNA is treated with these chemicals very briefly (less than 5 minutes) at 15°C so that not all bases on a DNA fragment are methylated. The treated DNAs are precipitated with ethanol and then reacted with 100 μl of 1 M piperidine at 90°C for 30 minutes to break the DNA at the modified bases. After removal of piperidine by vacuum drying, the DNA in the reaction mixture is dissolved in a loading reagent (80% formamide, 10 mM NaOH, 1 mM EDTA, 0.1% xylene cyanol, and 0.1% bromphenol blue) and electrophoresed in 4 separate lanes marked G, G+A, T+C, and C in a polyacrylamide gel (Fig. 11-2). The gel is then transferred to a filter paper and dried, and an X-ray film is placed in the dark on the dried gel. After an overnight exposure, the X-ray film is developed to visualize bands. The sequence on the X-ray film is then read. All the bands in the G lane are read as G, and the bands that appear in the G+A lane but not in the G lane are read as A. Similarly, bands that appear in the C lane are read as C, and those in the T+C lane but not in the C lane are read as T. The sequence is read from bottom up in the order of appearance in all four lanes (Fig. 11-2).

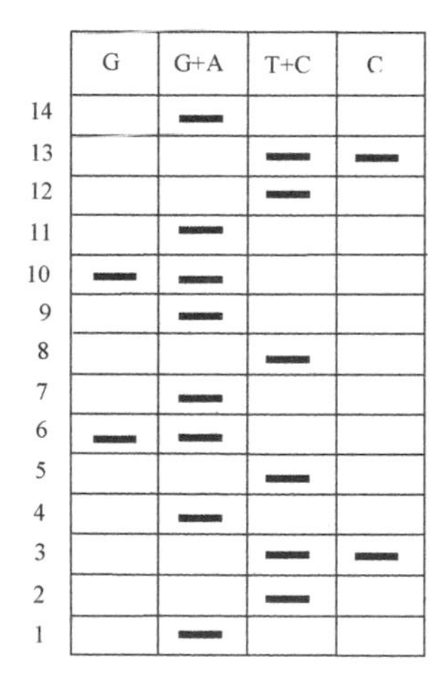

5'-ATCATGATAGATCA-3'

Fig. 11-2: Electrophoregram of Maxam and Gilbert DNA Sequencing Reactions.

11.3: Sanger's DNA Sequencing Method

Sanger's DNA sequencing method also follows the 3 basic principles mentioned above. The major difference is that it uses DNA synthesis to obtain DNA fragments with known 3' ends. During DNA synthesis, dideoxynucleotide triphosphate (ddNTP) in addition to regular deoxynucleotide triphosphate (dNTP) is used. ddNTPs are modified dNTPs in which a hydrogen (H) instead of hydroxyl (OH) group is present in both 2' and 3' positions of the nucleotides. A ddNTP can be joined to a regular dNTP during DNA synthesis but cannot accept another nucleotide when it is incorporated into DNA because it lacks the 3' OH group to which a second nucleotide is joined (Fig. 11-3).

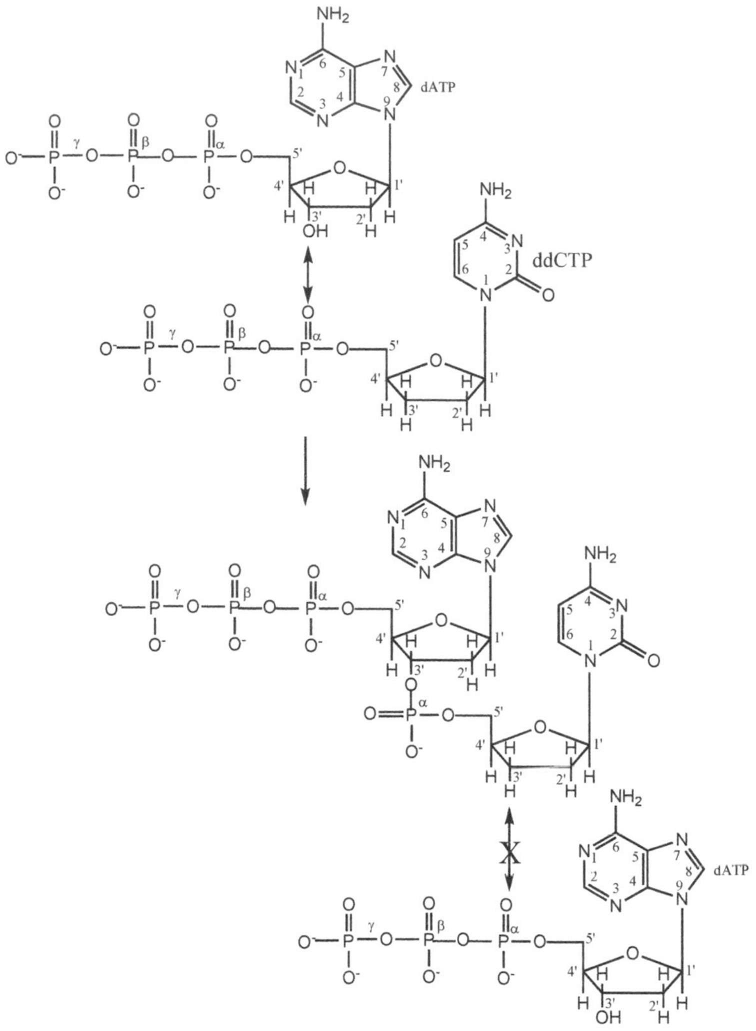

Fig. 11-3: Termination of DNA Synthesis by ddNTP.

For Sanger's sequencing, the following components are required: a DNA template (the DNA to be sequenced), a primer, a DNA polymerase (e.g., the Klenow fragment of *E. coli* DNA polymerase I), dNTP (dATP, dCTP, dTTP, and dGTP), and ddNTP (ddATP, ddCTP, ddTTP, or ddGTP). When a primer is annealed to the template, DNA polymerase will start to synthesize DNA. Similar to the Maxam and Gilbert method, four separate reactions G, A, T, and C are performed for every sequencing run (Fig. 11-4). In the G reaction, ddGTP (~

0.1 mM) in addition to dNTP (~ 0.25 mM) is present so that the DNA synthesis reaction will stop at the place where the template is a C, and the last base (3' end) of the newly synthesized DNA is G. To prevent DNA synthesis with different templates from stopping at the same place, a very low concentration (~ 0.1 mM) of ddGTP is used. If the template has 5 C's, some of the DNA synthesis reactions stop at the first C, and others terminate at the second, third, fourth, or fifth, resulting in five newly synthesized DNA fragments of different lengths with G as the last base of all of these fragments. In the same manner, A, C, and T reactions are performed with ddATP, ddCTP, or ddTTP in each respective reaction (Fig. 11-4). The DNA molecules in these

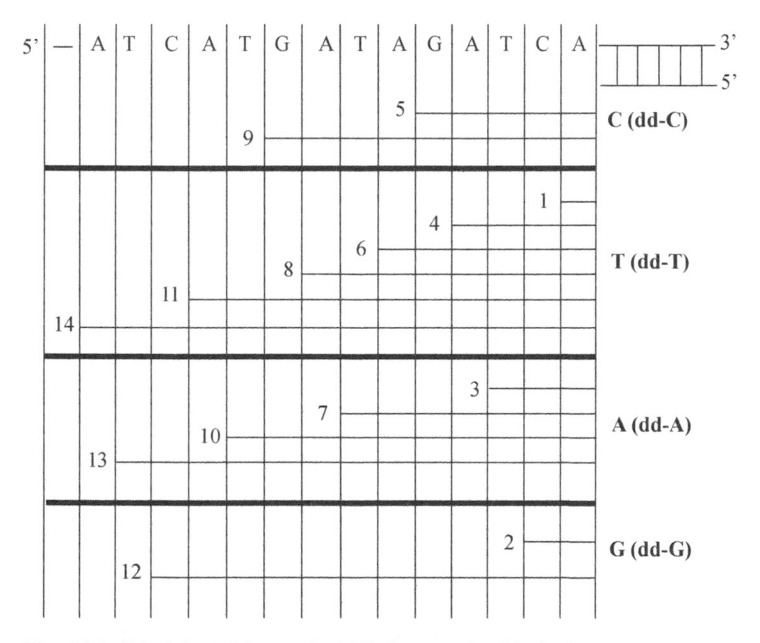

Fig. 11-4: Principles of Sanger's DNA Sequencing Method.

four reactions are then denatured and electrophoresed in a denaturing polyacrylamide gel to separate the newly synthesized DNA fragments. The sequence of the DNA is then read starting from the bottom of the gel according to the locations of the bands in the gel (Fig. 11-5). All the bands in the G lane are read as G, and those in A lanes are read as A. Likewise, bands in the T lane are read as T, and those in the C lane are read as C. A major difference between the Sanger's method and the Maxam and Gilbert method is that all bands generated in the Sanger's method are in different positions after electrophoresis. Another difference is that the sequence read from the gel of Sanger's method is complementary to that of the template, whereas that of the Maxam and Gilbert method is the same as that of the template.

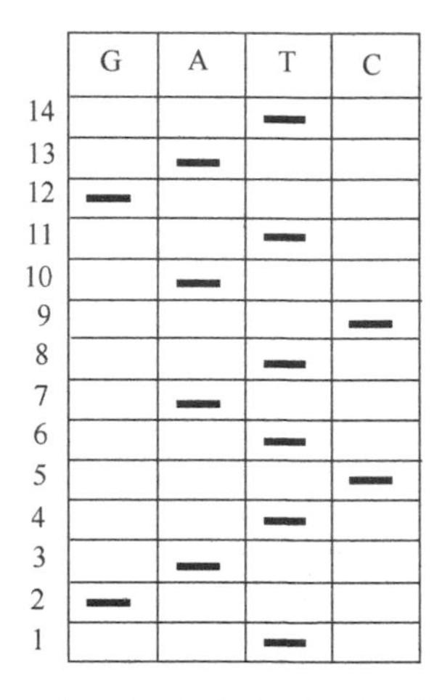

5'- tgatctatcatgat -3'
3'- ACTAGATAGTACTA -5'

Fig. 11-5: Electrophoregram of Sanger's DNA Sequencing Reactions.

11.4: DNA labeling in Sanger's Sequencing Method

The newly synthesized DNA fragments in the original Sanger's sequencing reaction are labeled by incorporating radioactive nucleotides, such as α-^{32}P-dATP. Since the radioactivity of ^{32}P is very strong, the bands seen on the X-ray film are not sharp and are not easy to discern if they are not well separated. With this method, less than 300 bp can be read per reaction. Therefore, ^{32}P is later replaced by ^{35}S or ^{33}P using α-^{35}S-dATP or α-^{33}P-dATP, and the number of readable bases is increased to approximately 400. Since radioactive nucleotides are expensive and troublesome to dispose of, they have been completely replaced by fluorescent nucleotides. The most commonly used ones are fluorescent ddNTP such as ddGTP-EO-dR110 in which the fluorescent dye EO-dR110 is linked to ddGTP. ddGTP-EO-dR110 releases blue light when it is irradiated with laser. ddATP-dR6G, ddUTP-EO-dROX, and ddCTP-EO-dTAMRA release green, red, and yellow light, respectively (Fig. 11-6). When these fluorescent ddNTPs are incorporated during DNA synthesis, the newly synthesized DNAs are colored. All DNA fragments synthesized in the G reaction will release blue light, and those in the A, T, and C reactions will emit green, red, and yellow light, respectively, when irradiated with laser. Since different ddNTPs release different color light, it is no longer necessary to perform G, A, T, and C reactions in four separate tubes and electrophorese the reactions in separate lanes. All four reactions are done in the same tube and electrophoresed in one lane in a gel. It is also possible to use fluorescent primers for sequencing, but four separate reactions are required, each with primers labeled with different color, e.g., G, blue; A, green; T, red; C, yellow. As with the original Sanger's method, the DNA fragments generated in the sequencing reactions are electrophoresed. When the DNA bands migrate to where they are irradiated with a laser, the color emitted is recorded by the computer associated with the sequencing equipment, and an electrophoregram is printed with peaks in four different colors.

Several different fluorescent ddNTP's are available. The most commonly used ones are those developed by Applied Biosystems, Inc. and PerkinElmer, Inc., including dRhodamine and BigDye ddNTP terminators. The ones mentioned above are dRhodamine terminators. Since each of these terminators absorbs different amounts of energy, they release different amounts of light. Therefore, the peak heights of the sequencing reaction are not even, making it difficult to read the sequence. BigDye terminators are modified from dRhodamine terminators and are composed of two parts (Fig. 11-6). The first part is a dRhodamine dye, and the second part is a fluorescein donor. When the molecule is irradiated with laser, the energy is absorbed by the fluorescein donor and then transferred to the dRhodamine dye. Since the fluorescein donor in different ddNTP transfers the same amount of energy to the dRhodamine dye, an equal amount of light is released by different terminators. Therefore, the peak heights are more even and the sequences are easier to read.

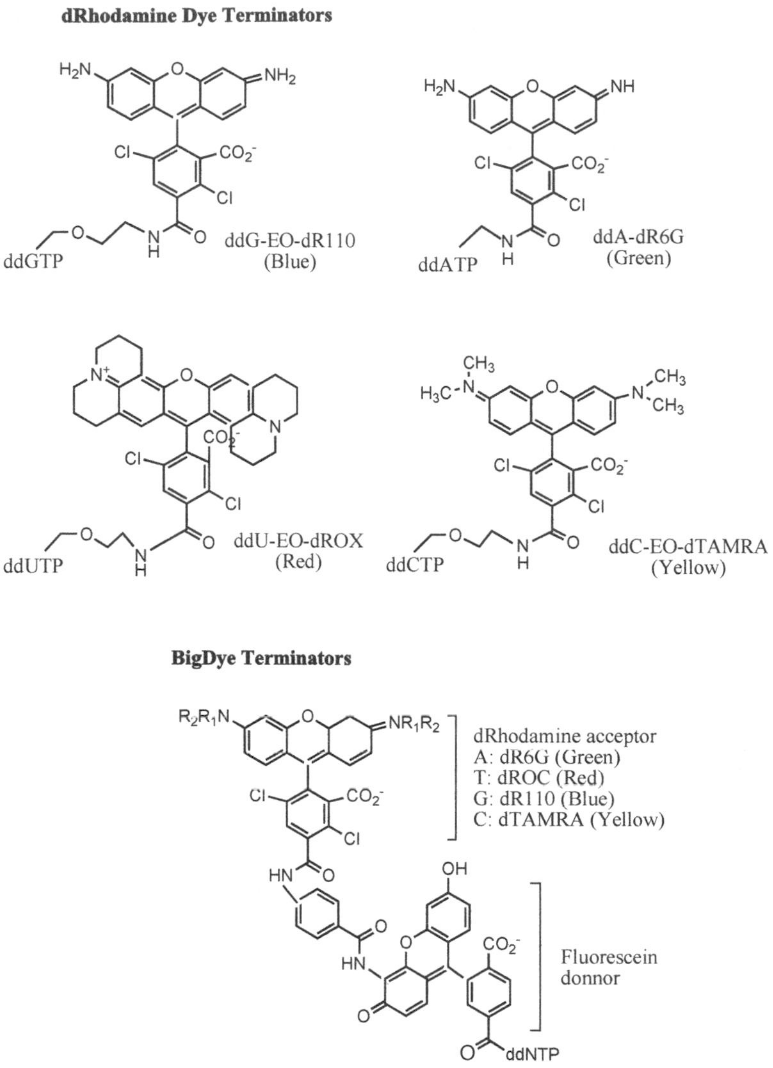
dRhodamine Dye Terminators
ddGTP
ddG-EO-dR110
(Blue)
ddATP
ddA-dR6G
(Green)
ddUTP
ddU-EO-dROX
(Red)
ddCTP
ddC-EO-dTAMRA
(Yellow)
BigDye Terminators
dRhodamine acceptor
A: dR6G (Green)
T: dROC (Red)
G: dR110 (Blue)
C: dTAMRA (Yellow)
Fluorescein
donnor
ddNTP

Fig. 11-6: Fluorescent ddNTP.

For Maxam and Gilbert sequencing reaction, a total of 40 µg of DNA, 10 µg for each G, A, T, and C reaction, is needed, and the entire sequencing reaction takes one day to complete. In contrast, Sanger's method requires less than 5 µg of DNA and no more than two hours to complete. Although these two methods were developed almost at the same time in early 70's, Sanger's method did not become popular until mid 80's. The major reason is that Sanger's method requires a single-stranded DNA template. Although it is possible to denture a double-stranded DNA to become single-stranded, it is very difficult to separate and isolate single-stranded DNA for sequencing. Preparation of single-stranded DNA was not straightforward until Joachim Messing developed M13 vectors in 1977.

11.5: Properties of M13 Bacteriophage

M13 is one of the commonly seen filamentous bacteriophages with a single-stranded DNA genome. Others are f1 and fd phages. They appear under electron microscope as filaments and thus are called filamentous bacteriophages. M13 infects only male *E. coli* because it enters *E. coli* through the sex pilus. Unlike lambda bacteriophage, M13 does not lyse infected *E. coli*. It only slows down the proliferation rate of infected *E. coli*.

When M13 infects *E. coli*, its single-stranded DNA genome is injected into the cell. This single-stranded (plus strand) DNA is then used as the template to synthesize the complementary (minus) strand, making the genome double-stranded. This double-stranded DNA is called the replicative form (RF) because it is used to replicate more M13 genome. The RF DNA is then nicked on the plus strand by the protein (gp II or 2P) encoded by gene II, and the 3' end of the nick is used as the primer to replicate DNA by the rolling-circle mechanism. As the replication proceeds counterclockwise, the plus strand is peeled off. This single-stranded region is used as the template to produce the minus-strand DNA. When one unit of the genome is produced, the double-stranded DNA is cleaved, and its two ends are joined to form a covalently closed circular DNA which is again nicked by the gene II protein (2P) and enters another round of replication.

The peeled off single-stranded (plus strand) DNA can also be packaged to become an infectious M13 viral particle. The 5P protein first binds to the DNA until one unit of the genome is peeled off. It is then replaced by 1P, 3P, 4P, 6P, 7P, and 8P proteins to form an infectious M13 particle which is released outside the cell without killing the infected cell. Therefore, there is double-stranded M13 DNA inside infected *E. coli*, and single-stranded M13 DNA outside the infected cells. Since double-stranded DNA can be cut by restriction enzymes, it is possible to insert a foreign DNA fragment into it. When the resulting recombinant DNA is introduced into *E. coli*, it will replicate according to the mode of M13 replication and release single-stranded recombinant M13 DNA outside the infected cells. This single-stranded M13 DNA contains the inserted DNA fragment. If a primer is annealed to a portion of the M13 genome adjacent to the insert, this DNA fragment can be sequenced. This was how M13 vectors were developed to produce single-stranded DNA template for sequencing.

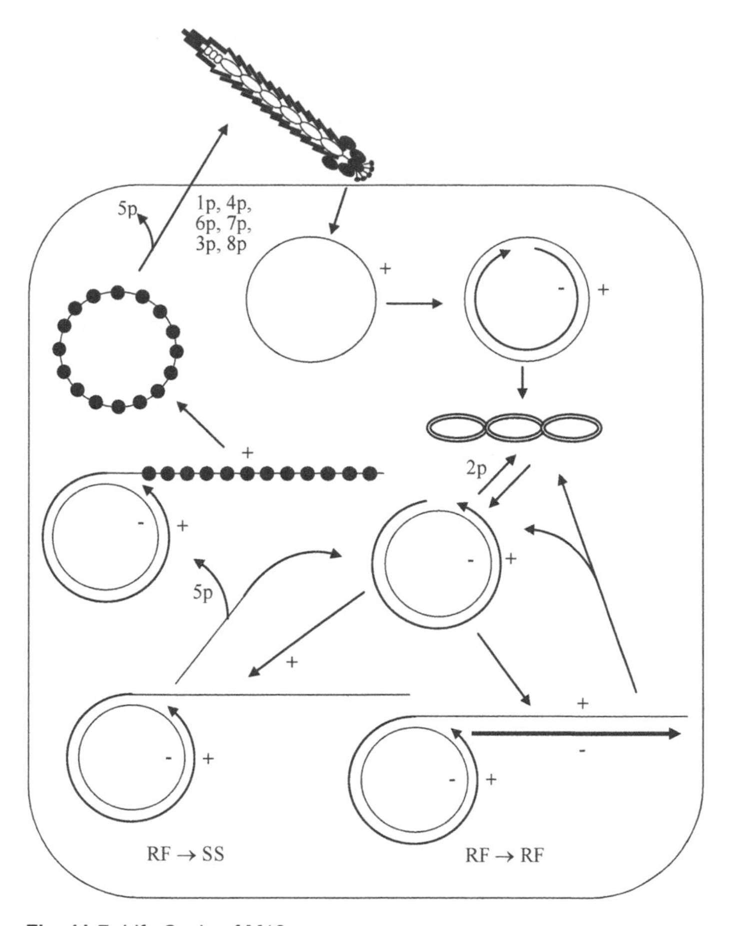

Fig. 11-7: Life Cycle of M13.

11.6. Development of M13 Vectors

As described previously, a DNA must have two selection markers and several unique cloning sites in order to become a useful cloning vector. M13 infects *E. coli* and slows down the proliferation rate of infected cells. This feature is a selection marker because infected cells grow slower than the surrounding uninfected cells and therefore form plaques when they are plated on an agar plate as a lawn. However, M13 does not have any other properties that can be used as a second selection marker. To solve this problem, Joachim Messing inserted the α portion of the *lacZ* gene into the M13 genome, making it able to produce β-galactosidase in an appropriate *E. coli* host. Therefore, the plaques of *E. coli* infected with this M13 are blue in the presence of X-Gal and IPTG. If a DNA fragment is inserted into the *lacZ* gene, *E. coli* infected with this recombinant M13 will not be able to produce β-galactosidase to convert X-Gal to a blue compound, and the plaques formed are colorless.

To create unique restriction sites, Joachim Messing performed chemical mutagenesis and selected M13 phages with an EcoRI site located at the 5' coding region of *lacZ*. This vector was named mp2 (Fig. 11-8), where mp stands for Max Planck Institute. To make the vector more versatile, more restriction sites were later added by inserting an oligonucleotide of approximately 50 bp containing multiple cloning sites into the EcoRI site, creating mp7. M13 mp7 then evolved to become mp8, mp9, mp10, mp11, mp18, and mp19 with different restriction sites. The regions that contain the multiple cloning sites of mp18 and mp19 were later used to construct pUC18 and pUC19 (Chapter 7).

With the development of the M13 vectors, it became very easy to produce single-stranded DNA for use as a template for Sanger's DNA sequencing. A primer, called universal primer, which anneals to a region of the *lacZ* gene immediately downstream from MCS was also developed. It is called universal primer because it can be used to sequence any DNA fragment that is inserted into the MCS of M13 vectors. With these developments, Sanger's method became enormously popular for DNA sequencing.

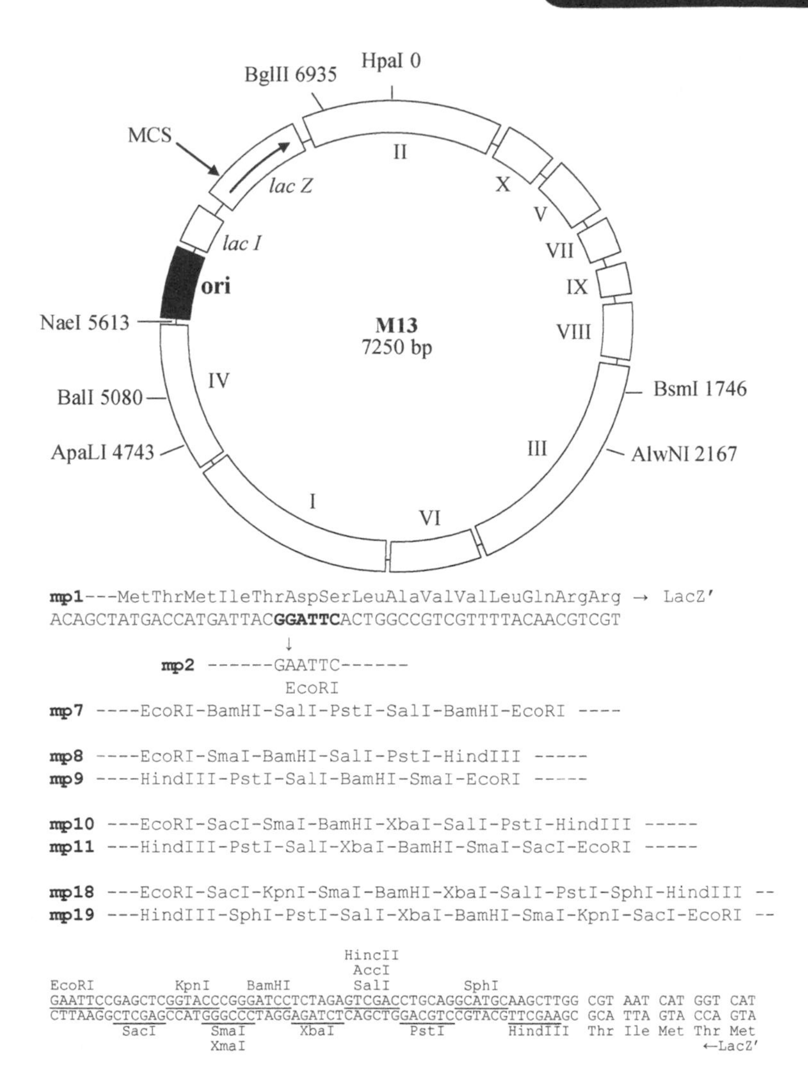

Fig. 11-8: M13 Vectors with Different Multiple Cloning Sites. The MCS sequence of mp18 is shown in the bottom of the figure. Underlined nucleotide sequences in the multiple cloning sites are recognition sequences of certain restriction enzymes.

11.7: Development of Phagemids

Although M13 vectors greatly simplify DNA sequencing using Sanger's method, they also have limitations. The major problem is that M13 has a rather large genome of greater than 7 kb. The average size of a foreign DNA fragment that M13 vectors can accept is only approximately 500 bp, although theoretically they can take larger DNA pieces. Therefore, a 5 kb DNA fragment to be sequenced would need to be cut into at least 10 smaller fragments of approximately 500 bp each. Each fragment is then separately cloned into an M13 vector to produce single-stranded DNA for use as template. In addition, M13 vectors are not useful for making RNA probes or proteins. Therefore, attempts were made to use a regular recombinant plasmid for DNA sequencing by directly denaturing it to become single-stranded. Although this is possible, the denatured DNA renatures back to double-stranded rather easily. Once it becomes double-stranded, it can no longer serve as template. Therefore, the number of bases that can be read with this method is very limited, usually less than 150 bases.

To solve this problem, phagemid was developed. The most important feature of phagemid is that it contains two replication origins. One is the regular colE1 replication origin, and the other is M13 or f1 replication origin. Since M13 and f1 replication origins are very similar, they are used interchangeably. If a phagemid is introduced into *E. coli* containing M13, it is replicated like M13 because the only sequence that M13 replicase recognizes is the M13 replication origin. Any plasmid containing an M13 replication origin is replicated as if it is M13 inside the cell. The single-stranded DNA thus produced is packaged by M13 proteins to become an infectious M13 particle and released outside the cell. These viral particles are then collected, and the DNA is isolated for use as template for sequencing by the Sanger's method. After phagemids became available, M13 vectors were no longer used. With phagemid, a cloned DNA fragment can be sequenced directly without subcloning. Phagemids can also be used to make RNA probes and to produce recombinant proteins.

To ensure that the sequence is correct, both strands of a DNA fragment are usually sequenced. To be able to sequence both strands, both plus and minus single-stranded templates need to be produced. If a single-stranded DNA template is made with a phagemid and found to be the minus strand, the plus-strand template can be made by cloning the same DNA fragment into the phagemid in an opposite orientation. Another way to make both plus and minus strand DNA templates is to clone the same DNA fragment into two different phagemids with the f1 replication origin in different orientations. A minus strand is made with a phagemid containing an f1 replication origin in clockwise orientation, and a plus strand DNA is produced using a phagemid with the f1 replication origin in the counterclockwise orientation.

Therefore, phagemids usually exist as a set of four. Two of them have the f1 replication origin in the same orientation, but multiple cloning sites (MCS) in different orientations. The other two have the MCS in the same orientation, but with f1 replication origins in different orientations. The pTZ series is one of the first phagemid sets developed (Fig. 11-9). In pTZ18U and pTZ19U or pTZ18R and pTZ19R, their f1 replication origins are in the same orientations, but their MCS are in different orientations. In contrast, pTZ18U and pTZ18R or pTZ19U and pTZ19R have their MCS in the same orientations but their f1 replication origins in opposite orientations.

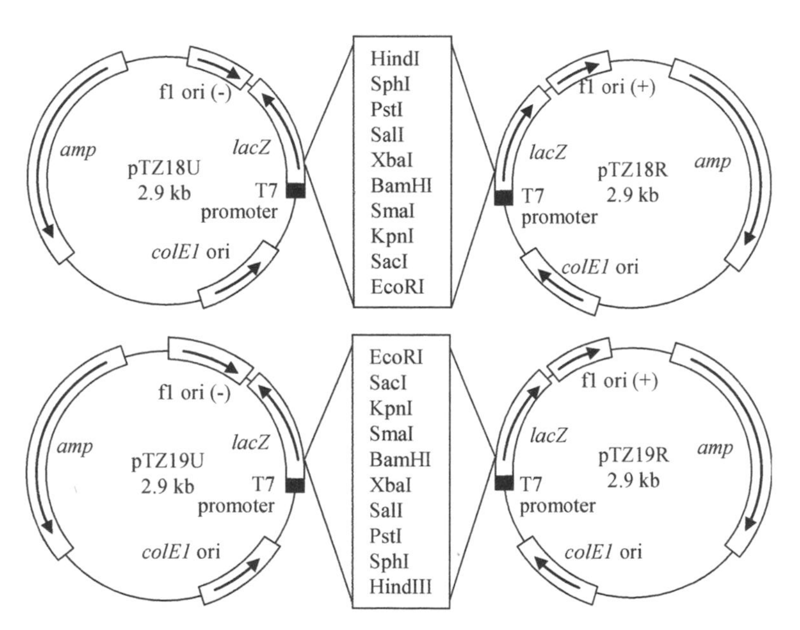

Fig. 11-9: pTZ phagemids.

At present, pBluescript KS(+/-) and pBluescript SK(+/-) are the most commonly used phagemids, where K represents KpnI and S represents SacI. KS and SK indicate different MCS orientations (Fig. 11-10). "+" indicates that the phagemid produces plus stranded DNA, and "-"

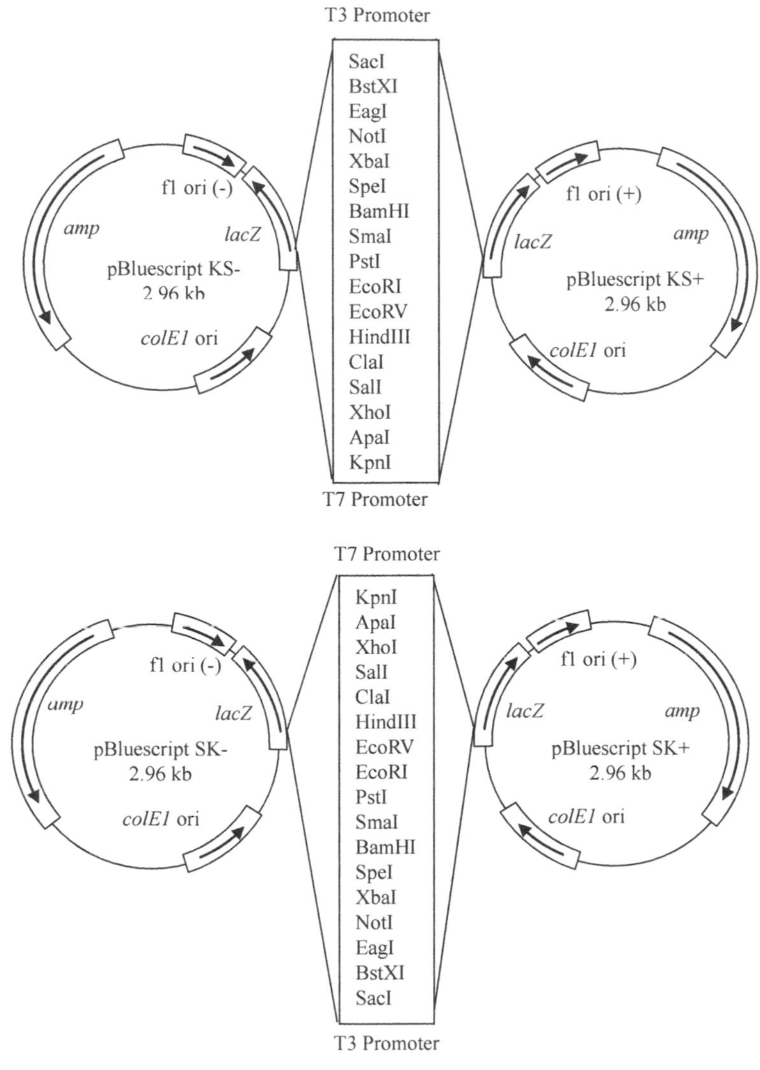

Fig. 11-10: pBluescript phagemids.

indicates that the phagemid produces minus stranded DNA. To produce single-stranded DNA using phagemid, a helper phage is required. A recombinant phagemid is first introduced into an *E. coli* that has an F or F' factor. The transformed cells are then infected with an M13 (M13KO7) or f1 (R408) helper phage. The helper phage M13KO7 is a derivative of M13 with a p15A replication origin. It has the Met40Ile mutation in gene II and a kanamymycin resistance gene from Tn903 inserted into its replication origin. M13KO7 can replicate independently of the gene II protein using the p15A replication origin. The mutated gene II protein interacts less efficiently with the mutated replication origin on its own genome than with the wild type M13 replication origin that is cloned into the phagemid. Therefore, the single-stranded DNA derived from the phagemid is preferentially packaged to become M13 phage particles. R408 is an f1 bacteriophage and a frequently used helper phage. It has a mutation in its packaging signal; therefore, its proteins package more efficiently the single-stranded DNA derived from the phagemid which has a normal packaging signal.

11.8: Common Sequencing Problems

Stop and compression are two problems commonly encountered in Sanger's sequencing reactions. Stop is due to the presence of secondary structures in the template, resulting in bands present in different lanes (G, A, T, and C lanes) in the same position after electrophoresis. The solution to this problem is to perform the sequencing reaction under a higher temperature (e.g., 60°C) that does not allow secondary structures to form. Since most DNA polymerases do not function at this temperature, a heat-resistant DNA polymerase, such as Taq DNA polymerase, is required.

Compression is due to the presence of GC rich sequences on the template. Therefore, DNA synthesized during the sequencing reaction remains annealed to the template during electrophoresis. The solution to this problem is to use modified nucleotides such as 7-deaza dGTP to replace dGTP for DNA synthesis in Sanger's sequencing reaction. In 7-deaza dGTP, its position 7 is a carbon (C) instead of a nitrogen

(N). Therefore, it pairs to cytosine with two, instead of three, hydrogen bonds. dITP (Fig. 11-11) can also be used to replace dGTP because inosine also pairs to cytosine with two hydrogen bonds. Currently, most sequencing reactions are performed using dITP to replace dGTP.

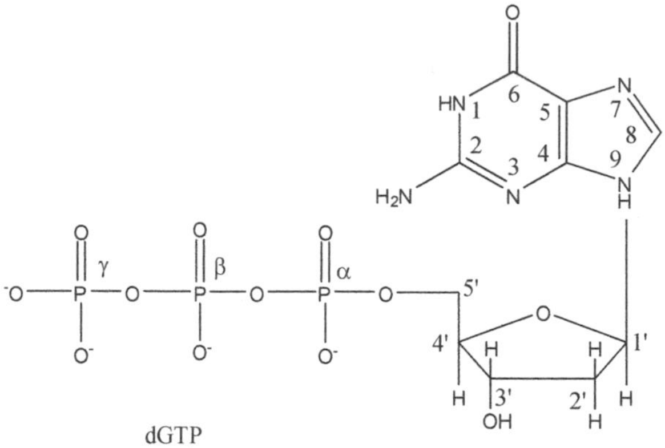

dGTP

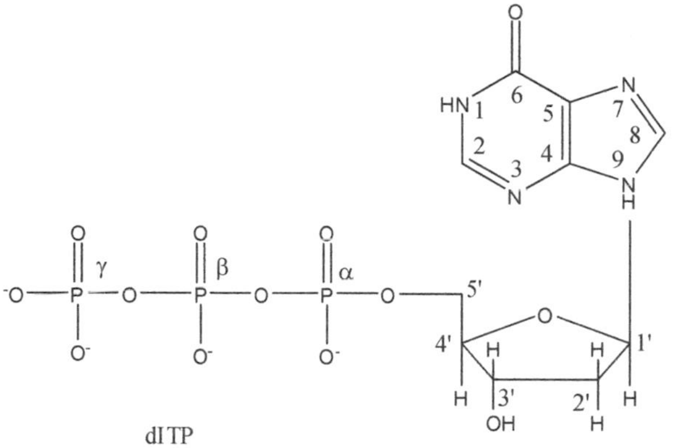

dITP

Fig. 11-11: dGTP and dITP.

11.9: Cycle Sequencing

In classical Sanger's sequencing method, the template is used only once. Therefore, a substantial amount (approximately 5 μg) of template DNA is required. The development of PCR (described in Chapter 13) also revolutionizes Sanger's sequencing method. With the use of Taq DNA polymerase, the reaction can be heated to 94°C to dissociate the newly synthesized DNA from the template without inactivating the DNA polymerase. The template is then reused. This method is therefore termed cycle sequencing. Components such as template, primers, dNTP, ddNTP, and Taq DNA polymerase required for a sequencing reaction are mixed together and then heated at 94°C for 10 minutes to ensure that the template is linear without secondary structures. The temperature is then lowered to approximately 60°C and held there for 30 seconds to 1 minute to allow primers to anneal to the template and Taq DNA polymerase to synthesize DNA. The temperature is then raised to 94°C for 1 minute to allow dissociation of the newly synthesized DNA from the template and then lowered to 60°C again for another round of primer annealing and DNA synthesis. This process is repeated for 25 – 30 cycle, after which an electrophoresis loading reagent containing formamide and NaOH is added, and the mixture is boiled in a water bath for 10 minutes before electrophoresis. In cycle sequencing, only approximately 50 ng of template DNA is required. Furthermore, double-stranded DNA can be heat-denatured to become single-stranded and used directly as template for sequencing because the entire reaction is done at a temperature greater than 60°C. Under such a high temperature, the two single-stranded DNAs cannot anneal back together. Therefore, it is no longer necessary to use M13 vectors or phagemids to produce single-stranded DNA to be used as templates for sequencing.

11.10: Automated DNA Sequencing

Because of the widespread use of heat-resistant Taq DNA polymerase and fluorescent ddNTP, DNA sequencing by Sanger's method is now mostly automated. After the sequencing reaction is completed, the reaction mixture is automatically loaded onto a gel or a capillary for electrophoresis. Because each ddNTP has a different color, the bands seen on the gel are colored. When each band migrates to the window to which a detector (usually a CCD camera) is affixed, the color is recorded by a computer and converted to a DNA sequence. At present, electrophoresis of sequencing reactions is done in a capillary which has a better resolution power than the conventional slab gel. Usually, 700-1000 bases, instead of 400 bp with slab gel, can be read with capillary gel electrophoresis. The diameter of a capillary is only approximately 50 microns. The capillary is packed with polyacrylamide based gel automatically by the equipment. After the electrophoresis, the capillary is automatically emptied and repacked to allow electrophoresis of another sample.

11.11: Pyrosequencing

In addition to Maxam and Gilbert and Sanger's methods, several other methods for DNA sequencing have also been developed. One such method is pyrosequencing which was developed by Ronaghi et al. in 1996. "Pyro" is derived from pyrophosphate. During DNA synthesis, the α phosphate of a dNTP is incorporated into DNA, while β and γ phosphates are released. The compound which comprises these two phosphate groups is called pyrophosphate (PPi). In pyrosequencing, the following components are required: a template, primers, dNTP, and a DNA polymerase. When a primer is annealed to the template, DNA synthesis will start. To determine the identity of a base, four different reactions (G, A, T, and C) are carried out. If pyrophosphate is released in the C reaction, the corresponding base on the template is G (Fig. 11-12).

To detect PPi, adenosine 5' phosphosulfate (APS) is added to the reaction. In the presence of sulfurylase and PPi, APS is converted to ATP. ATP in turn activates luciferase which converts luciferin to oxyluciferin and releases light that can be detected with a luminometer (Fig. 11-12). Therefore, the DNA sequence can be elucidated by determining which of the four reactions releases light. Because dATP can also activate luciferase, pyrosequencing uses α-S-dATP instead of dATP in the A reaction. If the template has two T's in a row, two A's are incorporated into the DNA in the A reaction, and twice as much PPi and thus twice as much light as a single T on the template is released.

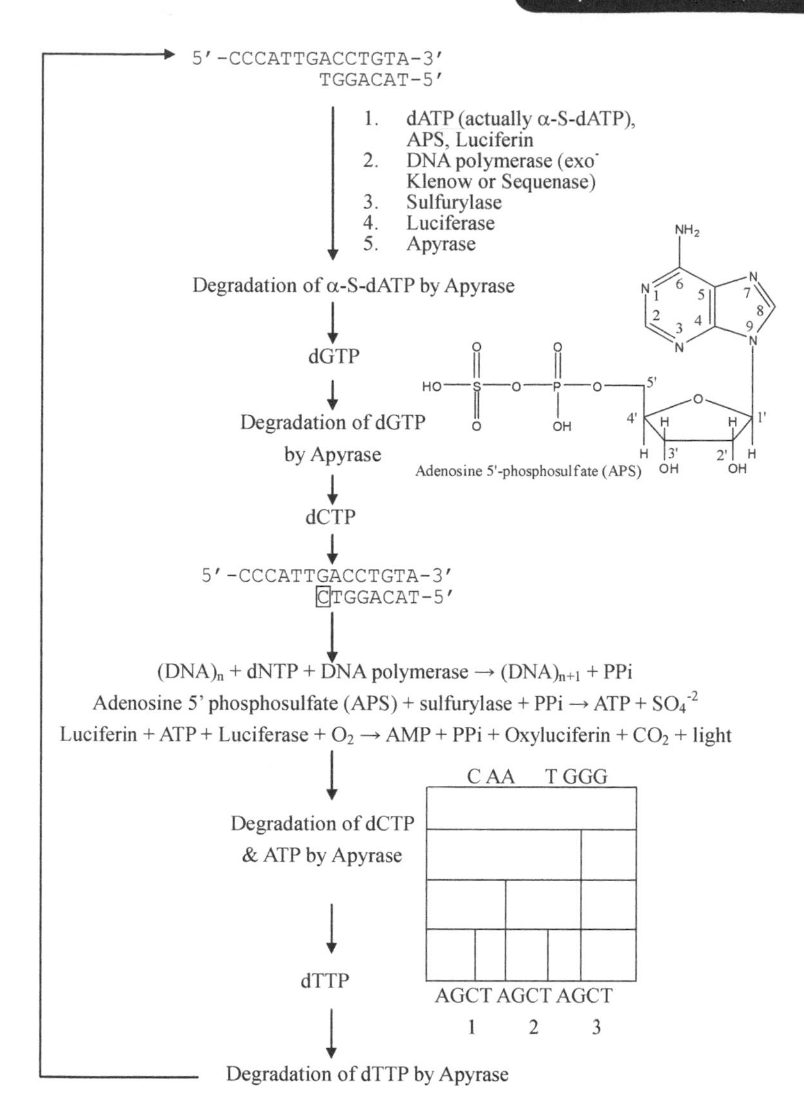

Fig. 11-12: Principles of Pyrosequencing.

Therefore, the amount of light released indicates the number of a certain residue on the template.

The DNA polymerase used for pyrosequencing is exo⁻ Klenow or Sequenase. Both enzymes lack the 3' to 5' exonuclease activity and do not remove the primer after it anneals to the template. In pyrosequencing, the G, A, T, and C reactions are carried out in sequence. In the G reaction, only dGTP is present, and only dATP, dTTP, or dCTP is present in the A, T, or C reaction, respectively. After each reaction, the unincorporated nucleotides and ATP that may be produced must be removed. This is achieved by using apyrase which converts ATP to ADP and dNTP to dNDP. However, apyrase is long lived. It is present when the next nucleotide is added to carry out the next reaction and will degrade the added nucleotides. Pyrosequencing works because DNA synthesis reaction by Klenow or sequenase is much faster than the degradation reaction by apyrase. Therefore, it is completed before apyrase starts to degrade the nucleotides added in each reaction. DNA synthesis is usually completed within 3 seconds, while apyrase reaction takes 5 seconds to complete. At present, each pyrosequencing reaction can sequence approximately 250 bp. The major advantage of using pyrosequencing is that it can be completely automated.

11.12: In vivo Excision of pBluescript from Lambda ZAP

The development of phagemid not only simplified preparation of single-stranded DNA template for sequencing, it also revolutionized isolation of genes from cDNA libraries. In cDNA library construction, a lambda phage is used as the vector because it can be introduced into *E. coli* by infection which has an efficiency close to 100%. However, lambda DNA is much harder to prepare and store than plasmid DNA. Therefore, once a recombinant lambda containing the desired gene is isolated, the DNA fragment containing the gene is cut out from the lambda vector and subcloned into a plasmid. The entire procedure, which takes several weeks to complete, consists of growing the lambda DNA, isolating lambda DNA, cutting out the insert, and cloning the insert into a plasmid. To simplify this process, the lambda ZAP cloning system was developed. This system allows conversion of a recombinant lambda to a recombinant plasmid without going through all the steps mentioned above.

This system utilizes the unique properties of the M13 or f1 replication origin, which are composed of two parts: initiator (I) and terminator (T). During replication, M13 or f1 replicase recognizes the initiator and DNA synthesis occurs until the terminator is reached. Although these two parts are located next to each other in native conditions, they can be separated by inserting a foreign DNA fragment without compromising their function. Therefore, it is possible to replicate only the portion located between the initiator and terminator in a recombinant lambda bacteriophage. In a lambda ZAP vector, a complete pBluescript plasmid DNA is inserted between the initiator and the terminator of the f1 replication origin (Fig. 11-13). During library construction, a cDNA fragment is inserted into the multiple cloning sites of the lambda ZAP vector. The recombinant DNA is then packaged and introduced into an appropriate *E. coli* host by infection to produce plaques. Plaques are screened to isolate the desired recombinant lambda ZAP. The isolated recombinant lambda ZAP is then introduced into an *E. coli* containing helper phage M13KO7 or R408. The replicase produced by the helper phage replicates the region including and between the initiator and terminator on the recombinant lambda ZAP. This region includes the entire pBluescript with the insert. The two ends of the newly synthesized DNA are then joined by the gene II protein of the M13 or f1 helper phage, and the DNA is packaged by phage proteins to become infectious viral particles.

Although the DNA in these viral particles is not f1 or M13 DNA, the viral particle can still infect *E. coli* and inject its DNA into the cells. Since the injected DNA is a recombinant pBluescript, it contains the *colE1* replication origin and can replicate in *E. coli* autonomously just like a regular plasmid. Thus, a recombinant lambda is converted to a recombinant plasmid without the conventional cloning process. This method works very well and greatly simplifies isolation of the gene of interest from a cDNA library.

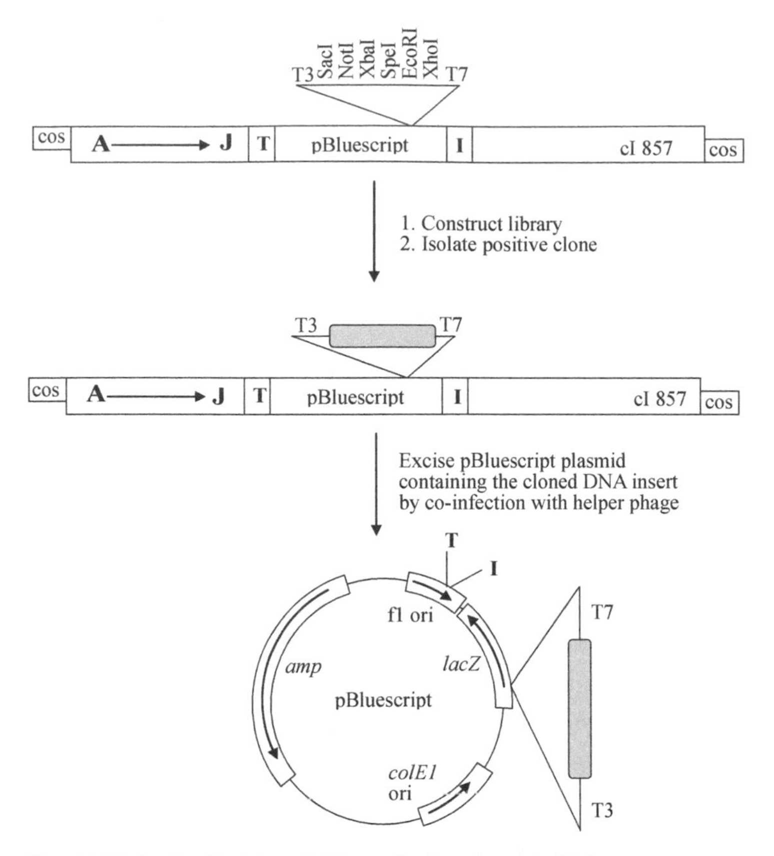

Fig. 11-13: In vivo Excision of pBluescript from Lambda ZAP.

Summary

In DNA sequencing, three problems need to be solved. The first is to be able to cleave the DNA at every base progressively from its 3' end. The second is to be able to separate these DNA fragments. The third is to be able to identify the last base of each DNA fragment generated. The second problem can be easily solved by using denatured polyacrylamide gel electrophoresis. The first and the third problems were solved by Maxam and Gilbert using a chemical approach and by Sanger using an enzymatic approach. To sequence a DNA fragment using the Maxam and Gilbert method, a DNA fragment is first labeled at one end, usually the 5' end. The DNA is divided into four aliquots. The first one is treated with dimethyl sulfate which methylates G. The second aliquot is treated with formic acid to methylate G and A. The third one is treated with hydrazine under no salt conditions to methylate T and C, and the last aliquot is treated with hydrazine under high salt conditions to methylate C. The reactions are then treated with piperidine at 90°C to break DNA at methylated bases. The four different reactions (G, G+A, T+C, and C) containing the DNA fragments thus generated are then electrophoresed in adjacent lanes in a denaturing polyacrylamide gel. The gel is then dried and exposed to an X-ray film to reveal radioactive bands. Since the DNA fragment is labeled at the 5' end, different sized bands seen on the X-ray film represent various DNA fragments with a common 5' end. The DNA sequence is then read in the order of appearance of bands starting from the bottom of the gel. All the bands in the G lane are read as G. The bands that appear in the G+A lane but not in the G lane are read as A. Those in the T+C lane but not in the C lane are read as T, and all bands in the C lanes are read as C.

Sanger's method uses dideoxynucleotide triphosphate (ddNTP) to terminate DNA synthesis at specific positions. It requires a template which is the DNA to be sequenced, primers, a DNA polymerase, dNTP, and ddNTP. Four different reactions are performed. The G reaction contains ddGTP so that DNA synthesis will stop at the place where the template is a C. The A reaction uses ddATP to stop DNA synthesis at the place where the template is a T. The T reaction contains ddTTP so that DNA synthesis will stop at the place where the template is an A, and the C reaction uses ddCTP to terminate DNA synthesis at the place where the template is a G. A radioactive nucleotide such as α-^{32}P-dATP is incorporated into the newly synthesize DNA so that the DNA can be visualized on the gel after electrophoresis. Very low concentration of ddNTP is used in each reaction; therefore, DNA synthesis reactions with different templates do not stop at the same place. If the template has 5 C's at different positions, five different fragments with G as their last base are produced. After electrophoresis, these five fragments can be seen on the gel as 5 bands in the G lane. There will be also bands in

A, T, and C lanes. The sequence of the DNA is read from the gel. Since the DNAs synthesized in Sanger's method are complementary to the template, the sequence read from the gel is that of the complementary strand.

A single-stranded DNA is required for Sanger's sequencing. Preparation of single-stranded DNA for sequencing was simplified by the development of M13 vectors. M13 is a bacteriophage of *E. coli*. Its genome is single-stranded DNA. However, it has a double-stranded DNA intermediate in its replication cycle. This double stranded DNA can be isolated from infected *E. coli*. Joachim Messing inserted a *lacZ* gene and an oligonucleotide with multiple cloning sites into the double-stranded genome, making it a useful cloning vector. A DNA fragment to be sequenced is first inserted into an M13 vector. The recombinant M13 DNA is then introduced into *E. coli*. The M13 replication system will replicate this double-stranded DNA and produce a recombinant single-stranded DNA which is then packaged by M13 proteins and released outside the cell. The single-stranded DNA of these viral particles is isolated and used as template for Sanger's DNA sequencing using a primer annealed to the M13 genome adjacent to the inserted DNA fragment.

Since M13 vectors can only accept DNA fragments of approximately 500 bp, a DNA fragment greater than 500 bp will need to be cut into smaller fragments. Each fragment is then cloned into M13 to produce single-stranded DNA for use as template in sequencing. This is very laborious and time consuming. Therefore, phagemids are developed to simplify the procedure. The goal was to use a regular plasmid to make single-stranded DNA. The most characteristic feature of a phagemid is that it has two replication origins. One is the regular *colE1* origin, and the other is M13 or f1 replication origin. When a phagemid is introduced into an *E. coli* containing an M13 or f1 helper phage, the phagemid is replicated as if it were M13, producing single-stranded DNA which is then packaged as an M13 viral particle and released outside the cell.

The invention of PCR also revolutionized Sanger's DNA sequencing method. With the use of heat-resistant Taq DNA polymerase, DNA sequencing can be done at high temperatures such as 60°C, which prohibits denatured DNA from being renatured. Because of this, it is no longer necessary to use M13 vectors or phagemids to produce single-stranded DNA for use as template in sequencing. Any double-stranded DNA can be denatured to become single-stranded and used directly as a template for sequencing. Furthermore, the newly synthesized DNA can be dissociated from the template, and the template can be reused. This process is termed cycle sequencing which requires only approximately 50 ng of template DNA. The development of fluorescent ddNTP further improves Sanger's DNA sequencing. Modified ddNTP such as ddGTP-EO-dR110, ddUTP-EO-dROX, ddATP-dR6G, ddCTP-EO-dTAMRA releases blue, red, green, and yellow light, respectively. When these ddNTPs are incorporated into DNA, fragments with G, A, T, or C as their last base will release blue, red, green, and yellow light, respectively. The use of capillary electrophoresis also greatly improves Sanger's DNA sequencing. It increases the number of bases that can be read in each reaction from 400 to approximately 1000.

Pyrosequencing is a different DNA sequencing method. Similar to the Sanger's method, it also requires a single-stranded DNA template, primer, a DNA polymerase, and dNTP. However, it does not require ddNTP. A primer is first annealed to the template, and four different reactions G, A, T, and C are performed to determine the identity of base on the template immediately adjacent to the 3' end of the primer. If the base on the template is C, only the G reaction will result in nucleotide incorporation. When a nucleotide is incorporated into DNA, its β and γ phosphates are released. The compound composed of these two phosphate groups is called pyrophosphate (PPi). Therefore, determining which of the four reactions releases PPi will identify the base on the template next to the 3' end of the primer. In the presence of PPi, sulfurylase converts adenosine 5' phosphosulfate (APS) to ATP which activates luciferase to convert luciferin to oxyluciferin and releases light. If light is detected in the G reaction, the base is C.

Although phagemids are no longer used to make single-stranded DNA for sequencing, their development greatly simplified isolation of genes from cDNA libraries. In cDNA library screening, a recombinant lambda phage containing the gene of interest is first isolated. The fragment containing the gene of interest is then isolated and subcloned into a plasmid since it is much easier to prepare plasmid DNA than lambda DNA. Lambda vectors are also not suitable for other purposes such as making RNA probes or recombinant proteins. Subcloning of a gene from lambda to plasmid is very time-consuming as it requires the following steps: isolation of lambda DNA, cutting out the DNA fragment, and cloning of the DNA into an appropriate plasmid vector. To simply this process, lambda ZAP vectors were developed. This development utilizes the unique properties of M13 or f1 replication origin which contains two parts: initiator and terminator. These two parts can be separated without compromising their function. In lambda ZAP, the entire pBluescript is inserted between the initiator and terminator, and a cDNA fragment is inserted into the MCS of the pBluescript. When a recombinant lambda ZAP is introduced into an *E. coli* containing helper phage M13 (M13KO7) or f1 (R408), the M13 or f1 replicase will replicate the region starting from the initiator and ending at the terminator. This region includes the pBluescript and the fragment inserted into it. The two ends of the single-stranded DNA thus produced are ligated by the gene II protein, and the circular DNA thus generated is packaged to become a phage particle which is then released outside the cell. The phage particles are then used to infect *E. coli*. Once the DNA is injected into *E. coli*, it replicates like a plasmid since it has the *colE1* replication origin. Therefore, a recombinant lambda is converted to a recombinant plasmid without going through the cloning processes. Since the entire pBluescript is excised, this technique is called "in vivo excision of pBluescript from lambda ZAP."

Sample Questions

1. In Maxam and Gilbert sequencing reaction, which of the following is used to methylate G: a) dimethyl sulfate, b) formic acid, c) hydrazine with no salt, d) hydrazine with high

salt.

2. In Maxam and Gilbert sequencing reaction, which of the following is used to methylate G and A: a) dimethyl sulfate, b) formic acid, c) hydrazine with no salt, d) hydrazine with high salt.
3. In Maxam and Gilbert sequencing reaction, which of the following is used to methylate T and C: a) dimethyl sulfate, b) formic acid, c) hydrazine with no salt, d) hydrazine with high salt.
4. In Maxam and Gilbert sequencing reaction, which of the following is used to methylate C: a) dimethyl sulfate, b) formic acid, c) hydrazine with no salt, d) hydrazine with high salt.
5. In Sanger DNA sequencing, which of the following is used to terminate a DNA synthesis reaction: a) dNTP, b) NTP, c) ddNTP, d) stop codons.
6. Which of the following apply to M13 bacteriophage: a) infects only female *E. coli*, b) the genome is single-stranded DNA, c) infects only male *E. coli*, d) the genome is double-stranded DNA.
7. Phagemid is a plasmid containing: a) P1 phage replication origin, b) both *colE1* replication origin and f1 replication origin, c) f1 replication origin only, d) lambda phage replication origin.
8. Which of the following are relevant for in vivo pBluescript excision from lambda ZAP: a) a pBluescript plasmid is inserted between the initiator and terminator of f1 replication origin, b) needs a helper M13 or f1 bacteriophage, c) gene II protein of M13 or f1 ligates newly synthesized DNA with M13 or f1 initiator and terminator flanking pBluescript, d) the circularized DNA is packaged by helper phage proteins to become infectious viral particles.

Suggested Readings

1. Blattner, F. et al. (1997). The complete genome sequence of *E. coli* K12. Science 277: 1453-1462.
2. Dotto, G. P. and Horiuchi, K. (1981). Replication of a plasmid containing two origins of bacteriophage f1. J. Mol. Biol. 153: 169-176.
3. Dotto, G. P., Horiuchi, K., and Zinder, N. D. (1984). The functional origin of bacteriophage f1 DNA replication: its signals and domains. J. Mol. Biol. 172: 507-521.
4. International Human Genome Sequencing Consortium (2001). Initial sequencing and analysis of the human genome. Nature 409: 860-921.
5. Maxam, A. M. and Gilbert, W. (1977). A new method for sequencing DNA. Proc. Natl. Acad. Sci. USA 74: 560-564.
6. Maxam, A. M. and Gilbert, W. (1980). Sequencing end-labeled DNA with base-specific chemical cleavage. Meth. Enzymol 65: 499-559.
7. Messing, J., Gronenborn, B., Muller-Hill, B., and Hofschneider, P. H. (1977). Filamentous coliphage M13 as a cloning vehicle: insertion of a HindII fragment of the lac regulatory region in M13 replicative form in vitro. Proc. Natl. Acad. Sci. 74: 3642-3646.
8. Sanger, F., Nicklen, S. and Coulson, A. R. (1977). DNA sequencing with chain-terminating inhibitors. Proc. Natl. Acad. Sci. USA 74: 5463-5467.
9. Short, J. M., Fernandez, J. M., Sorge, J. A., and Huse, W. D. (1988). λ ZAP: a

bacteriophage λ expression vector with in vivo excision properties. Nucleic Acids Res. 16: 7583-7600.

10. Venter, J. C. et al. (2001). The sequence of the human genome. Science 291: 1304-1350.

11. Vieira, J. and Messing J. (1987). Production of single-stranded plasmid DNA. Methods Enzymol. 153: 3-11.

12. Waterson et al. (2002). Initial sequencing and comparative analysis of the mouse genome. Nature 420: 520-562.

Chapter 12

Cloning Vectors

Outline

Cloning potential of various vectors
EMBL3 and EMBL4 vectors
Cosmid
Yeast artificial chromosome
Bacterial artificial chromosome
P1 artificial chromosome

12.1: Cloning Potential of Various Vectors

Both bacteriophage and plasmid can be used as cloning vectors. Regardless of the type of cloning vector, the size of the insert that each vector can accept is limited. The insert size of plasmid vectors is usually inversely proportional to the copy number of the plasmid. For example, the F plasmid exists in *E. coli* as a single copy and can accept DNA fragments of approximately 100 kb. In contrast, both pBR322 and pACYC184 have a copy number of approximately 20 and can accept DNA fragments of only approximately 10 kb. Although fragments bigger than 10 kb may be ligated with the vector, the recombinant DNA may not be able to get in *E. coli* by transformation or may not replicate well inside *E. coli* cells. pUC, pTZ, and pBluescript vectors have a copy number greater than 200. The average size of DNA fragments that these vectors can accept is approximately 4 kb. The insert size of bacteriophage vectors is limited by the packaging potential of the bacteriophage. For example, the head of lambda phage can package a DNA fragment of approximately 50 kb. If the size of the lambda vector is 40 kb, the vector can accept an insert of approximately 10 kb. The cloning potential of λgt10 and λgt11 is 7.6 kb, and that of lambda ZAP vectors is 10 kb.

12.2: EMBL3 and EMBL4 Vectors

In genomic library construction, the DNA fragments to be cloned are usually greater than 10 kb. Therefore, vectors such as EMBL3 and EMBL4 capable of accepting bigger fragments of DNA are used. Both EMBL3 and EMBL4 are derivatives of lambda bacteriophage (Fig. 12-

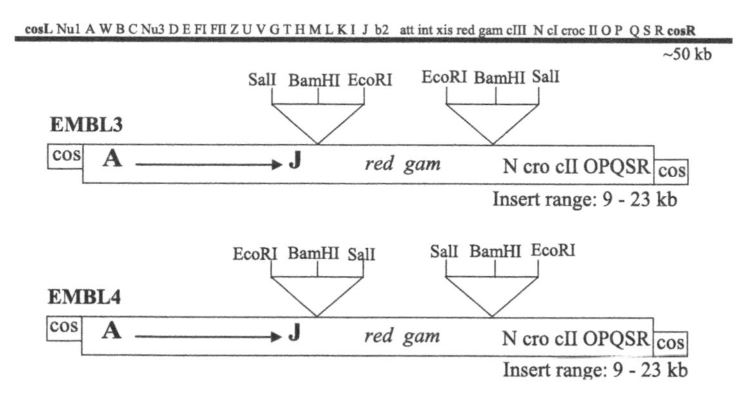

Fig. 12-1: EMBL3 and EMBL4 Vectors.

1). EMBL stands for European Molecular Biology Laboratory. These two vectors can accept DNA fragments ranging from 9 to 23 kb. For lambda bacteriophage, the region of the genome located between the J gene and N gene is not essential for its replication. Therefore, this region is referred to as stuffer fragment and can be replaced with a foreign DNA fragment. To use a lambda vector for the construction of a genomic library, the vector is first digested with a certain restriction enzyme to release the stuffer fragment. The resulting left and right arms are isolated, usually with a sucrose gradient, so that the stuffer fragment will not be ligated back to generate the original wild type lambda. This is a rather labor intensive process. With EMBL3 and EMBL4 vectors, this step can be bypassed. Both vectors have a pair of SalI, BamHI, and EcoRI sites located in two clusters 23 kb apart. The order of these 3 sites in the left cluster of EMBL3 is SalI, BamHI, and EcoRI, and that in the right cluster is EcoRI, BamHI, and SalI (Fig. 12-2). If the DNA fragments to be cloned are generated by BamHI digestion, EMBL3 is first digested with BamHI and then with EcoRI, so that the stuffer fragment has EcoRI ends instead of BamHI ends and cannot be ligated back to the two arms that have BamHI ends. Since the distance between BamHI and EcoRI sites is only approximately 10 bp, the 10 bp fragments generated can be removed by spin dialysis. The genomic DNA fragments generated by BamHI digestion are then ligated to the BamHI and EcoRI-digested EMBL3. The recombinant EMBL3 is then packaged by the lambda packaging system to become phage particles. The collection of these phage particles is a genomic library. If the DNA fragments to be cloned are generated by EcoRI digestion, EMBL4 vector is used (Fig. 12-2). EMBL4 is first digested with EcoRI and BamHI so that the stuffer fragment will not be religated back to the lambda genome.

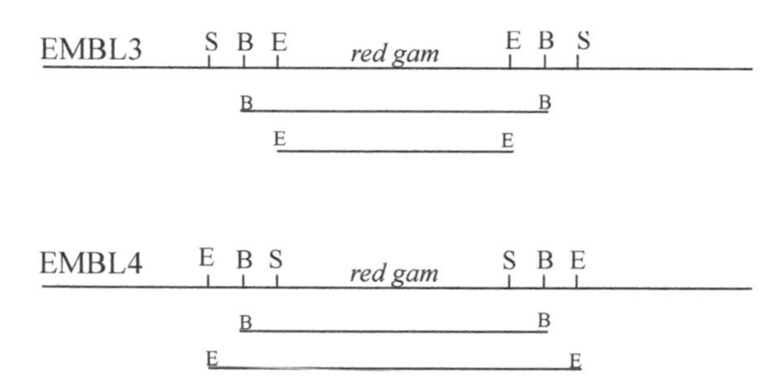

Fig. 12-2: Application of EMBL3 and EMBL4 vectors. S, B, and E represent SalI, BamHI, and EcoRI sites, respectively. To clone BamHI fragments, use EMBL3 which is cut with BamHI and EcoRI. To clone EcoRI fragments, use EMBL4 which is cut with EcoRI and BamHI.

If EcoRI or BamHI digestion is not complete in the preparation of EMBL3 or EMBL4 vector, wild type EMBL3 or EMBL4 lambda phages will be produced during the cloning process. Therefore, a method to distinguish wild type from recombinant is needed. This can be achieved by taking advantage of a unique property of lambda bacteriophage called sensitivity to P2 interference (spi). Although the mechanisms are unknown, lambda bacteriophages that have the *red* (recombination deficiency) and *gam* (gamma) genes cannot replicate in a P2 lysogen. The *red* gene encodes the Red protein which is composed of exo and β subunits. The exo protein is a 5' to 3' exonuclease analogous to the small fragment of the *E. coli* DNA polymerase I. The β protein is a single-strand DNA binding protein. The lambda Red protein is functionally equivalent to the RecA protein of *E. coli*. The product of the *gam* gene is the Gamma protein which is a potent inhibitor of the *E. coli* RecBCD exonuclease. When EMBL3 or EMBL4 is used to construct a genomic library, the stuffer fragment on which both the *red* and *gam* genes are located is replaced by the insert. Therefore, recombinant EMBL3 or EMBL4 phages do not have *red* and *gam* genes and are able to replicate in P2 lysogen. An example of P2 lysogen is *E. coli* strain Q359 [F^- *hsdR* (r_k^- m_k^+) *glnV44 fhuA* ($\Phi 80^r$) (P2)]. Theoretically, all plaques formed with this host contain recombinant EMBL3 or EMBL4 phages.

12.3: Cosmid

If a DNA fragment to be cloned is greater than 23 kb, none of the cloning vectors described above will work. Other types of vectors such as cosmid are used. A cosmid is a plasmid containing the cohesive site (*cos*) of lambda bacteriophage. pHC79 is one example of a cosmid (Fig. 12-3). It can accept a fragment of approximately 45 kb. The genome of lambda bacteriophage is a linear double-stranded DNA with a 12-base overhang at the 5' end of both upper and lower strands. The sequence of the overhang on the upper strand is GGGCGGCGACCT, and that of the lower strand is AGGTCGCCGCCC. Since these two sequences are complementary, the two ends of the lambda genome can readily anneal to each other. Therefore, the ends of lambda bacteriophage genome are referred to as cohesive sites (*cos*). The cohesive site is the packaging signal of lambda bacteriophage. During lambda DNA replication, a concatemer containing many units of lambda genome is formed. The region located between two *cos* sites on the concatemer is a unit of lambda genome, and *cos* signifies that one unit of the lambda genome is packaged. The concatemer is cut when *cos* is detected by the lambda A protein sitting at the opening of lambda head while the DNA is being pulled into the head to be packaged; therefore, the A protein is also called terminase.

To clone a DNA fragment into pHC79, pHC79 is first linearized by cutting with a certain restriction enzyme such as BamHI. DNA fragments that are generated by BamHI digestion are then ligated with the linearized pHC79. During ligation, a concatemer may be formed, in which a DNA fragment is flanked by two linear pHC79 DNA followed by another DNA fragment and pHC79 (Fig. 12-3). Since pHC79 has a *cos* site, some of the DNA fragments in the concatemer are flanked by two *cos* sites. This is a perfect substrate for the lambda packaging system if the distance between the two *cos* sites is approximately 50 kb. When the lambda phage particles formed are used to infect *E. coli*, the packaged DNA is injected into *E. coli*. The two ends of the injected DNA are then joined to form a circular DNA. Since this circular DNA has the *colE1* replication origin and ampicillin-resistance gene from pHC79, it is a typical plasmid and will replicate autonomously. The lambda genome is approximately 50 kb and pHC79 is 6 kb; therefore, pHC79 can accept a DNA fragment of approximately 45 kb.

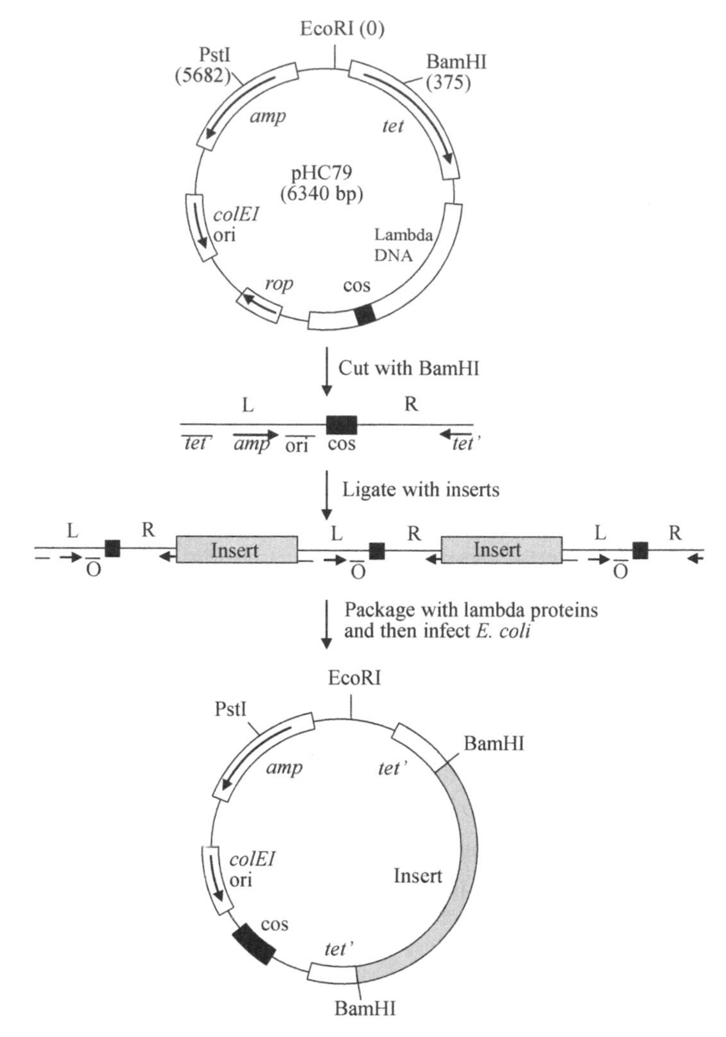

Fig. 12-3: Cosmid pHC79 and its applications. L, R, and O represent left arm, right arm, and replication origin. The filled black box denotes the cos site.

12.4: Yeast Artificial Chromosome

If the DNA fragment to be cloned is greater than 45 kb, artificial chromosome vectors are used for cloning. pYAC2 was the first vector developed for construction of a yeast artificial chromosome (Fig. 12-4). It is a derivative of pBR322 containing a centromere (CEN), two telomeres (TEL), an autonomous replication sequence (ARS1) which is a replication origin of yeast, and the *SUP4* gene from yeast. It also contains yeast *TRP1* and *URA3* genes. Any DNA with telomeres at both ends and a centromere in between is a chromosome.

To use pYAC2 to construct an artificial chromosome, it is first digested with BamHI and then with SmaI to generate two DNA fragments: the left and right arms (Fig. 12-4). Each arm has a telomere

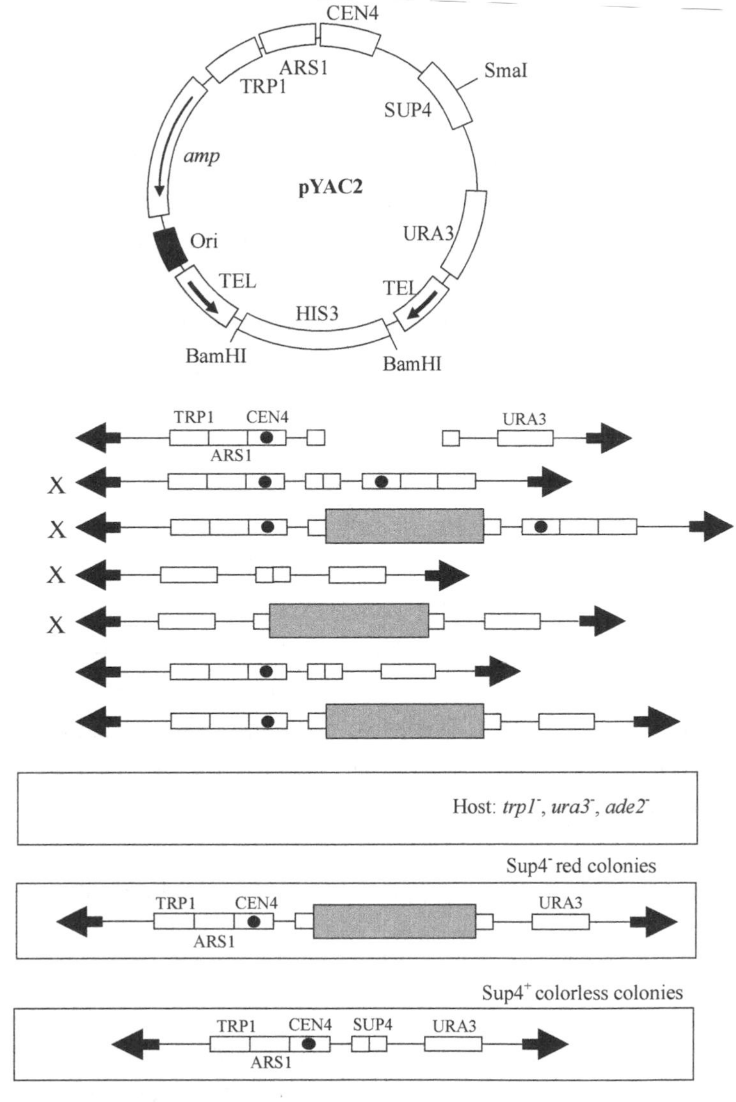

Fig. 12-4: pYAC2 and its application.

at one end. The centromere (CEN), *ARS1*, and the *TRP1* gene are located on the left arm, and the *URA3* gene is located on the right arm. If a DNA fragment is inserted between the left and right arms, an artificial chromosome is formed. However, many different combinations may be formed during ligation. Two left arms as well as two right arms may be ligated. It is also possible that a DNA fragment is ligated between two left arms or between two right arms. Another possibility is that the left and right arms are ligated together. If the DNA is ligated between two left arms or if two left arms are ligated together, the recombinant DNA will have two centromeres and will not be able to replicate in yeast cells. If the DNA fragment is ligated between two right arms or if two right arms are ligated together, the recombinant DNA will have no centromere and will also not be able to replicate in yeast cells. Only the recombinant in which the DNA fragment is ligated between the left and right arms or the left and right arms are ligated together can replicate in yeast cells. Since yeast chromosomes range between 225 to 1900 kb in size, the YAC vector has a cloning potential of approximately 2000 kb.

Once recombinant YACs are formed, they are introduced into yeast cells to replicate. Three different types of yeast cells may result. The majority of the cells are those that do not take up any DNA. The second type is those that take up the wild type YAC containing only the left and right arms. The third type of yeast cells are those that take up the desired recombinant with a DNA fragment inserted between the left and the right arms. Since the genotype of the yeast cells used as host cells is [*trp1, ura3, ade2*], cells that do not take up wild type or recombinant YAC cannot grow in minimum medium. In contrast, yeast cells containing the wild type YAC will grow and form colorless colonies, and those containing recombinant YAC will form red colonies on agar plates containing minimal medium.

The mechanism for selection of yeast cells containing recombinant YAC are described as follows. The symbol *trp1* indicates that the gene responsible for production of the *TRP1* enzyme required for biosynthesis of tryptophan is defective (Fig. 12-5). Unless the growth

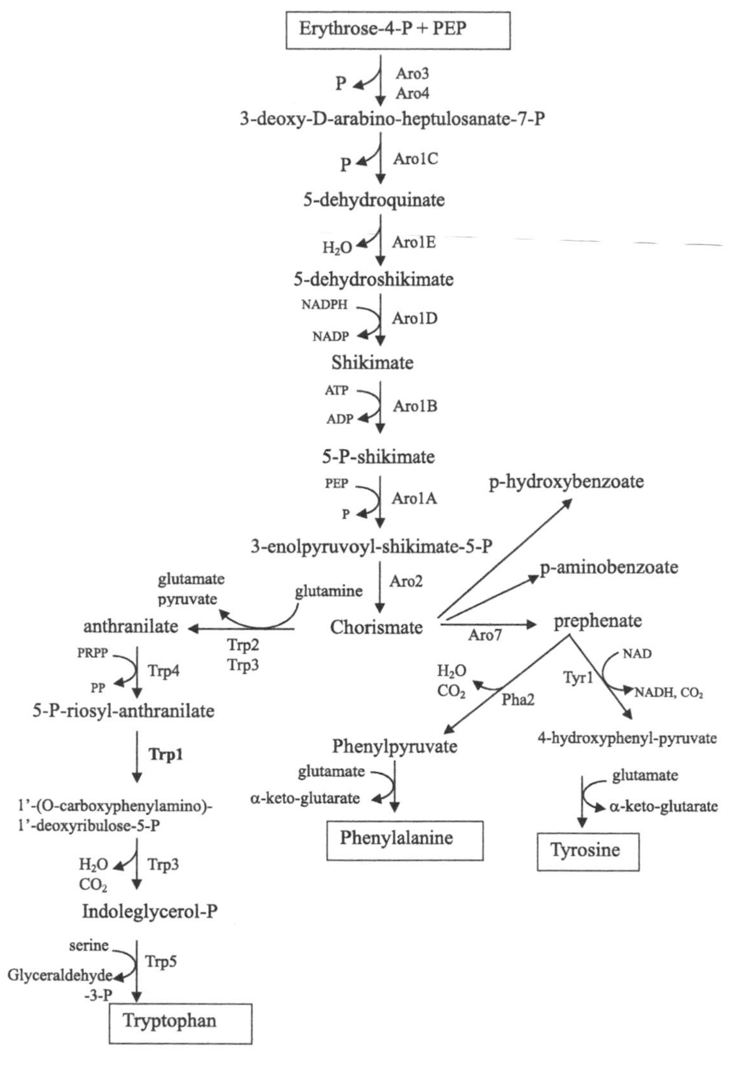

Fig. 12-5: Biochemical Pathway for Conversion of Erythrose-4-P to Tryptophan, Phenylalanine, and Tyrosine

medium is supplemented with tryptophan, the yeast cells cannot grow in minimal medium. This strain of yeast also has the *ura3* mutation and does not make the Ura3 protein which is one of the enzymes required for conversion of glutamine to CTP. Therefore, uracil or CTP must be supplemented to allow the cells to grow in minimum medium. Since the

YAC vector contains normal *TRP1* and *URA3* genes, cells containing either the wild type or recombinant YAC are able to grow in minimum medium without supplementation of tryptophan and uracil.

Ade2 is one of the enzymes required for conversion of p-ribosyl-pp to AMP or GMP (Fig. 12-7). The yeast strain used in this system contains a defective *ade2* gene; therefore, the pathway stops after the formation of p-ribosylamino imidazole. Since this compound is red, colonies of yeast cells without a functional *ADE2* gene is red. The *ade2* mutation in this yeast is due to the change of a sense codon to Amber stop codon. Therefore, a suppressor tRNA is needed to suppress this mutation to allow the cells to produce functional Ade2 protein. Since the YAC vector used in this system has the *SUP4* gene, cells containing the wild type YAC will be able to produce the Ade2 protein to allow the pathway to continue all the way to the end without accumulating p-ribosylamino imidazole which makes the colonies red. In recombinant YAC, the DNA fragment is inserted into the SUP4 gene. Therefore, yeast cells containing recombinant YAC do not make Ade2 protein and will accumulate p-ribosylamino imidazole, making their colonies red.

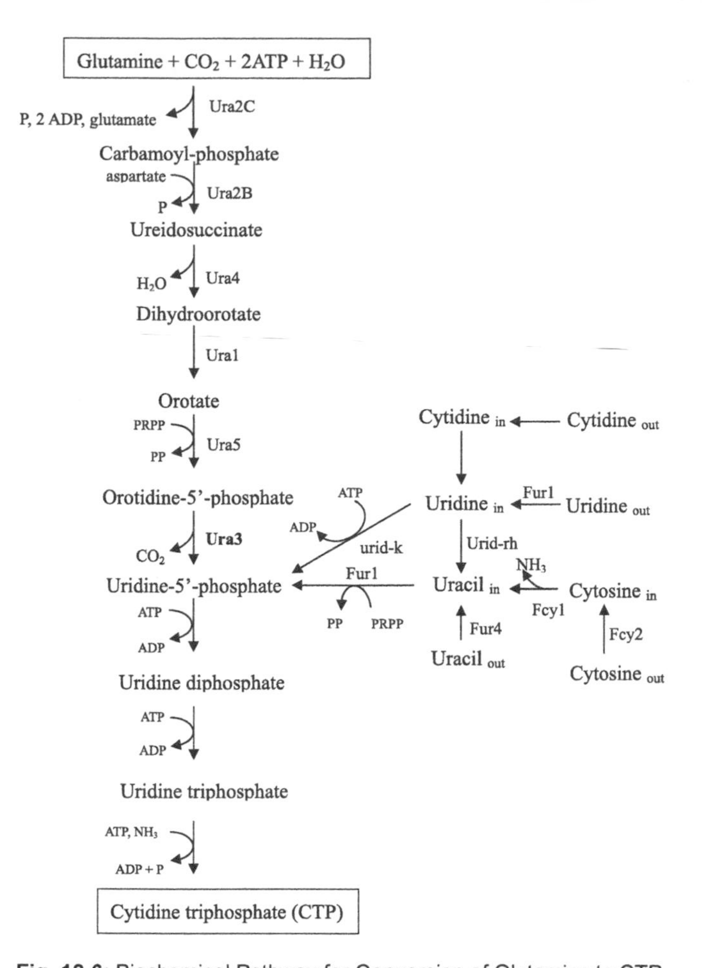

Fig. 12-6: Biochemical Pathway for Conversion of Glutamine to CTP.

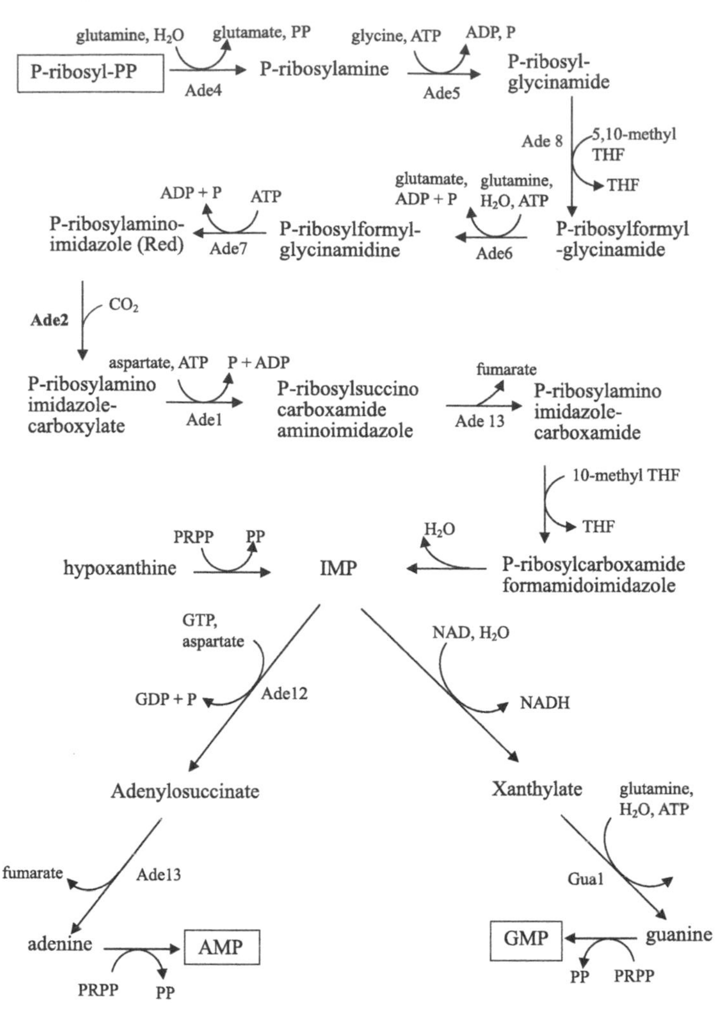

Fig. 12-7: Biochemical Pathway for Conversion of P-ribosyl-PP to AMP and GMP.

12.5: Bacterial Artificial Chromosome

Although YAC vectors can accept very large DNA fragments, recombinant YACs are unstable and may undergo deletions or rearrangements in yeast cells. Therefore, other artificial chromosome vectors such as BAC were developed. Although BAC stands for bacterial artificial chromosome, the BAC vector is not used to construct an *E. coli* artificial chromosome. Rather, it is used to construct a recombinant F plasmid. The original BAC vector is pBAC108L (Fig. 12-8). Its replication origin is *oriS* which is the replication origin of F

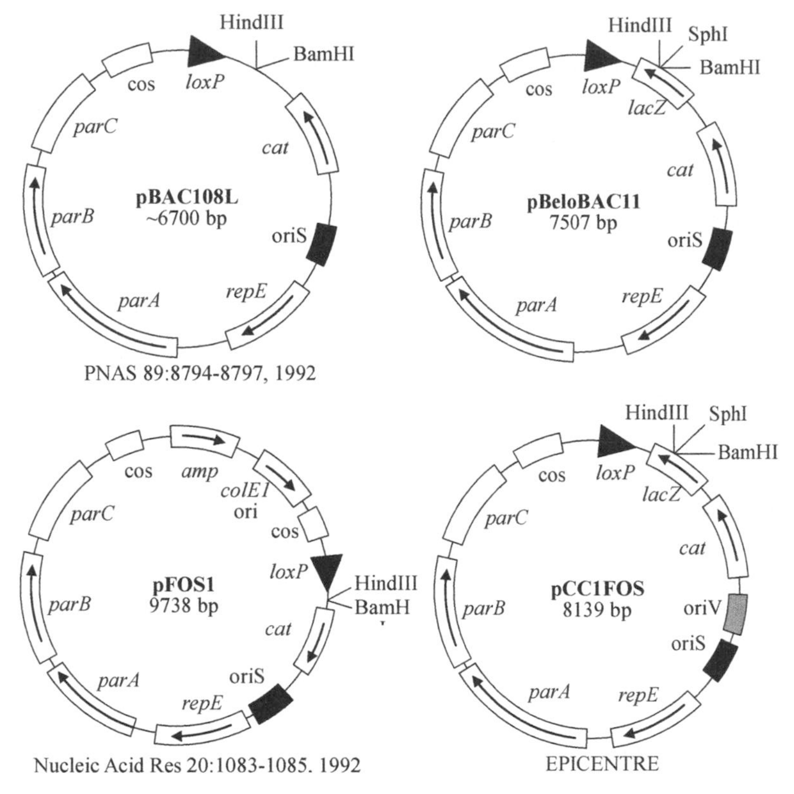

Fig. 12-8: BAC and Fosmids.

plasmid. It also contains the chloroamphenical resistance (*cam*) gene which encodes chloroamphenical acetyl transferase. Therefore, the *cam* gene is also referred to as *cat*. The *repE* gene on pBAC108L encodes the replication initiation protein of F plasmid. The symbol "*par*" stands for partition. The ParA protein is an ATPase, and ParB is a DNA binding protein. ParA and ParB interact with *parC* to regulate partition of F plasmid during cell division so that each daughter cell gets a copy of the F plasmid.

To use pBAC108L for cloning, it is first digested with HindIII or BamHI. DNA fragments to be cloned are then ligated with the linearized pBAC108L. The recombinant plasmids generated are introduced into *E. coli* by electroporation. Although the F plasmid is approximately 100 kb in size, BAC vectors can accept DNA fragments as large as 300 kb. Since pBAC108L does not have a selection marker for recombinant clones, pBeloBAC11 (Fig. 12-8) is constructed. It has the *lacZ* gene, and therefore, *E. coli* cells containing recombinant pBeloBAC11 can be selected based on their ability to produce β-galactosidase.

BAC plasmids are also called fosmids because they have the F plasmid replication origin. Since each *E. coli* cell contains only one copy of F plasmid, it is very difficult to prepare sufficient amounts of BAC plasmid for cloning. Therefore, pBeloBAC11 is modified by the insertion of a DNA fragment containing *oriV*, which is the replication origin of RK2 plasmid, creating pCC1FOS. In the presence of the TrfA protein (transacting function of fragment A), pCC1FOS uses *oriV* to replicate to approximately 50 copies per cell, thus greatly simplifying preparation of BAC vector DNA for cloning.

12.6: P1 Artificial Chromosome

PAC (P1 artificial chromosome) is another vector for construction of artificial chromosomes. It has a cloning potential of approximately 100 kb. The original PAC vector is pAD10sacBII (Fig. 12-9). P1 is another bacteriophage of *E. coli* and also has both lytic and lysogenic life cycles. Unlike lambda bacteriophage, P1 phage genome is not integrated into *E. coli* chromosome during lysogenic cycle but is integrated into many different sites in *E. coli* genome during lytic cycle. When these integrated P1 genomes are excised, the *E. coli* genome is broken into pieces, and infected *E. coli* cells are killed. During lysogenic cycle, P1 replicates in infected *E. coli* like a plasmid and does not harm the host cells.

Important components of pAD10sacBII include Plasmid rep, Lytic rep, *kan*, *sacB*, two *loxP* sites, and *pac* (Fig. 12-9). Plasmid rep is the

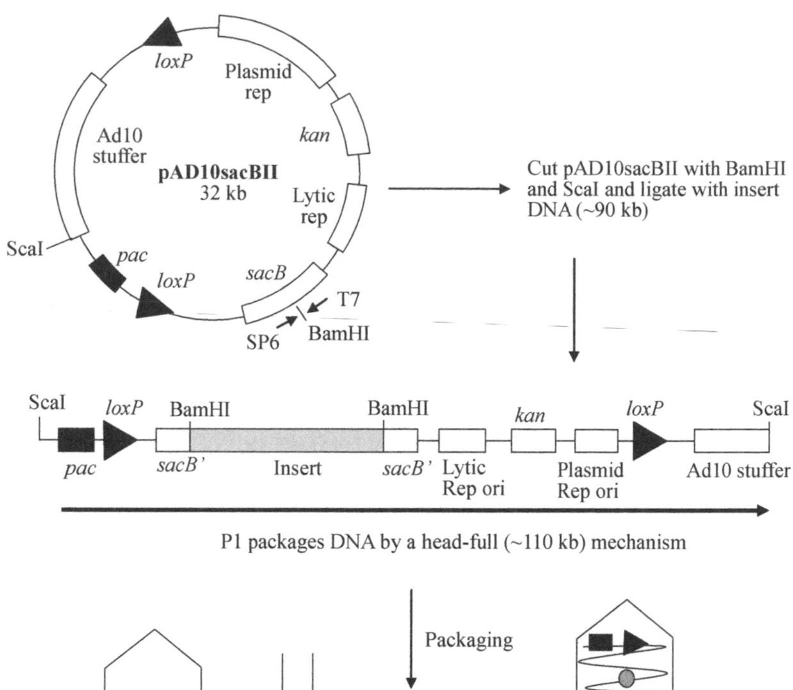

Fig. 12-9: PAC Vector and Its application. The shaded circle represents the plasmid replication origin of P1.

replication origin of P1 used during the lysogenic cycle, and Lytic rep is the replication origin used during the lytic cycle. *kan* represents the kanamycin-resistance gene, and *loxP* stands for locus of crossing over. It is a very important recombination signal of P1 phage. *pac* is the packaging signal of P1 and is functionally equivalent to the *cos* site of lambda phage. The *sacB* gene encodes sucrase which metabolizes sucrose to levan. Since levan is toxic to *E. coli*, *sacB* can be used as a selection maker to screen for desired recombinants if the foreign DNA fragment is inserted into *sacB*. *E. coli* containing the wild type pAD10sacBII will not be able to grow in a medium containing sucrose because levan converted from sucrose will kill the cell. Ad10 Stuffer is a DNA fragment derived from adenovirus. It serves to make recombinant DNA large enough for packging by the P1 packaging system. To use pAD10sacBII for cloning, it is first digested with ScaI and BamHI. DNA fragments generated by BamHI digestion are then ligated with the linearized vector. The recombinant DNA generated is packaged to become P1 phage in a manner similar to the in vitro packaging of lambda DNA. Since the only sequence required for P1 packaging is *pac*, any DNA with *pac* can be packaged. P1 is packaged by a head-full mechanism, i.e, the DNA is pulled into P1 head until the head is full. Since the P1 head can package a DNA fragment of approximately 110 kb, the PAC vector has a cloning potential of approximately 100 kb. When these P1 phage particles are used to infect *E. coli*, the packaged DNA is injected into *E. coli* cells. If the cells contain Cre (cyclization recombinase), the region between the two *loxP* sites are cut and then ligated to become circular DNA. Since the circular DNA generated contains the P1 plasmid replication origin and a kanamycin-resistance gene, it will replicate just like a plasmid in *E. coli*.

Summary

Every cloning vector has a certain cloning potential, meaning that the size of DNA fragment which a vector can accept is limited. The cloning potential of plasmids is usually inversely proportional to their copy numbers. The copy number of F plasmid is one; it can accept a DNA fragment of approximately 100 kb. The copy number of pBR322 or pACYC184 is approximately 20. These two plasmids can accommodate fragments of approximately 10 kb, whereas pUC derived plasmids can accept fragments of only approximately 4 kb. For bacteriophage vectors, the insert size is limited by the packaging potential of the phage. The lambda genome is 50 kb. Its packaging potential is 75 –

105% the size of the genome. λgt10 and λgt11 can accept fragments up to 7.6 kb, λZAP vectors can take DNA fragments of approximately 10 kb, and EMBL3 and EMBL4 can accept fragments up to 23 kb. When EMBL3 or EMBL4 is used, the stuffer region is replaced by a foreign DNA fragment. Since wild type lambda phage is sensitive to P2 interference (spi) and cannot replicate in P2 lysogen due to the presence of the *red* and *gam* genes, this property can be used to select desired recombinants. Recombinant EMBL3 or EMBL4 can replicate in P2 lysogen because the stuffer fragment containing the *red* and *gam* genes is replaced by a foreign DNA fragment.

If a DNA fragment greater than 23 kb is to be cloned, a different type of vector such as cosmid is used. Cosmid is a regular plasmid containing the *cos* sequence of lambda bacteriophage. During lambda packaging, the region located between two *cos* sites of a lambda concatemer is packaged. Since *cos* is the only signal recognized by the packaging system, any DNA of approximately 50 kb flanked by two *cos* sites can be packaged. pHC79 is the original cosmid. It is a derivative of pBR322 and contains a small portion of lambda DNA with a *cos* site. If a DNA fragment is ligated between two linearized pHC79, the DNA fragment is flanked by two *cos* sites and can be packaged by the lambda packaging system. The packaged DNA is injected into *E. coli* cells when the phage particles thus produced are used to infect *E. coli.* The two ends of the injected DNA are then joined to form a circular DNA. Since this circular DNA contains the *colE1* replication origin from pHC79, it will replicate as a plasmid. Cosmids have a cloning potential of approximately 45 kb.

To clone DNA fragments greater than 45 kb, YAC vectors may be used. YAC stands for yeast artificial chromosome. Since yeast chromosomes are 225 to 1900 kb in size, the cloning potential of YAC is approximately 2000 kb. The original YAC vector is pYAC2. It is a derivative of pBR322 containing two telomeres and a centromere of yeast. When pYAC2 is digested with BamHI and SmaI, two arms with a telomere are generated. The left arm also contains the *TRP1* gene and the centromere. The right arm contains the *URA3* gene in addition to a telomere at the end. An artificial yeast chromosome is formed if a foreign DNA fragment is inserted between these two arms. The recombinant DNA generated is introduced into an appropriate yeast host such as the one with the genotype of *ade2*, *trp1*, *ura3*. In this cloning system, the foreign DNA fragment is inserted into the *SUP4* gene on pYAC2. Since the *SUP4* gene encodes a suppressor tRNA which can suppress the Amber mutation on the *ade2* gene, yeast cells containing a recombinant YAC do not produce suppressor tRNA. Therefore, the pathway which converts p-ribosyl-pp to AMP and GMP will stop after the step which produces p-ribosylamino-imidazole. Since this compound is *red*, colonies of yeast cells containing a recombinant YAC are *red*. Yeast cells that do not take up either wild type or recombinant YAC are eliminated because they do not have functional *TRP1* and *URA3* genes that are located on YAC arms and cannot grow on minimum medium due to *trp1* and *ura3* mutations.

BAC is another vector for construction of artificial chromosomes. Although BAC stands for bacterial artificial chromosome, it actually allows construction of an artificial F plasmid. pBAC108L is the original BAC vector. Its replication origin is *oriS* which is the replication origin of F plasmid. The DNA fragment to be cloned is inserted into the HindIII or BamHI site of pBAC108L, and the recombinant plasmid is introduced into *E. coli* by electroporation. Since the replication origin of BAC vectors is from F plasmid, BAC vectors are also called fosmids. F plasmid exists as single copy in *E. coli*. Therefore, it is very difficult to prepare sufficient BAC vector for cloning. One improvement in the system was the introduction of another replication origin such as *oriV* from a high copy number plasmid RK2 into the vector. pCC1FOS is one such plasmid. In the presence of the TrfA (transacting function of fragment A) protein, pCC1FOS can replicate to approximately 50 copies per cells. The cloning potential of BAC is approximately 300 kb.

PAC (P1 artificial chromosome) is another artificial chromosome cloning system. The original PAC vector is pAD10sacBII. It contains both plasmid and lytic replication origins, *pac*, and two *loxP* sites of P1. It also contains a kanamycin-resistance gene (*kan*) and the *sacB* gene as the two selection markers of the plasmid. To use it for cloning, pAD10sacBII is linearized by digestion with SacI and BamHI and then ligated with foreign DNA fragments to generate a recombinant DNA containing *pac*, the insert, *kan*, and the two replication origins flanked by two *loxP* sites. This recombinant DNA is then packaged by the P1 packaging system which will package any DNA with the *pac* signal by a head-full mechanism. The P1 phage particles produced are used to infect *E. coli*. If the *E. coli* cells contain the Cre (cyclization recombinase) protein, Cre will cut the DNA at the two *loxP* sites and ligate the two ends together, making the recombinant DNA circular. Since this circular DNA has the P1 replication origin and the kanamycin resistance gene, it will replicate autonomously as a plasmid. The P1 packaging system can package a DNA fragment of approximately 110 kb; therefore, PAC has a cloning potential of approximately 100 kb.

Sample Questions

1. *Cos* is: a) replication origin of lambda phage, b) cohesive site of lambda phage, c) packaging signal of lambda phage, d) integration site of lambda phage
2. Cosmid is a plasmid containing: a) P1 phage replication origin, b) the *cos* sequence of lambda phage, c) f1 replication origin, d) lambda phage replication origin
3. Cosmid allows cloning of a DNA fragment approximately: a) 40 kb, b) 400 kb, c) 100 kb d) 1000 kb
4. YAC vector allows cloning of a DNA fragment approximately: a) 40 kb, b) 400 kb, c) 100 kb, d) 2000 kb
5. BAC vector allows cloning of a DNA fragment approximately: a) 40 kb, b) 5 kb, c) 100 k, d) 1000 kb
6. PAC vector allows cloning of a DNA fragment approximately: a) 40 kb, b) 400 kb, c) 100 kb, d) 1000 kb
7. BAC vector is also called: a) cosmid, b) phagemid, c) fosmid, d) bacmid.

Suggested Readings

1. Bertani, G. (2004). Lysogeny at mid-twentieh century: P1, P2, and other experimental systems. J. Bacteriol. 186: 595-6000.
2. Blackburn, E. H. and Szostak, J. W. (1984). The molecular structure of centromeres

and telomeres. Annu. Rev. Biochem. 53: 163-194.

3. Burke, D. T., Carle, G. F., and Olson, M. V. (1987). Cloning of large segments of exogenous DNA into yeast by means of artificial chromosome. Science 236: 806-812.
4. Clarke, L. and Carbon, J. (1985). The structure and function of yeast centromeres. Annu. Rev. Genet. 19: 29-56.
5. Figurski, D. H. and Helinski, D. R. (1979). Replication of an origin-containing derivative of plasmid RK2 dependent on a plasmid function provided in trans. Proc. Natl. Acad. Sci. USA 76: 1648-1652.
6. Frischauf, A. M., Lehrach, H., Poustka, A., and Murray, N. (1983). Lambda replacement vectors carrying polylinker sequences. J. Mol. Biol. 170: 827-842.
7. Hohn, B. and Collins, J. (1980). A small cosmid for efficient cloning of large DNA fragment. Gene 11: 291-298.
8. Kim, U. -J., Shizuya, H., de Jong, P. J., Birren, B. and Simon, M. I. (1992). Stable propagation of cosmid sized human DNA inserts in an F factor based vector. Nucleic Acids Res. 20: 1083-1085.
9. Kittell, B. L. and Helinski, D. R. (1991). Iteron inhibition of plasmid RK2 replication in vitro: evidence for intermolecular coupling of replication origins as a mechanism for RK2 replication control. Proc. Natl. Acad. Sci. USA 88: 1389-1393.
10. Murray, A. and Szostak, J. W. (1983). Construction of artificial chromosome in yeast. Nature 305: 189-193.
11. Pierce, J. C. and Sternberg, N. L. (1992). Using bacteriophage P1 system to clone high molecular weight genomic DNA. Meth. Enzymol. 216: 549-574.
12. Shizuya, H., Birren, B., Kim, U. -J., Mancino, V., Slepak, T., Tachiiri, Y., and Simon, M. (1992). Cloning and stable maintenance of 300-kilobase-pair fragments of human DNA in *Escherichia coli* using an F-factor-based vector. Proc. Natl. Acad. Sci. USA 89: 8794-8797.
13. Sternberg, N. (1990). Bacteriophage P1 cloning system for the isolation, amplification, and recovery of DNA fragments as large as 100 kilobase pairs. Proc. Natl. Acad. Sci. USA 87: 103-107.
14. Yarmolinsky, M. B. (2004). Bacteriophage P1 in retrospect and in prospect. J. Bacteriol. 186: 7025-7028.
15. Zakian, V. A. (1989). Structure and function of telomeres. Annu. Rev. Genet. 23: 579-604.

Chapter 13

Polymerase Chain Reaction

Outline

Invention of polymerase chain reaction
PCR primer design
Types of PCR
PCR contamination problems
PCR controls
Prevention of non-specific reactions
PCR applications
Cloning of PCR products
Enzymes used for PCR
Other nucleic acid amplification methods

13.1: Invention of Polymerase Chain Reaction

Polymerase chain reaction (PCR) was invented by Kary B. Mullis in 1983 to amplify a segment of a DNA molecule in vitro. Its principle is similar to that of primer extension described in Chapter 8. The major difference is that two primers are used for PCR, whereas only one is used for primer extension. To perform PCR, a double-stranded DNA is first denatured to become two single-stranded DNA molecules. A primer is then annealed to the upper (plus) strand at the right hand end of the region to be amplified, and another primer is annealed to the lower (minus) strand at the left hand end of the region to be amplified (Fig. 13-1). In the presence of dNTP (dATP, dCTP, dGTP, and dTTP) and a DNA polymerase, the primers are extended to form two double-stranded DNAs. When these two double-stranded DNAs are denatured, each of the four single-stranded DNAs can be used as a template to synthesize

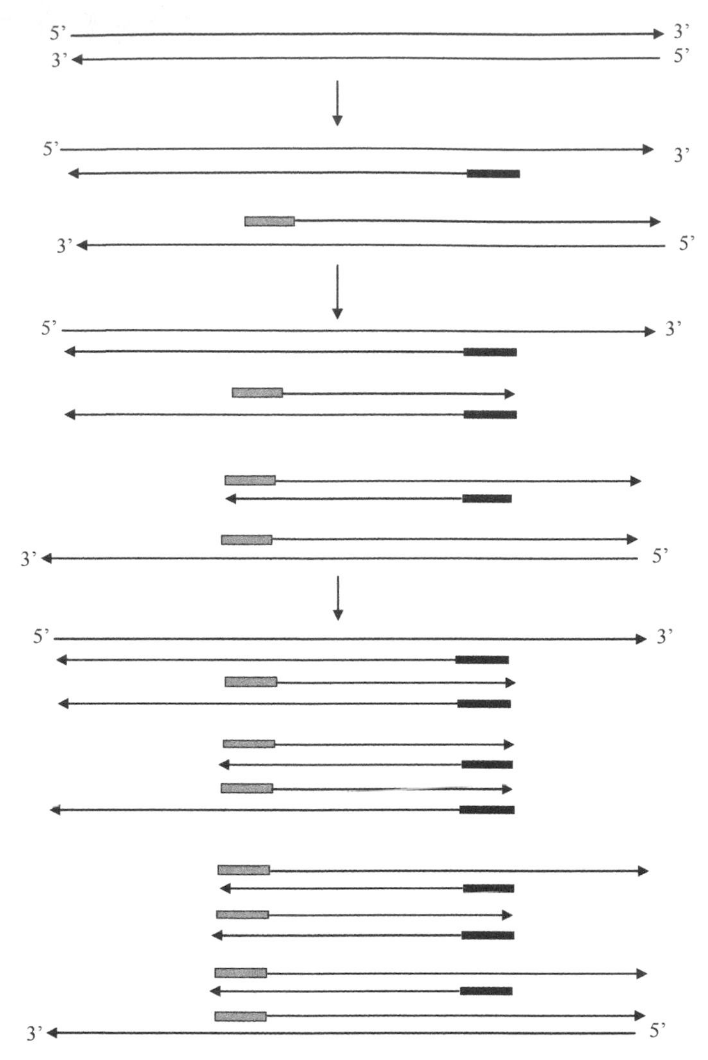

Fig. 13-1: Polymerase Chain Reaction. Gray and black bars represent primers.

DNA, thus increasing the number of DNA strands from the original two to a total of eight. If the process is repeated thirty times, billions of DNA strands are produced.

There are three different steps in PCR. The first step is to denature the template DNA to single-stranded by heating the DNA at a temperature of approximately 95°C, usually 94°C. The second step is to anneal the two primers to the template; this is done at a temperature of 50 – 65°C. The third step is DNA synthesis through extension of the primers. The temperature required for this step depends on the type of DNA polymerase used. If the Klenow fragment of *E. coli* DNA polymerase I is used, the required temperature is 37°C. When PCR was first developed, three water baths at three different temperatures 94°C, 50°C, and 37°C were used. DNA template, primers, and dNTP are mixed in a test tube. The tube is placed in the 94°C water bath for 30 - 60 seconds to denature DNA and then transferred to the 50°C water bath for 30 seconds to allow the primers to anneal to the template. DNA polymerase such as the Klenow fragment is then added to the tube, and the tube is incubated in the 37°C water bath for 30 – 60 seconds to synthesize DNA. After which, the tube is transferred back to the 94°C water bath to denature the newly synthesized DNA so that they can be used as the template for a second round of the reaction. Since the Klenow fragment is not heat resistant, it is inactivated in this heating step. Therefore, additional Klenow fragment is added in the third step of every cycle.

Since it is a very labor intensive process to repeat this for 30 – 35 cycles, two improvements were made to simplify the process. The first was to replace Klenow fragment with Taq DNA polymerase. This enzyme is isolated from *Thermus aquaticus* which is a thermophilic bacterium found in hot springs. Taq DNA polymerase is heat resistant and is not inactivated even after 20 minutes of boiling. Its optimal reaction temperature is 72°C. The other improvement is to develop an automatic temperature cycler. With these improvements, it is no longer necessary to add additional DNA polymerase in every cycle. All reagents required for PCR are mixed together in a test tube, and the tube is placed in the thermocycler which automatically performs the reaction for the set number of cycles. An example of temperature cycling is 94°C for 60 seconds, 50°C for 30 seconds, and 72°C for 60 seconds for 30 – 35 cycles.

13.2: PCR Primer Design

Components needed for PCR include a DNA template, two primers, Taq DNA polymerase, dNTP, and a buffer to maintain the pH of the reaction mixture. DNA template is the DNA isolated from a certain type of cells or a DNA fragment containing the gene to be amplified. Taq DNA polymerase, buffer, and dNTP can be purchased from various vendors. Since amplification of different genes requires different primers, most primers are not commercially available and need to be custom made. Two primers are required to perform a PCR reaction. One primer anneals to the 5' end, and the other anneals to the 3' end of the region to be amplified. The one that anneals to the 5' end is called 5' primer, and the one that anneals to the 3' end is called 3' primer. The length of each primer must be at least 18 nucleotides, usually 20 – 30 nucleotides. If the primer is too short, it may not anneal to the template. If it is too long, it would take more time to anneal to the template, and the total time required for a PCR reaction would be increased.

The first step in performing PCR is to determine which region of the target is to be amplified. Since GC-rich regions are more difficult to denature, these regions are avoided. The length of the region to be amplified varies depending on the purpose for DNA amplification. For diagnostic purpose such as detection of a certain microorganism, a region smaller than 200 bp is amplified.

To make primers, the target sequence, which is usually written in single-stranded form (Fig. 13-2), is examined to find two 20-base

Fig. 13-2: 5' and 3' Primers for PCR.

regions whose sequences can be used. Once these two regions are selected, the primer sequences are written in 5' to 3' direction and give to vendors to synthesize the primers. The sequence of the 5' primer thus written is identical to that of the first 20 bases at the 5' end, and the sequence of the 3' primer thus written is complementary to that of the last 20 bases at the 3' end of the region to be amplified in this single-stranded form target sequence (Fig. 13-2 and Fig. 13-3).

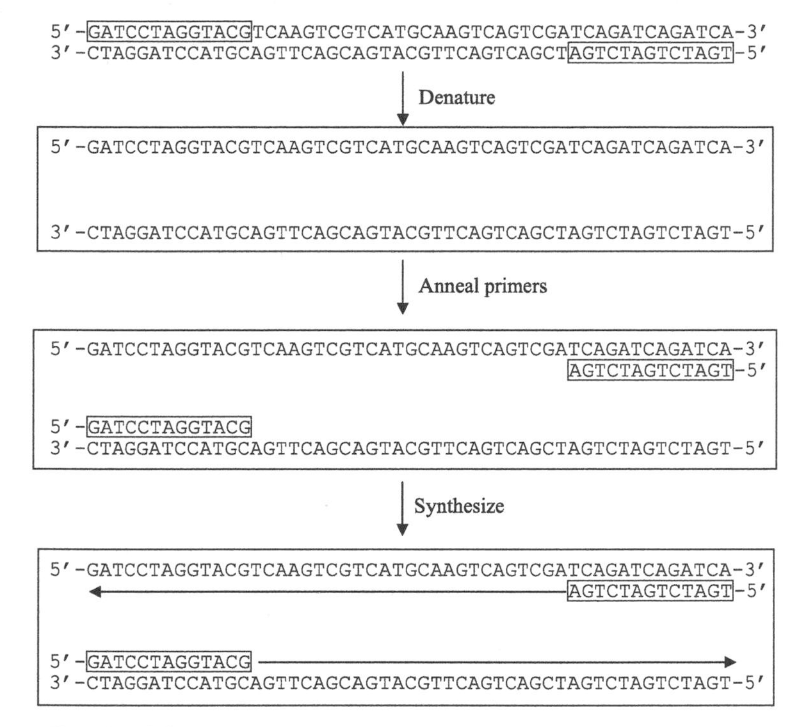

Fig. 13-3: PCR primers and Principles.

Since the plus (upper) strand of a double-stranded DNA is the sense strand, the 5' primer is also called sense primer, and the 3' primer is also called the anti-sense primer. The 5' primer leads DNA synthesis from left to right which is conventionally considered as forward direction; hence, it is also called the forward primer. The 3' primer leads DNA synthesis in an opposite direction; therefore, it is also called the reverse primer. Not all sequences can be used to make primers. Certain rules are followed in order to select sequences that are appropriate. The number of G and C bases should be approximately 50% of the total bases in a primer, and the G and C bases in a primer should not be in clusters. In general, primers should not contain more than 3 of the same base in a row. The 3' end of the primer should be GG, CC, GC, or CG because G pairs to C with three hydrogen bonds, making the primers anneal to the template more tightly and allowing for more efficient priming in DNA synthesis. The 5' end of a primer is not critical. Therefore, extra bases that do not pair with the template can be added (Fig. 13-4). These extra bases are used as the template to synthesize

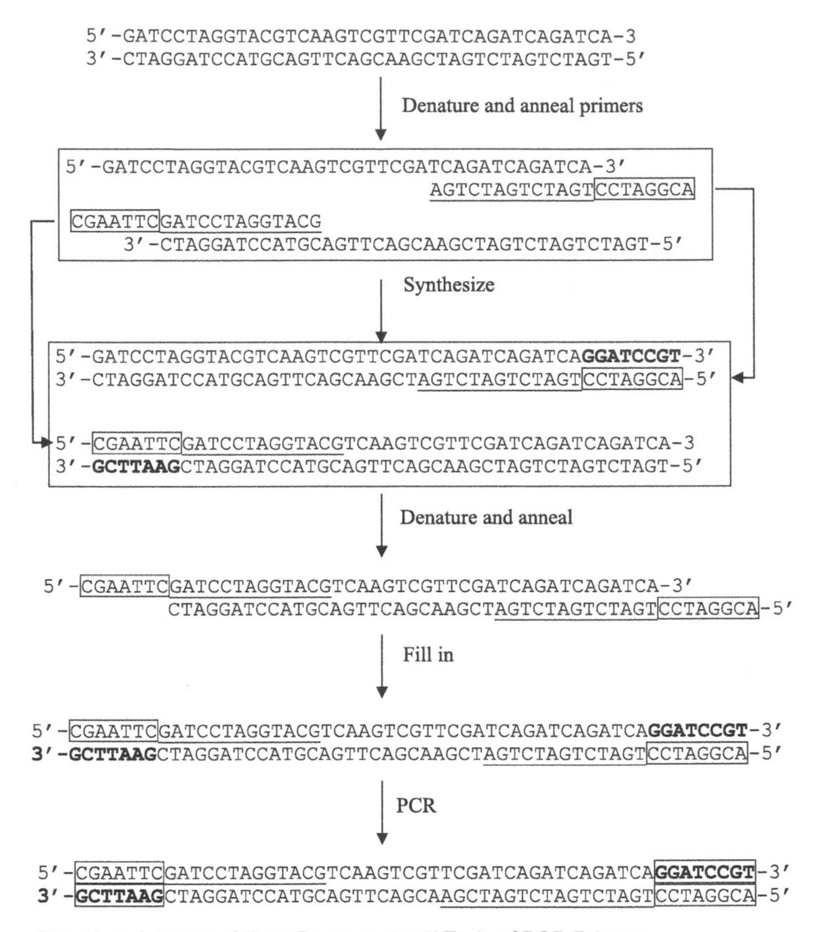

Fig. 13-4: Addition of Extra Bases to the 5' Ends of PCR Primers.

the complementary strand, resulting in a double-stranded DNA with the added bases. The extra bases added to the 5' end of a primer are usually the recognition sequence of a certain restriction enzyme so that the resulting PCR product can be cut with the restriction enzyme to make ends compatible with that of a cloning vector.

The primer must not form a stem-loop structure, i.e., the primer sequence cannot have inverted repeats. In addition, the 3' ends of the two primers cannot be complementary; otherwise, they will anneal to each other not to the template, and all PCR products generated would be primer dimers. The primer sequence must be unique to the target so that non-specific reactions do not occur. To ensure the uniqueness of a primer sequence, the sequence of the designed primer is compared with sequences in the DNA sequence banks. This can be done by using the program BLAST on http://www.ncbi.nlm.nih.gov or BLAT on http://genome.ucsc.edu/. BLAST is basic local alignment search tool, and BLAT is BLAST-like alignment tool. A primer sequence that has more than 30% homology with other sequences cannot be used; otherwise, non-specific reactions will occur. The melting temperatures (Tm) of the two primers should be identical or very close to each other. Tm of a certain primer can be calculated using the formula: Tm = 4x (# of G + # of C) + 2 x (# of A + # of T). If the sequence of a primer is GATCGTAGC, its Tm is (4x5)+(2x4) = 28°C since the total number of G and C bases is 5, and that of A and T bases is 4. If it is not possible to make the Tm's of the two primers identical, the difference should be less than 10°C.

PCR can also be used to amplify RNA, usually mRNA. However, the RNA to be amplified must first be converted to DNA by reverse transcription. Therefore, the method is called RT-PCR for reverse transcription-PCR. Since mRNAs do not have introns, primers must anneal to exons. However, if both primers anneal to the same exon, it would be impossible to distinguish whether the PCR product generated is derived from mRNA or from contaminating DNA. For example, an RT-PCR is performed to amplify a portion of the mRNA of a gene containing two exons of 200 bp and 500 bp, respectively separated by an intron of 1000 bp. If the 5' primer anneals to the 5' end of exon 2 and the 3' primer anneals to the 3' end of the same exon, the PCR product derived from the mRNA or form the DNA is the same size, 500 bp. In contrast, if the 5' primer anneals to the 5' end of exon 1 and the 3' primer anneals to the 3' end of exon 2, the PCR product derived from mRNA is 700 bp, but that derived from DNA is 1700 bp. Therefore, for RT-PCR, the two primers are designed to anneal to different exons.

The amount of each primer required for PCR is 10 – 20 pmol. Primers are chemically synthesized oligonucleotides. Their concentrations are usually expressed in OD_{260} units. To convert OD_{260} units to molar concentration, the following formula can be used: Total OD_{260} ÷ (10 x length of primer) = μmol/ml. Since the molecular weight of a nucleotide is approximately 330 g/mole, the molecular weight of a 20-base oligonucleotide is 330 g/mole x 20. One OD260 unit of oligonucleotides is approximately 33 μg/ml. Therefore, the molar concentration of a 20-base primer with OD_{260}=1 is 33 μg/ml ÷ (330 g/mole x 20), which equals to [1 ÷ (10 x 20)] μmole/ml. If the concentration of another 20-base primer is OD_{260}=2, its molar concentration would be 66 μg/ml ÷ (330 g/mole x 20) which is equal to [2 ÷ (10 x 20)] μmole/ml. Similarly, if the concentration of the third 20-base primer is OD_{260}=3, its molar concentration would be 99 μg/ml ÷ (330 g/mole x 20) which equals to [3 ÷ (10 x 20)] μmole/ml. Therefore, the formula for converting OD_{260} units to molar concentration is: Total OD_{260} ÷ (10 x length of primer) = μmole/ml.

For unknown reasons, oligonucleotides are susceptible to degradation when subjected to repeated freeze and thaw cycles. Therefore, primers are divided into many aliquots for storage, and each aliquot is used only once. Several factors may affect the efficiency of PCR. Among them, Mg^{++} is most critical. The required Mg^{++} concentration is 1 to 10 mM, and the most common concentration used is 1.5 mM. Primer annealing temperature is another factor and is dependent on its Tm. The appropriate annealing temperature is Tm – 5. For example, the annealing temperature for a primer with a Tm of 65°C is approximately 60°C. Primer annealing temperature is usually between 42 to 70°C. The following is one example of a PCR condition: 50 ng of chromosomal DNA, 20 pmol of each primer, 1.5 mM Mg^{++}, 0.2 mM dNTP, 1 - 2 units of Taq DNA polymerase, and annealing temperature

50°C. If this condition fails to produce expected PCR product, the first variable to be adjusted is Mg^{++}, followed by annealing temperature and primer concentrations. If none of the adjustments work, there must be problems with the primer design.

13.3: Types of PCR

Many types of PCR exist. Nested PCR amplifies the same target twice with two overlapping sets of primers. If the copy number of the target is very low, a simple PCR may not be sufficient. The product of the first PCR reaction is used as a template for amplification by the second set of primers that anneal to regions slightly internal to where the first set of primers binds. Multiplex PCR amplifies more than one target in the same reaction. To perform multiplex PCR, all primers used must have similar Mg^{++} requirements and Tm's.

Quantitative PCR is used to determine the copy number of a certain target. If the increase in the amount of product during PCR is graphed, the curve is similar to the growth curve of bacteria with lag, log (exponential), and stationary phases. If two samples containing the same target in different concentration are amplified, the curves in the log phase are parallel, but those in the stationary phase overlap, indicating that the amounts of PCR products of these two samples are the same in the stationary phase. Since most PCR reactions must reach stationary phase to produce sufficient amounts of product to be visualized in a gel after electrophoresis, it is not possible to use regular PCR to quantify the copy number of a target. Competitive PCR is a modified PCR for this purpose. In this method, the same set of primers is used to amplify two different targets with the same primer binding sites. One of the targets is the real target to be quantified; the other is a fake target constructed in vitro. The fake target is made slightly longer or shorter than the real target. Since it is made in vitro, its concentration is known. To perform a competitive PCR, a series of 5 - 6 reactions with 10, 20, 30, 40, 50, or 60 ng of the fake target and the same volume (e.g., 10 μl) of the real target are performed. After PCR, the products are electrophoresed in a gel. Two bands representing the real and fake targets should be seen. The intensities of the PCR product bands are then compared. If the intensities of the two bands in the lane with 30 ng of the fake target are the same, the amount of the real target in this sample is also 30 ng.

If only the fold increase, not the absolute amount, is to be determined, semi-quantitative PCR can be performed. An example of this application is for determining whether a chemical or physical treatment of cells affects the expression of a certain gene. The RNAs isolated from untreated and treated cells are reverse transcribed to cDNA. The cDNA samples are 10-fold serially diluted, and each dilution is amplified by PCR to detect the mRNA of interest. If the highest dilution that yields a positive PCR reaction is 10^{-4} for the untreated sample but is 10^{-5} for the treated sample, the treatment induces a 10 fold increase in the expression of the gene.

The most commonly used quantitative PCR is real-time PCR. This method makes PCR products fluorescent or PCR reactions emit fluorescence. Since some spectrofluorometers can detect as few as one photon, fluorescent PCR products generated can be monitored from the beginning to the end. Therefore, the cycles in which PCR is in the exponential phase can be determined. During exponential phase, the amount of PCR product is directly proportional to the copy number of the target. In real-time PCR, the fluorescence emitted is measured every 3-5 seconds while the PCR reaction is going on. The most important aspect of real-time PCR is the cycle number at which the fluorescence becomes detectable or reaches a certain level. This cycle number is referred to as the cycle threshold (C_T). The C_T value is inversely proportional to the copy number of the target, the higher the target copy number the lower the C_T value. To make fluorescent PCR products, SYBR Green can be added to the PCR reaction mixture. SYBR Green fluoresces brighter when it binds to double-stranded DNA. PCR products are double-stranded DNA. The more PCR products produced the higher the amount of light emitted.

Real-time PCR is very commonly used to determine changes in the expression of a certain gene after induction. For example, if the C_T value of the real-time RT-PCR of IL-1β mRNA of a sample is 30 before and 20 after stimulation of the cells with a certain drug, it is clear that the expression of IL-1β is increased after the drug treatment. However, it is not known how much increase in expression the drug treatment caused. The most commonly used formula for calculating fold change is $2^{-\Delta\Delta CT}$. Since a fold change observed may be due to technical errors such as pipetting, an internal control is needed. If the levels of the IL-1β mRNA are to be determined, another cellular gene such as the β-actin gene is co-amplified to serve as the internal control. Therefore, two sets of primers, one for IL-1β and the other for β-actin, are used in the same RT-PCR. This reaction will result in two different C_T values. The difference between these two C_T values is called ΔC_T which is usually the value of target C_T minus that of control C_T (target C_T - control C_T.). If the target C_T is 30 and the control C_T is 40, the ΔC_T value is -10. To determine the effect of the drug on IL-1β gene expression, the same two sets of primers are used to perform the real-time RT-PCR on the RNA sample from drug treated cells. If the target C_T is 20 and the control C_T is 40, the ΔC_T value of this drug treated sample is -20. With these values, $\Delta\Delta C_T$ can be determined. It is the ΔC_T value of the treated sample minus that of the untreated sample; therefore, it is (-20) – (-10) = -10, and $2^{-\Delta\Delta CT}$ is 2^{10} = 1024. This indicates that the drug treatment caused a 1024 fold increase in the expression of the IL-1β gene. It is customary that each real-time PCR is performed in triplicate. Therefore, each of the C_T values used in the calculation is the average C_T value of the three reactions.

The major disadvantage of using SYBR Green is that it binds to any double-stranded DNA and really cannot distinguish primer dimers and nonspecifically amplified product from the intended product. To determine whether a real-time PCR is specific, melting-curve analysis of the PCR product is performed. A melting curve traces the number of double-stranded DNA molecules that become single-stranded at different temperatures (Fig. 13-5). If the PCR is specific, only one

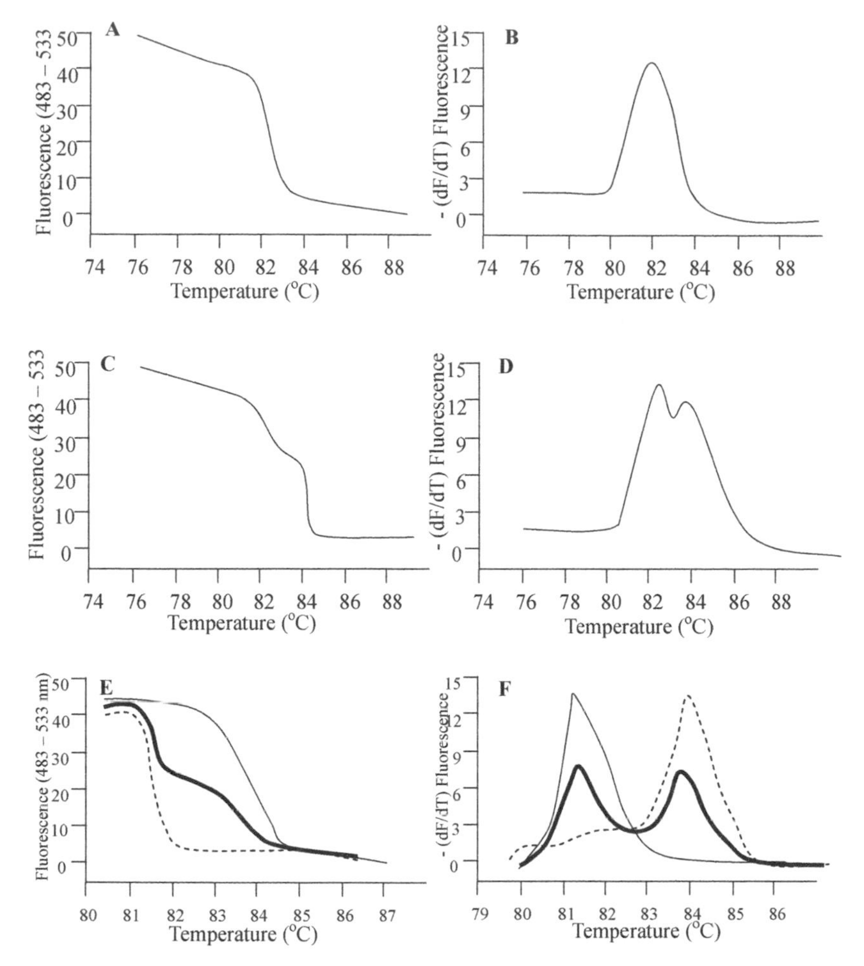

Fig. 13-5: Melting-curve analysis of real-time PCR products. (A) and (B): single PCR product. (C) and (D): two PCR products. (E) and (F): PCR products of homozygous normal (thin solid line), homozygous mutant (dash line), and heterozygous (thick line). (B), (D), and (F): first negative derivative plots of (A), (B), and (C).

product is generated. Since every molecule of this PCR product is the same, all molecules will melt at the same temperature. Therefore, only one peak on the melting curve is observed. If two or more PCR products are present, multiple peaks will appear on the melting curve.

Melting curve analysis of a real-time PCR performed with SYBR Green is commonly used for detection of single nucleotide polymorphisms (SNP). If a genomic DNA sample has no mutation on the gene to be examined, its PCR product will have a single peak on the melting curve. If there is a single-base mutation on the gene, the resulting PCR product will peak at a different temperature on the melting curve. If the DNA sample from an individual who is heterozygous for the gene to be examined, e.g., the paternal chromosome is normal and the maternal chromosome has a single-base mutation on the gene, two different PCR products will be generated. Therefore, two peaks will appear on the melting curve, and the height of each peak will be half of that of homozygous normal or mutant (Fig. 13.5).

Since SYBR Green binds to all double-stranded DNA, it is not readily applicable to multiplex PCR because it is not possible to determine which fluorescence is emitted from which target. A more commonly used method for real-time PCR is the TaqMan PCR, also referred to as the 5' nuclease assay. TaqMan PCR is based on a phenomenon called fluorescence resonance energy transfer (FRET). If two fluorescent molecules are located very close to each other, the energy released from one molecule when it is excited is absorbed by the other molecule; therefore, no emitted light is detected. These two fluorescence molecules are referred to as donor and quencher, respectively. For TaqMan PCR, a probe of approximately 20 bases, in addition to the two primers, is used. This probe is labeled with the donor molecule such as TET (tetrachloro-6-carboxyfluorescein) or FAM (5,6-carboxyfluorescein) at its 5' end and the quencher molecule such as TAMRA (tetramethyl-6- carboxyrhodamine) or ROX (6-carboxyl-X-rhodamine) at its 3' end. The probe is annealed to a template 2-3 bases down stream from where the 5' or 3' primer binds. When Taq DNA polymerase starts to synthesize DNA by extending the primers, it will degrade the probe with its 5' to 3' exonuclease activity. When the base labeled with the donor molecule is degraded, it is no longer associated with the quencher molecule and will emit light. The higher the copy number of the target, the more light is released. Since different probes can be labeled with different donor molecules, TaqMan PCR can be easily applied to multiplex PCR.

Molecular beacons can also be used as probes for a real-time PCR. A molecular beacon is an oligonucleotide of 15 to 30 bases long with the sequence of the first and the last 5 – 7 bases occurring in an inverted orientation. Therefore, the molecule can form a hairpin structure when it is free in solution. A fluorophore is conjugated to each end of the molecular beacon probe. When the molecules form a hairpin structure, the two fluorophores, the donor and quencher molecules, are brought together and do not fluoresce. When they hybridize with the target, the two fluorophores are separated. Therefore, the light emitted by the donor fluorophore is no longer quenched and can be detected by the spectrofluorometer of the real-time PCR system. The length of the PCR product in this system is usually 75 – 250 bp, and the Tm of a molecular beacon is usually 7-10°C above the annealing temperature of the PCR.

Differential PCR is another form of quantitative PCR. It is a multiplex PCR amplifying simultaneously the target and a reference gene. An example of its application is determination of the copy number of the *N-MYC* gene. The copy number of *N-MYC* is increased (amplified) in neuroblastoma, and the higher the *N-MYC* copy number the worse the prognosis is. To detect amplified *N-MYC*, a multiplex PCR amplifying both the *N-MYC* gene and a single-copy gene such as the *IFN*-γ gene of the patient and a normal individual is performed. The PCR products are labeled with isotopes or fluorescence. After electrophoresis, the radioactivity or fluorescence in both the *N-MYC* and *IFN*-γ bands is measured. If the ratio of *N-MYC* PCR product to that of *IFN*-γ of the normal individual is 100/100=1, but that of the patient is 300/100=3, the *N-MYC* copy number of the patient is 3 times that of the normal individual.

Allele-specific PCR uses different primer sets to amplify different copies of a gene. An example of its application is detection of the

ΔF508 mutation of the gene encoding cystic fibrosis transductant receptor (CFTR). Approximately 70% of cystic fibrosis patients have this mutation in which codon 508 encoding phenylalanine is deleted. The sequence of the region containing the 508 codon of a normal CFTR gene is GAAAATATCA<u>CTT</u>GGT, and that of the mutated gene is GAAAATATCAGGT, without the underlined CTT bases. To diagnose cystic fibrosis, an oligonucleotide with the sequence GAAAATATCA<u>CTT</u>GGT can be used as the 5' primer to amplify the normal CFTR gene, and the one with the sequence GAAAATATCAGGT can be used as the 5' primer to amplify the mutated gene. This type of PCR is called allele-specific PCR.

13.4: PCR Contamination Problem

Since the number of molecules of a PCR product can be very high, some molecules may become aerosolized when the tube containing the PCR product is opened. The aerosolized PCR products may drop into the tube containing PCR reagents, such as Taq DNA polymerase, primers, or dNTP. If contaminated reagents are used to perform another PCR for the same target, a false-positive reaction will result. Several methods may be used to minimize or prevent contamination. A common practice is to perform template DNA preparation, PCR reaction set up, and analysis of PCR products in three different rooms. In addition, the barrel of the pipetman used to pipet PCR reagents may be removed and soaked in 0.2 N HCl to degrade any PCR products that may be present. Pipet tips with a filter, referred to as barrier tips, are used to prevent PCR products from entering the barrel of a pipetman. Another approach is to irradiate the reagents with UV before use. Since UV irradiation may cross link TT or TC between strands, the irradiated DNA can not be denatured for use as a template. However, UV irradiation may damage dNTPs causing poor PCR product yield. The most effective method to prevent PCR contamination is the use of dUTP and uracil-N-glycosylase (UNG). Since most DNA polymerases, including Taq DNA polymerase cannot distinguish dUTP from dTTP, dTTP can be replaced by dUTP in PCR. The resulting PCR products will have U in every position where it is supposed to be a T. For simplicity, this DNA is referred to as U-DNA. If U-DNA contaminates a PCR reaction mixture, this mixture will have two types of DNA: the contaminating U-DNA and the target DNA. If this mixture is treated with UNG, all the U's in the U-DNA are removed, thus breaking the contaminating U-DNA into pieces that are too small to serve as template. Therefore, a false-positive reaction cannot occur. Since UNG is not heat resistant, it is inactivated when the reaction mixture is heated at 94°C for 10 minutes prior to initiation of PCR. If restriction enzyme analysis is to be used to characterize the PCR products, a mixture of dUTP and dTTP at 1:1 ratio is used as most restriction enzymes do not cut U-DNA.

13.5: PCR controls

Several different controls are usually included in each PCR reaction. Like all other enzymatic reactions, both positive and negative controls are needed. If PCR is performed to detect cytomegalovirus (CMV) in a blood sample, a sample known to contain CMV is used as the positive control, and another one that is negative for CMV is used as the negative control. A very important control is reagent control which is performed with a mixture containing all reagents required for the PCR except the template DNA. If this control becomes positive in a PCR run, one or more of the reagents are contaminated. The sensitivity control is critical for diagnostic purposes. A positive control sample is 10-fold serially diluted, and each dilution is used to perform PCR. If this sample can normally be diluted 1000 fold to yield a positive reaction, but can only be diluted to 100 fold in a new reaction, the sensitivity of the new reaction is 10 times lower than the previous one. This type of change in sensitivity is not acceptable in clinical diagnosis. Another very important control is the internal control. This is used to ensure that a negative PCR result is due to the absence of target and not inhibition of the reaction. Many clinical samples may contain substances such as heparin or hemoglobin that may inhibit Taq DNA polymerase, resulting in false-negative reactions. To control for this, a cellular gene such as the β-actin gene is amplified together with the intended target such as CMV during the multiplex PCR reaction. Since CMV resides in white blood cells (WBC), CMV DNA is present in the DNA isolated from infected cells. If the β-actin gene PCR is negative, there must be inhibitors in the reaction that inhibit the PCR reaction. A reaction is considered negative for CMV PCR only when the β-actin gene PCR is positive and the CMV PCR is negative. For a real positive CMV PCR reaction, both CMV and β-actin gene PCRs must be positive.

13.6. Prevention of Non-specific PCR

Non-specific PCR is due to primers that bind to places where they are not supposed to bind. Double-stranded DNA may undergo strand breathing in which the DNA transiently becomes single-stranded under ambient temperature. This may allow primers to bind non-specifically to the template and to cause DNA synthesis to start as soon as Taq DNA polymerase is added, resulting in a non-specific reaction. To prevent non-specific reactions, the PCR mixture may be heated at 94°C for 10 minutes before Taq DNA polymerase is added to dissociate all primers that bind to the template. When the temperature is decreased to the intended temperature, such as 60°C, for primer annealing, the primers can only anneal to their specific binding sites; therefore, non-specific reactions will not occur. This process is called hot start.

To simplify the hot start process, one approach is to place a layer of wax on top of the reagent mixture containing template DNA, primers, dNTP, and buffer, and then place the Taq DNA polymerase on top of the wax layer. When the tube is heated at 94°C for 10 minutes, the wax dissolves and Taq DNA polymerase becomes mixed with other components of the PCR mix, starting the PCR. With this approach, it is no longer necessary to open the tube to add Taq DNA polymerase after 10 minutes of heating. Another method is to use modified Taq DNA polymerase such as AmpliTaq Gold (PerkinElmer). This DNA polymerase is inactive under ambient temperature. It becomes active only when it has been heated to 94°C for 10 minutes. A similar approach is to use the Platinum Taq DNA polymerase from Invitrogen, Inc. This is a mixture of Taq DNA polymerase and anti-Taq antibody. The antibody binds to and inactivates the Taq DNA polymerase. When the reaction mixture is heated at 94°C for 10 minutes, the antibody dissociates, and Taq DNA polymerase is activated.

13.7: PCR Applications

Before the invention of PCR, it was very difficult to isolate a DNA fragment if no restriction sites are available. BAL 31 or combination of Exo III and S1 nucleases are used to delete the portions that are not needed. Since it is not possible to direct BAL 31 or Exo III to cut at precise locations, this process is very tedious and required a tremendous effort to screen the digested products. With PCR, the region of the desired portion of the DNA can be amplified if DNA sequence is available to design appropriate primers. Since the genomes of humans and many different species of animals have been sequenced, most genes can be amplified and used as probes for hybridizations.

Many traditional molecular biology reactions can also be done by PCR. For example, PCR can be used to join two DNA fragments together without using ligase. This method is called recombinant PCR (Fig. 13-6). To achieve this, PCR is used to amplify the two fragments to be joined separately using specially designed primers with the last 25 – 30 bp of the 3' end of the first fragment identical to that of the first 25 – 30 bp of the 5' end of the second fragment. When the resulting PCR products are denatured, the upper strand of the first fragment will anneal to the lower strand of the second fragment at this 25 – 30 bp region because their sequences are complementary. When the single-stranded regions of these recombinant molecules are filled in, a DNA fragment with the sequence the same as that of the two fragments joined together is produced. To make the sequences of the 3' end of the first fragment and the 5' end of the second fragment identical, additional bases are added to the 3' PCR primer of the first fragment and the 5' PCR primer of the second fragment. The additional bases added to the 3' PCR primer of the first fragment are complementary to the sequence of the 3' end of the lower strand of the first fragment. The bases added to the 5' primer of the second fragment is also complementary to the 5' sequence of the lower strand (Fig. 13-6).

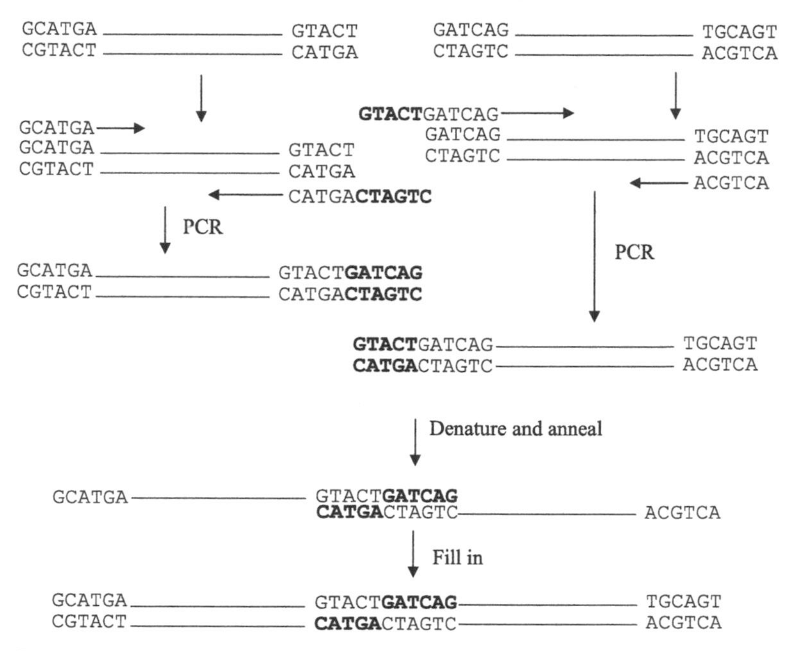

Fig. 13-6: Recombinant PCR.

RACE (rapid amplification of cDNA ends) is a very important application of PCR. To isolate the gene of interest, a cDNA library is screened. If the selected clone is found to contain only a portion of the gene, the insert of this cloned is isolated, labeled, and used as the probe to screen the same or a different cDNA library to isolate other clones that may contain the missing portion of the gene. This process may need to be repeated many times in order to isolate the entire gene. With RACE, the missing portion of the gene can be amplified by PCR without re-screening the libraries. The insert of the clone containing a portion of the gene of interest is sequenced, and the sequence obtained is used to design a primer to amplify the 5' portion of the gene. The primer with the sequence complementary to a portion of the mRNA is annealed to the mRNA (Fig. 13-7). Reverse transcription is then performed using the mRNA as the template to synthesize the missing 5' portion of the gene. The mRNA is then degraded with alkali, and the resulting single-stranded cDNA is tailed with poly-C at its 3' end by using terminal deoxynucleotide transferase. Oligo-dG is then used as the 5' primer, and the primer used for reverse transcription is used as the 3' primer to amplify the cDNA fragment. The resulting double-stranded DNA is the missing portion of the 5' end of the gene. This method is called 5' RACE (Fig. 13-7).

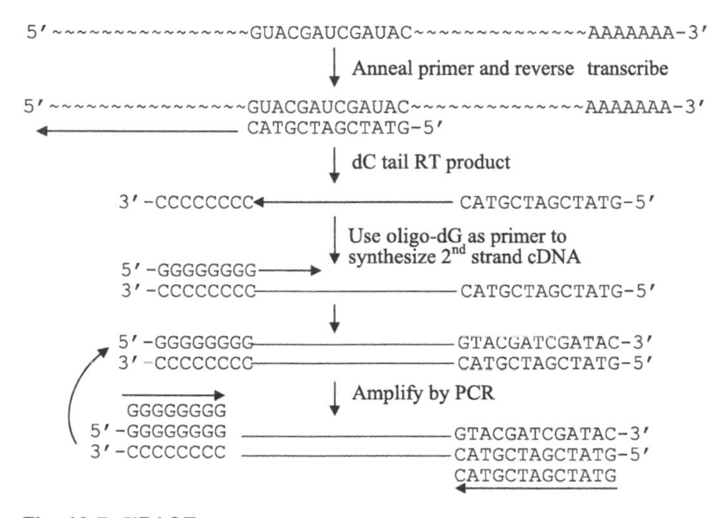

Fig. 13-7: 5'RACE.

3' RACE is used to synthesize the missing 3' portion of the gene (Fig. 13-8). Oligo-dT is first used as the primer to convert the mRNA of the gene to cDNA. The mRNA is then degraded with alkali, and the resulting single-stranded DNA is used as the template to amplify the missing 3' portion of the gene. A primer designed based on the insert sequence of the cDNA clone containing a portion of the gene of interest is annealed to the cDNA and then extended to the 3' end to synthesize the 3' portion of the gene. This primer is then used as the 5' primer and oligo-dT is used as the 3' primer to amplify the fragment. The resulting

PCR product is then joined with other portions of the gene to construct the entire gene.

A very important application of PCR is mutagenesis including single-base mutation, deletion, and insertion. These applications are described in Chapter 15.

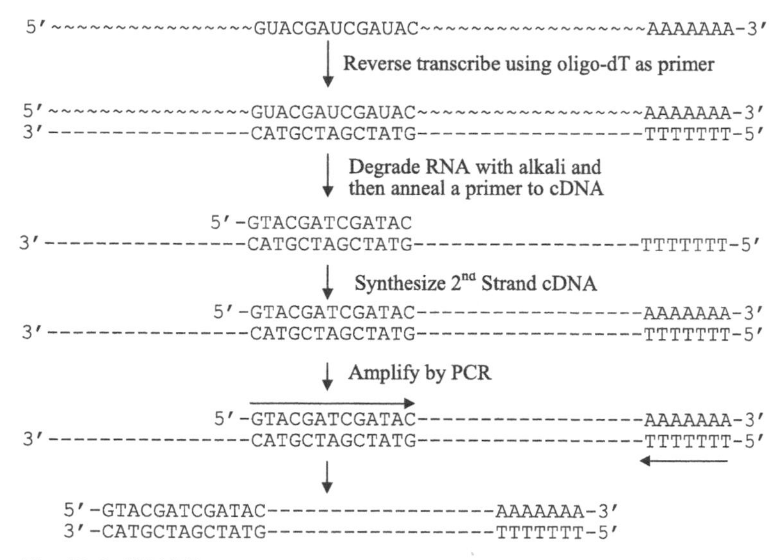

Fig. 13-8: 3'RACE.

13.8: Cloning of PCR Products

To express a gene, PCR products are cloned into a plasmid vector. The PCR product is first treated with T4 DNA polymerase to ensure that they are blunt ended and then cloned into the SmaI or HincII site of a vector such as pBluescript. This is how PCR products were cloned before the TA-cloning vector was developed by Invitrogen. A very interesting property of PCR is that PCR products generated by Taq DNA polymerase have an extra A at the 3' ends of both upper and lower strands. The TA-cloning vector developed by Invitrogen has an extra T at the 3' ends of the linearized vector to pair with this extra A at the 3' ends of the PCR products. The PCR product is then ligated to the vector after the ends are paired to each other. The TA-cloning system was improved with the conjugation of a topoisomerase I molecule to the extra 3' T of the TA-cloning vector. This modified vector is called pCRII-

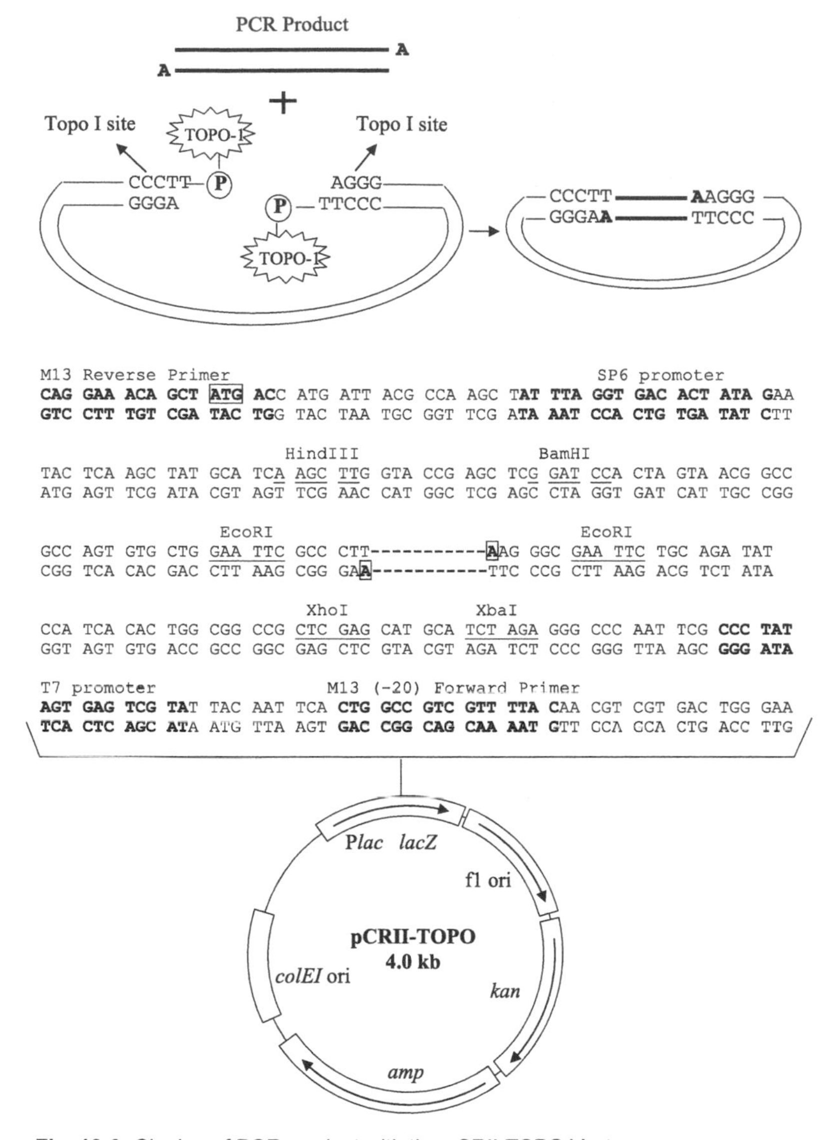

Fig. 13-9: Cloning of PCR product with the pCRII-TOPO Vector.

TOPO (Fig. 13-9). The topoisomerase I used in this system recognizes the sequence CCCTT. It will cut any DNA with this sequence and religate the DNA back together. This serves to relieve the topological strain of a DNA molecule so that it does not break. Topoisomerase I cuts only one strand of the DNA, whereas topoisomerase II cuts both strands of the DNA. After cutting, the DNA is unwound and religated back together. TOPO cloning takes advantage of the ligation function of topoisomerase I to ligate PCR product to the vector. Since this ligation reaction takes less then 5 minutes to complete and is more efficient than that of T4 DNA ligase, this cloning system is currently the method of choice for cloning PCR products.

13.9: Other DNA Polymerases for PCR

Taq DNA polymerase is not the only enzyme that can be used for PCR. Many different heat-resistant DNA polymerases can also be used. Tth DNA polymerase is isolated from *Thermus thermophilus*. This enzyme has both reverse transcriptase and DNA polymerase activities. It functions as a DNA polymerase in the presence of Mg^{++} and a reverse transcriptase in the presence of Mn^{++}. Tth DNA polymerase is commonly used to perform RT-PCR such as the one used to detect the hepatitis C virus. Since its genome is single-stranded RNA, it is first converted to cDNA and then amplified by PCR. With Tth DNA polymerase, these two reactions can be done simultaneously by adding both Mg^{++} and Mn^{++} to the reaction. Tfl DNA polymerase isolated from *Thermococcus litoralis* is another enzyme with both reverse transcriptase and DNA polymerase activities similar to Tth DNA polymerase.

A major problem with Taq DNA polymerase is that it does not have proofreading function since it lacks 3' to 5' single-stranded exonuclease activity. Therefore, the PCR products synthesized may have errors. It has been estimated that Taq DNA polymerase mis-incorporates at a rate of one in ten thousand nucleotides. Therefore, Pfu DNA polymerase isolated from *Pyrococcus furiosus* is used when high fidelity DNA synthesis is required because it has the 3' to 5' exonuclease activity and can correct mistakes made during PCR. However, the activity of Pfu DNA polymerase is only approximately 10% of Taq DNA polymerase and was not commonly used until modified Pfu DNA polymerase, such as Pfu Ultra II, was developed. A high efficiency double-stranded DNA binding domain was fused to Pfu DNA polymerase so that it binds to DNA template more efficiently to carry out DNA synthesis reactions. In addition, dUTPase is added to the reaction to remove any dUTP that are converted from dCTP. When dUTP is incorporated into DNA, Pfu DNA polymerase cannot correct mistakes and therefore loses its proofreading function. With these improvements, the modified Pfu DNA polymerase is as efficient as Taq DNA polymerase in DNA synthesis and can amplify DNA fragments as big as 20 kb.

13.10: Other Methods for Nucleic Acid Amplification

In addition to PCR, several other methods can also be used to amplify DNA or RNA. The ligase chain reaction (LCR) is one of them. A double-stranded DNA is first denatured at 94°C for approximately 60 seconds to form single-stranded DNA. Four oligonucleotide probes of approximately 30 bases each are then annealed to the single-stranded DNA at approximately 50°C. Two of the probes anneal to the upper strand juxtaposed with each other, and the other two anneal to the lower strand, also juxtaposed with each other. If the sequences of the probes are complementary to that of corresponding regions of the template, the two probes that anneal to the upper or lower strand of the template can be ligated together by ligase to produce a product the size of two probes combined, i.e, 60 bases long. If a heat-resistant ligase, such as the ligase from *Thermus aquaticus* or *Aeropyrum pernix*, is used, the reaction can be cycled like PCR, resulting in the production of numerous 60-base long products. If there is a mismatch between the probe and the template, the ligation reaction will not occur and no 60-base long products are generated. Therefore, LCR can also be used to detect mutations in a target gene.

Self-sustained sequence replication (3SR) is another amplification method (Fig. 13-10). Its target is RNA. A primer is first annealed to a

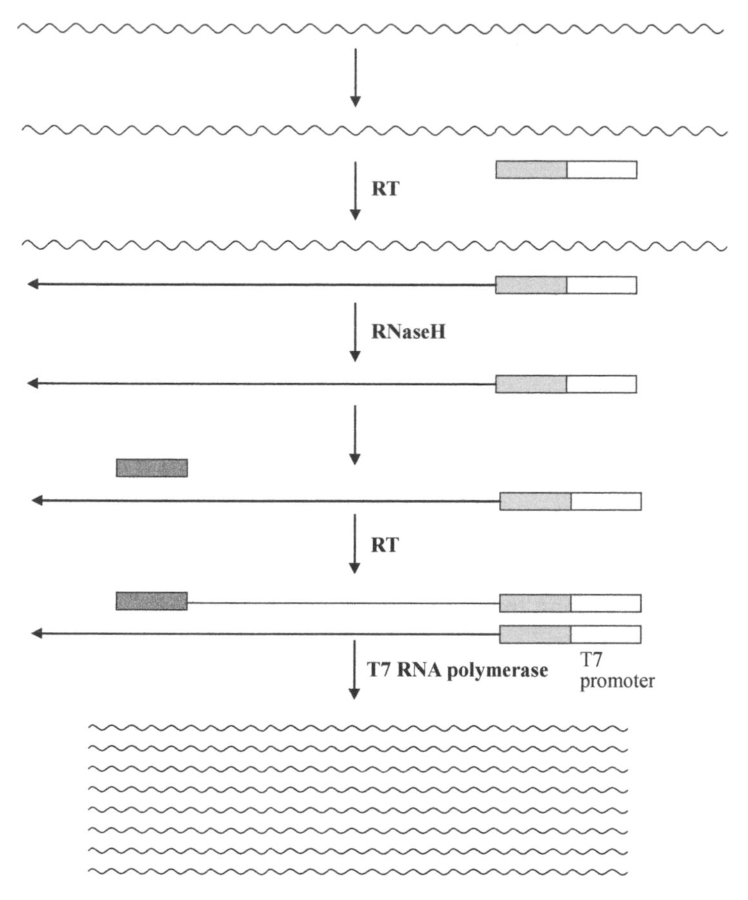

Fig. 13-10: Self-sustained Sequence Replication (3SR).

region near the 3' end of the RNA to perform a reverse transcription. This primer is composed of two parts. The 5' portion is the sequence of a T7 promoter, and the 3' portion is the sequence required for the primer to anneal to the RNA target. After reverse transcription, the RNA is digested away with RNase H. A second primer is then annealed to the single-stranded cDNA generated. Since reverse transcriptase can also use DNA as its template, the second primer is extended by reverse transcriptase in the presence of dNTP to produce a double-stranded DNA, generating a functional T7 promoter. T7 RNA polymerase then binds to the promoter and transcribes thousands of RNAs from each template. This method uses three different enzymes including reverse transcriptase, RNase H, and T7 RNA polymerase. Since the optimal reaction temperature of all three enzymes is 37°C, 3SR is isothermal and temperature cycling is not required. 3SR is also called nucleic acid sequence based amplification (NASBA) or transcription-mediated amplification (TMA).

Strand displacement amplification (SDA) is another method for DNA amplification (Fig. 13-11). The DNA polymerase used in this method is the Klenow fragment or other enzymes with similar activities such as the Bca DNA polymerase isolated from *Bacillus caldotenax*. A double-stranded DNA is first denatured to become two single-stranded molecules. Four primers are then annealed to the template. Two of them anneal to the 3' end of the upper strand next to each other, and the other two anneal to the 3' end of the lower strand also next to each other. The first primer in each primer set is composed of two parts. Its 5' portion contains the HincII recognition site GTTGAC, and the 3' portion contains bases required for primer annealing. When this primer is extended, the newly synthesized DNA is displaced by another DNA synthesized with the second primer which anneals behind the first primer. This phenomenon is called strand displacement.

The displaced single-stranded DNA is then used as the template for amplification. Another molecule of the first primer with the HincII recognition site is annealed to the 3' end of this single-stranded DNA. When this primer is extended and the single-strand region of the lower strand (the template) is filled in, a double-stranded DNA with a HincII site at both ends is produced (Fig. 13-11). If α-S-dATP instead of dATP is used in the DNA synthesis reaction, these HincII sites can be recognized by HincII, but can only be cut by HincII on the strand that does not contain thioated A bases. When this cutting occurs, a small fragment of DNA which anneals behind the bigger piece on the template is generated. This small DNA fragment is then used as the primer to synthesize DNA, displacing the DNA synthesized by the primer annealed in front of it. Since the sequences of the DNAs displaced from the two strands of the template are complementary, the displaced DNAs will anneal to each other to form a double-stranded

DNA which is the product of this SDA reaction. The DNA synthesis and HincII cutting reactions continue to occur until all dNTP are used or the DNA polymerase loses its activity. The DNA synthesized using the lower strand of the original template will undergo the same cycling reaction and produce more DNA. SDA can achieve a 10^7 fold amplification.

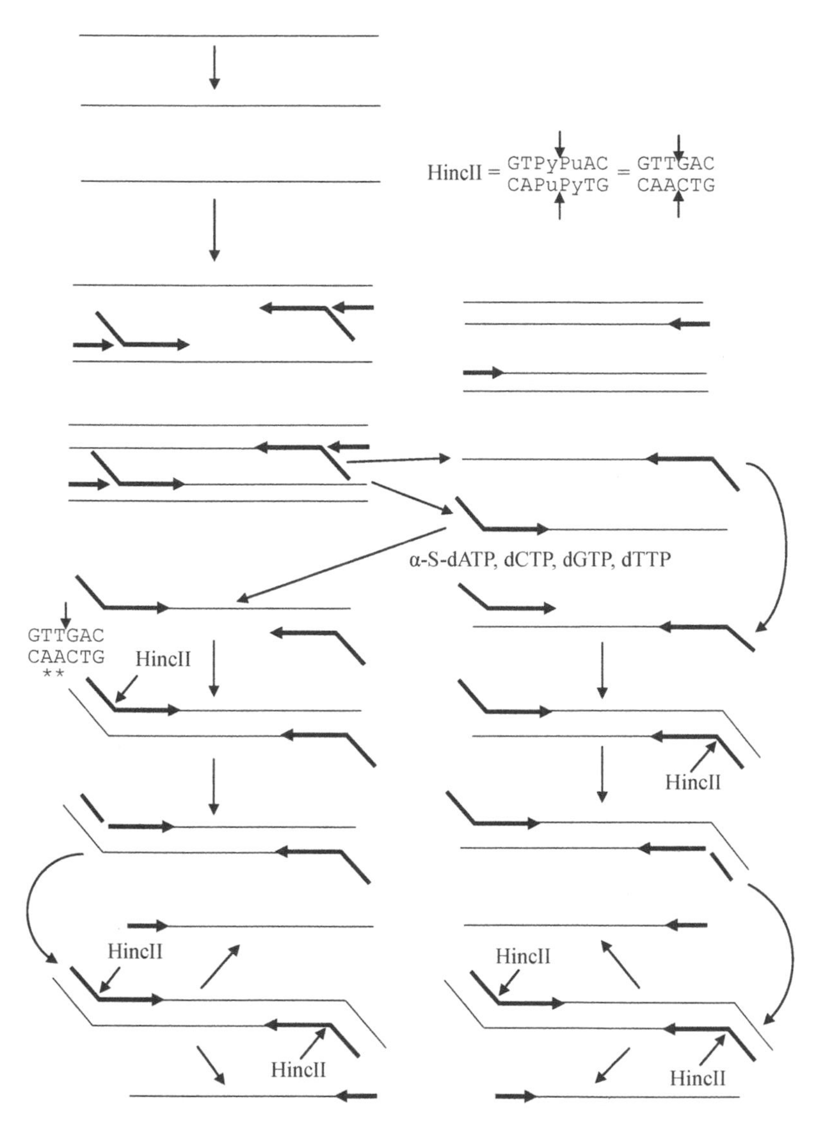

Fig. 13-11: Strand Displacement Amplification (SDA). Asterisks indicate thioated A bases.

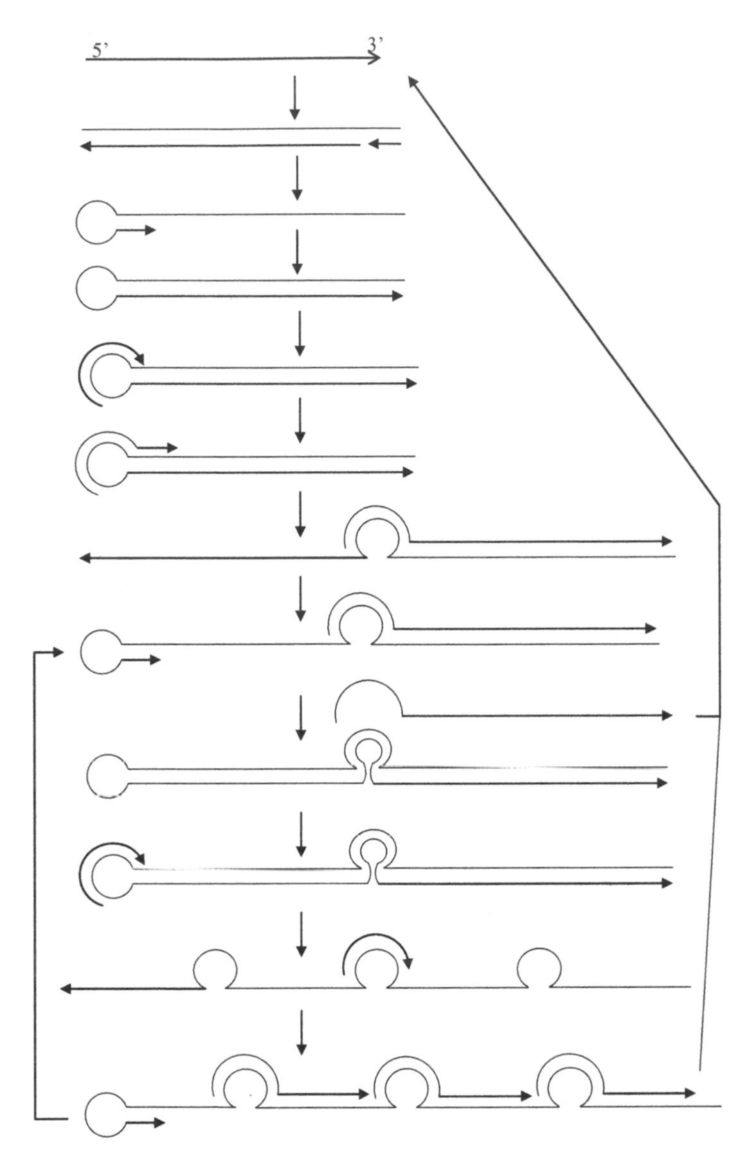

Fig. 13-12: Principle of Loop-mediated Isothermal DNA Amplification (LAMP).

Loop-mediated isothermal DNA amplification (LAMP) is a derivative of SDA (Fig. 13-12). Two primers are annealed to a single-stranded DNA or RNA template next to each other so that the DNA synthesized with the second primer will displace the one synthesized with the first primer. The basic principle is to make the displaced DNA form a loop at its 3' end. The end of the loop is then used as the primer to synthesize a double-stranded DNA. Since the loop region is single-stranded, a primer can anneal to it to prime a DNA synthesis reaction. When DNA synthesis is initiated, the double-stranded region will open to become single-stranded, generating a DNA twice as long as the original template. The 3' end of this DNA is again looped back and extended to form a double-stranded DNA. A primer again anneals to the single-stranded loop region to initiate another DNA synthesis. This again opens the double-stranded DNA which is now four times as long as the original template. This looping and opening cycle continues until all dNTP are used up or the DNA polymerase loses its activity, thus amplifying the original DNA or RNA molecule.

Similar to SDA, LAMP also uses two sets of primers, two in each set. In the original LAMP protocol, these primers are named FIP (forward inner primer), BIP (backward inner primer), F3, and B3 (Fig. 13-13). Both FIP and BIP primers are composed of three parts. The central portion of FIP is TTTT which is flanked by F1c and F2, 24 bases each. Similarly, primer BIP is composed of B1c-TTTT-B2. Both B1c and B2 are 24 bases long. When primer FIP anneals to the template, its F2 portion binds to the F2c region of the template. The DNA synthesized with primer BIP is then displaced by the one primed by F3 which anneals to the template behind FIP [(Fig. 13-13(c)]. Since the sequence of the B2c region located at the 3' end of the displaced DNA is complementary to that of the B2 region of BIP, BIP will anneal to it and prime the synthesis of another DNA which is then displaced by the DNA primed by B3 annealed behind BIP [(Fig. 13-13(g)]. Since the sequences of the F1c and F1 regions at the 3' end of this displaced DNA are complementary, these two regions will anneal to each other to form a loop [(Fig. 13-13(h)], and the F1 region of the loop is used as a primer to synthesize DNA [(Fig. 13-13(i)]. Another FIP primer will then anneal to the F2c region of the loop through an F2c and F2 interaction [(Fig. 13-13(i)]. When this FIP primer is extended, the double-stranded

DNA will open to become single-stranded [(Fig. 13-13(j)]. This single-stranded DNA will form another loop at its 3' end because the sequences of its B1 and B1c regions are complementary. The B1 region is then used as the primer to synthesize another DNA and displace the one primed by FIP. Since this displaced DNA is similar to the one displaced by F3 primed DNA [(Fig. 13-13(d)], primer BIP will anneal to it to prime DNA synthesis. Therefore, a cycle shown in Fig. 13-13(e) and Fig. 13-13(k) is repeated continuously to produce numerous DNA molecules shown in Fig. 13-13(l).

Since the loop region in the lower part of the DNA shown in Fig. 13-13(l) is B2c, primer BIP will anneal to it through B2-B2c interaction. When this BIP primer is extended, this double-stranded DNA [(Fig. 13-13(l)] will open to become single-stranded. Its 3' end will then form a loop again because B1 and B1c regions are complementary [(Fig. 13-13(m)], and its B1 region is used as the primer to synthesize DNA to displace the one primed by BIP [(Fig. 13-13(o)]. Since the 3' end of the displaced DNA is B1, BIP primer will bind to it. When this BIP is extended, a double-stranded DNA is formed [(Fig. 13-13(p)]. The B1 region of the molecule shown in Fig. 13-13(m) will also be used as primer to synthesize the DNA shown in Fig. 13-13(n) which is very similar to, but twice as long as the one shown in [(Fig. 13-13(l)]. Another BIP is then annealed to the single-stranded loop region of this molecule. Therefore, the reactions in Fig. 13-13(l) to Fig. 13-13(n) are repeated until the dNTP supply is exhausted, producing numerous DNA molecules similar to the one shown in Fig. 13-13(m), with the length of new product doubled in every cycle.

LAMP is isothermal and does not require a thermal cycler. The temperature used for the reaction is dependent on the DNA polymerase used. The most commonly used one is the Bst DNA polymerase from *Bacillus stearothermophilus*. Its optimal reaction temperature is 65°C.

Among the various methods for amplification of nucleic acids, PCR is most straightforward to perform and most efficient. The other methods require more adjustments in order to work properly; therefore, they are not commonly used for research. These other methods, however, are widely used for clinical diagnosis because the same protocol can be used for tests that are performed routinely once it has been established.

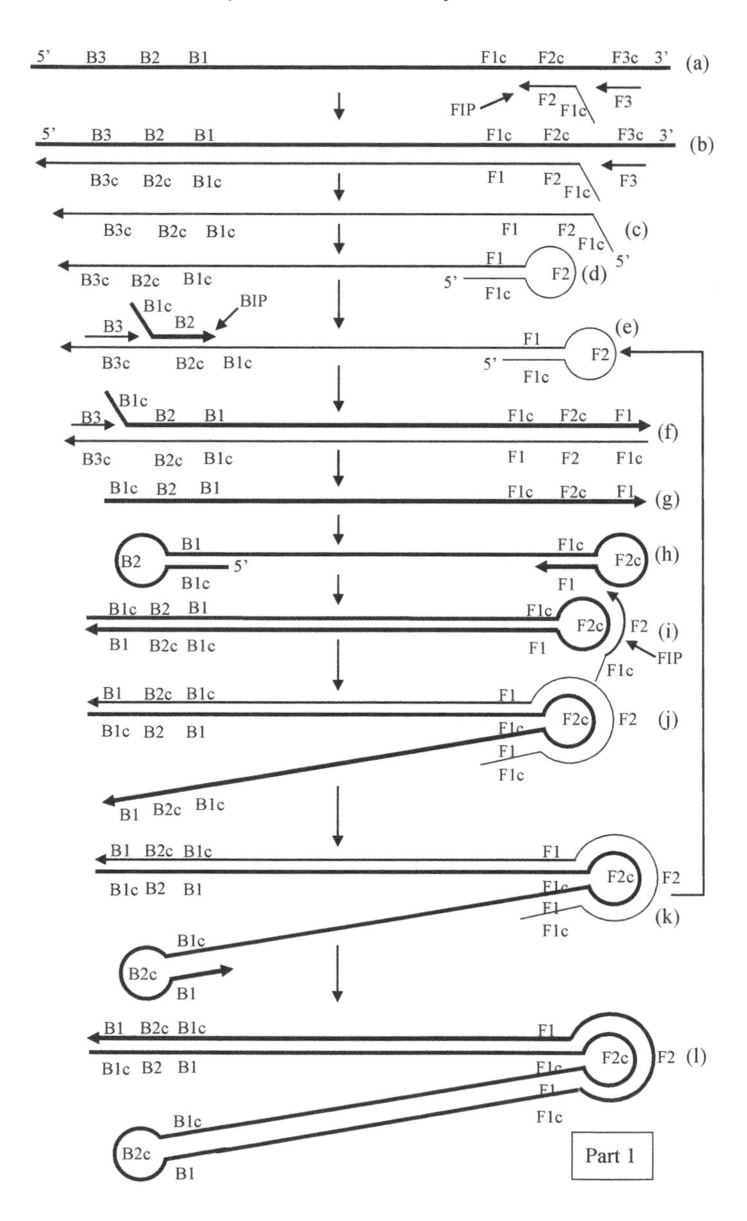

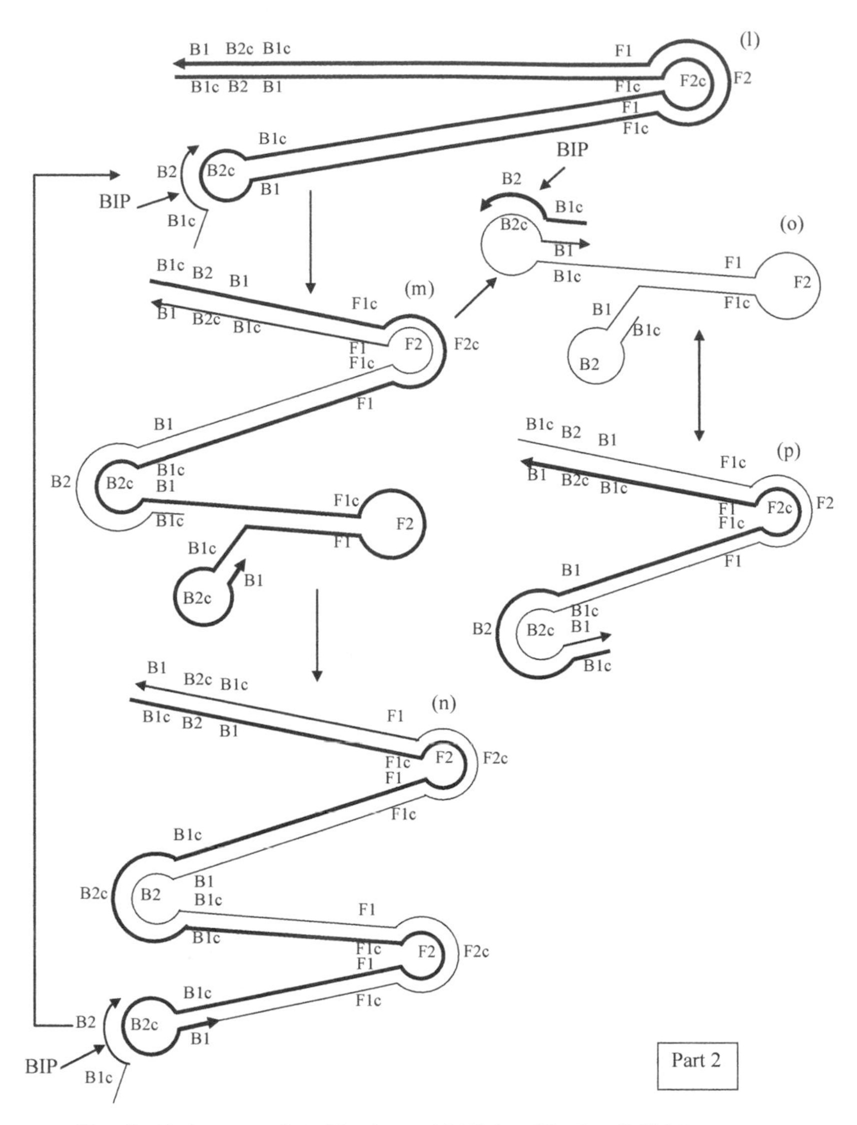

Fig. 13-13: Loop-mediated Isothermal DNA Amplification (LAMP).

Summary

PCR is a method used to amplify a DNA in vitro. It is performed in a reaction mixture containing a DNA template, two primers, a heat-resistant DNA polymerase such as Taq DNA polymerase, and dNTP in three steps. In the first step, the double-stranded DNA is heated at 94°C to form single-stranded DNA to be used as a template for DNA synthesis. In the second step, the temperature is decreased to approximately 50°C to allow primers to anneal to the templates. In the third step, the temperature is raised to 72°C to allow Taq DNA polymerase to synthesize DNA by extending the primers. When these three steps are repeated for 30 cycles, billions of copies of the region bordered by the two primer binding sites are produced.

The most critical part in PCR is primer design, in which two sequences of approximately 20 bases each are selected to make oligonucleotides for use as primers. These two primers anneal to two different regions on the template by base pairing, and the region between the two primer binding sites is amplified. Therefore, one of the primers anneals to the 5' end, and the other anneals to the 3' end of the region to be amplified. The one annealing to the 5' end is called the 5' primer, sense primer, or forward primer; and the one annealing to the 3' end is called the 3' primer, anti-sense primer, or reverse primer. The sequence of the 5' primer is identical to that of its binding site, and that of the 3' primer is complementary to that of its binding site. The 5' end of the primers is not critical; therefore, extra bases can be added. However, the 3' end of the primers must match perfectly with the template and usually end with GG, CC, CG, or GC because these bases allow the primers to anneal to the template more tightly. The bases within the primer must not be able to anneal to each other to form a stem-and-loop structure; otherwise, the primer will not anneal to the template. In addition, the sequence of a primer must be unique to the target gene so that non-specific amplification does not occur. The number of GC bases should be approximately 50% of the total number of bases of the primer, and the GC bases should not be clustered together. The Tm of the primer can be calculated using the formula 4x(# of G+C) + 2x(# of A+T) and should be approximately 65°C. The Tm's of the two primers should be the same or differ by less than 10°C.

To use PCR to amplify RNA, the RNA must first be converted to DNA by reverse transcription. This type of PCR is called RT-PCR. The two primers for RT-PCR must anneal to two different exons across at least one intron, so that the products derived from RNA or from contaminating DNA can be distinguished. Mg^{++} is a very important component of a PCR reaction; its concentration is usually 1.5 mM although some primers may require a higher concentration. The primer concentration for PCR is approximately 10 – 20 pmol. The annealing temperature is usually Tm-5°C. A typical PCR reaction may contain 50 ng of template DNA, 10 pmol of each primer, 1.5 mM Mg^{++}, 0.2 mM dNTP, and 1 - 2 units of Taq DNA polymerase.

Several PCR methods can be used to quantify the copy number of a certain target. In competitive PCR, a fake target with the same primer binding sites is first constructed. Several reactions with different amounts of the fake target and the same amount of the real target are performed. If there is a reaction in which the amount of PCR product from the fake and real targets are the same, the amount of the real target is the same as that of the fake target. A more accurate method to quantify the copy number of a target is real-time PCR in which fluorescent PCR products are made. The cycle number at which the fluorescence becomes detectable is called cycle threshold (C_T). The higher the target copy number, the lower the C_T value is. To make fluorescent PCR products, SYBR Green, which does not fluoresce unless it binds to a double-stranded DNA, can be added to the reaction. Since PCR products are double-stranded, the more PCR products produced the greater the fluorescence is released. This method, however, does not work for multiplex PCR if only the C_T value is determined. A better alternative to the SYBR Green method is TaqMan PCR. This method requires a probe of approximately 20 bases long labeled at both ends with different fluorescent dyes such as FAM and TET in addition to the two primers. This probe anneals 2-3 bases in front of one of the primers. Since the two fluorescent molecules on the probe are located very close to each other, the energy released from one upon activation is quenched by the other. This phenomenon is called fluorescence resonance energy transfer (FRET). When Taq DNA polymerase starts to synthesize DNA, it will degrade the probe because it has 5' to 3' single-strand exonuclease activity. When the first base of the probe is degraded, the light emitted from the dye attached to the first base will fluoresce because it is no longer quenched. If more targets are present, more probes are bound and more fluorescence is released. Since several different dyes can be used, different probes can be labeled with different dyes. Therefore, TaqMan PCR can be used in multiplex PCR to quantify multiple targets.

In PCR, enormous amounts of PCR product is generated. When the tube containing PCR product is opened, some of the PCR products may be aerosolized. The aerosolized PCR products may contaminate PCR reagents, resulting in a false-positive result. One approach to minimize contamination problem is to perform template DNA isolation, PCR set up, and analysis of PCR products in three different rooms. The most effective way is to substitute dTTP with dUTP. Therefore, the PCR products will have U incorporated where the base is supposed to be T. If the U containing PCR products contaminate a PCR reaction, it can be degraded by uracil-N-glycosylase.

Several different controls are usually included in each PCR run. In addition to the regular positive and negative controls, a very important control is reagent control which is done with no template DNA. If the reagent control reaction is positive, one or more of the PCR reagents are contaminated. Sensitivity control which ensures consistent efficiency of different PCR runs is critical for clinical uses of PCR. Since inhibitors may be present in samples, an internal control is needed. This control amplifies two different targets. One is the real target such as cytomegalovirus, and the other is a cellular gene such as the β-actin gene from the cell in which the real target resides. If the internal control is negative, a negative PCR reaction is due to inhibition of the reaction and may not be due to the absence of the real target.

Hot start is a process to minimize non-specific PCR reactions. This is done by heating the reaction mixture at 94°C for 10 minutes before Taq DNA polymerase is added to remove all primers that bind to the template. The temperature is then lowered to the temperature intended for primer annealing such as 50°C so that the primers will not bind non-specifically. To simplify this procedure, modified Taq DNA polymerase such AmpliTaq Gold or Platinum Taq can be used. AmpliTaq Gold DNA polymerase is not active until it has been heated at 94°C for 10 minutes. The Platinum Taq DNA polymerase has an antibody bound to it and is not active unless the antibody is dissociated which can be done by heating. With these enzymes, it is not necessary to heat other PCR components before Taq DNA polymerase is added.

There are many different applications of PCR. RACE (rapid amplification of cDNA ends) allows isolation of missing portions of a gene without re-screening cDNA libraries. The insert containing a portion of the gene of interest is sequenced, and the sequence obtained is used to design a primer to anneal to the mRNA of the gene. Reverse transcription is then performed to synthesize the missing 5' portion of the gene. The resulting product is then amplified by PCR. This method is called 5'RACE. To obtain the missing 3' portion of the gene, cDNA is first made using mRNA as the template and oligo-dT as the primer. The primer designed based on the available sequence of the gene is annealed to cDNA to synthesize the missing 3' portion of the gene. The product obtained is amplified by PCR. Since this method allows isolation of the 3' end of the gene, it is referred to as 3'RACE.

PCR products generated by Taq DNA polymerase have an extra A at their 3' ends. Therefore, they can be easily cloned by using a vector containing an extra T at its 3' ends. This vector is called T-A cloning vector. A modification of this method is the TOPO-TA cloning in which a topoisomerase I molecule is conjugated to the extra T residues of the vector. When a PCR product is linked to the vector through TA pairing, the topoisomerase ligates the two molecules together to form a covalently closed circular DNA. This ligation reaction is much more efficient than that using the T4 DNA ligase.

Although Taq DNA polymerase is the most commonly used enzyme for PCR, it is not the only heat-resistant DNA polymerase that can be used for PCR. Tth DNA polymerase isolated from *Thermus thermophilus* is commonly used for RT-PCR because it has both reverse transcriptase and DNA polymerase activities in the presence of both Mg^{++} and Mn^{++}. The Pfu DNA polymerase isolated from *Pyrococcus furiosus* has proofreading ability. Unlike Taq DNA polymerase which mis-incorporates at a rate of one in 10,000 nucleotides, Pfu DNA polymerase does not make mistakes and is commonly used to amplify genes for expression. DNA fragments amplified by Pfu DNA polymerase are blunt ended.

PCR is not the only method capable of amplifying DNA or RNA. LCR (ligase chain reaction) is another method. A double-stranded DNA is first denatured to become two single-stranded molecules. Two probes then anneal to the upper strand of the target, and another two probes anneal to the lower strand of the target juxtaposed to each other. If the sequences of the probes match perfectly with that of the target, ligase will join the two primers together, creating a DNA fragment the size of the two probes combined. The newly generated DNA molecules can also serve as template for probe binding. This reaction can be repeated like PCR if a heat-resistant ligase is used.

3SR (self-sustained sequence replication) is a method used for amplifying RNA targets. It is also called TMA (transcription mediated amplification). A primer is designed to anneal to a region near the 3' end of the target RNA. This primer is composed of two parts. The 3' portion contains bases required for the primer to anneal to the template, and the 5' portion contains the sequence of the T7 promoter. This primer is used to reverse transcribe RNA to DNA. The RNA is then degraded with RNase H, and the single-stranded cDNA generated is used as the template to synthesize a second strand of cDNA. This reaction creates a functional T7 promoter which is recognized by the T7 RNA polymerase to produce numerous copies of RNA. This method does not require temperature cycling as all three enzymes can function at the same temperature (37°C).

SDA (strand displacement amplification) is another method for DNA amplification. Two primers are annealed to the same template next to each other. When both primers are extended, the DNA primed by the first primer is displaced by the one primed by the second primer. This phenomenon is called strand displacement. The first primer is composed of two parts. The 3' portion contains sequence required to anneal to the template, and the 5' portion contains a HincII recognition site GTTGAC. When the displaced DNA is used as the template, the double-stranded DNA formed will have a HincII site at both ends. If dATP is substituted with α-S-dATP, the HincII site can only be cut on the strand that does not contain thioated A. This is the strand that contains the primer. When this double-stranded DNA is nicked at the HincII site, two DNA molecules annealed to the template in tandem are generated. The DNA molecule annealed behind the first will prime DNA synthesis and displace the DNA molecule in front of it. Both strands of the original template can be used as templates. Since the products derived from these two strands are complementary, they will anneal to each other and become double stranded.

LAMP (loop-mediated isothermal DNA amplification) is derived from SDA. It also uses two primers to anneal next to each other on a single-stranded DNA or RNA template. The first primer is designed so that the newly synthesized DNA will form a loop at its ends. When the DNA synthesized with this primer is displaced by the DNA primed with the second primer, the 3' end of the loop can be used as primer to synthesize DNA. Since the loop region is single-stranded, a primer can also anneal and prime the synthesis of another DNA, thus opening the double-stranded DNA to become single-stranded. This DNA is twice as long as the original template. Since this DNA will also form a loop at its 3' end. The same reaction will repeat creating a product that is increased in length two fold in every cycle.

Sample Questions

1. PCR is actually primer extension using: a) one specific primer, b) random primers, c) two specific primers, d) oligo-dT primers
2. Which of the following are critical for PCR: a) Mg^{++} concentration, b) Ca^{++} concentration, c) primer annealing temperature, d) primer concentration
3. Which of the following would not be a good PCR primer: a) GGATTTTTGCCCCCAAAAATTT, b) GATCCTAGCTGGCTAGCTAGG, c) GGACGATGACGTAGGATTCC, d) GGATTCGATGATGTGCAAAA
4. The two primers for RT-PCR must anneal: a) within the same intron, b) within the same exon, b) across at least one intron, d) across at least one stop codon
5. The Tm of the two PCR primers should be: a) differ by 20°C, b) the same or very close, c) differ by 30°C, d) differ by 40°C.
6. The most effective way to prevent false-positive PCR due to contamination by previous PCR products is to use: a) ddNTP instead of dNTP, b) NTP instead of dNTP, c) combination of dUTP and UNG, d) UV irradiation of PCR reagents
7. The most common method to prevent a non-specific PCR reaction is: a) false start, b) hot start, c) addition of ddNTP, d) treatment of PCR mix with UNG before PCR
8. 3SR (or TMA) is used to amplify: a) DNA, b) RNA, c) protein, d) polysaccharide
9. Taq DNA polymerase has: a) very good proofreading function, b) no proofreading function, c) 5' to 3' single-strand exonuclease activity, d) 3' to 5' single-strand exonuclease activity
10. Pfu DNA polymerase has: a) very good proofreading function, b) no proofreading function, c) 5' to 3' single-strand exonuclease activity, d) 3' to 5' single-strand exonuclease activity
11. PCR products generated by Taq DNA polymerase usually have: a) one extra T at 5' end, b) one extra A at 3' end. c) one extra G at 3' end, d) one extra C at 5' end

Suggested Readings

1. Champoux, J. J. (2001). DNA topoisomerases: structure, function, and mechanism. Annu. Rev. Biochem. 70: 369-413.
2. Wang, J. C. (2002). Cellular roles of DNA topoisomerases: a molecular perspective. Nat. Rev. Mol. Cell. Biol. 3: 430-440.
3. Savva, R., McAuley-Hecht, K., Brown, T. and Pearl L. (1995). The structural basis of specific base-excision repair by uracil-DNA glycosylase. Nature 373: 487-493.
4. Gilliland, G., Perrin, S., Blanchard, K., and Bunn, H. F. (1990). Analysis of cytokine mRNA and DNA: detection and quantitation by competitive polymerase chain reaction. Proc. Natl. Acad. Sci. 87: 2725-2729.
5. Longo, M. C., Berninger, M. S., and Hartley, J. L. (1990). Use of uracil DNA glycosylase to control carry-over contamination in polymerase chain reactions. Gene 93: 125-128.
6. Saiki, R. K., Scharf, S., Faloona, F. A., Mullis, K. B., Horn, C. T., Erlich, H. A., and Arnheim, N. (1985). Enzymatic amplification of β-globin genomic sequences and restriction site analysis for diagnosis of sickle cell anemia. Science 230: 1350-1354.

7. Mullis, K. B. and Faloona, F. A. (1987). Specific synthesis of DNA in vitro via a polymerase-catalyzed chain reaction. Methods Enzymol. 155: 335-350.

8. Frohman, M. A., Dush, M. K., and Martin, G. R. (1988). Rapid production of full-length cDNAs from rare transcripts: amplification using a single gene-specific oligonucleotide primer. Proc. Natl. Acad. Sci. USA 85:8998-9002.

9. Eyal, Y., Neumann, H., Or, E., and Frydman, A. (1999). Inverse single-strand RACE: an adapter-independent method of 5′ RACE. BioTechniques 27:656-658.

10. Park, D. J., Park, A. J., Renfree, M. B., and Graves, J. A. M. (2003). 3′ RACE walking along a large cDNA employing tiered suppression PCR. BioTechniques 34:750-756.

11. Fahy, E., Kwoh, D. Y., and Gingeras, T. R. (1991). Self-sustained sequence replication (3SR): an isothermal transcription-based amplification system alternative to PCR. PCR Methods and Applications, Cold Spring Harbor Laboratory Press, p25-32.

12. Notomi, T., Okayama, H., Masubuchi, H., Yonekawa, T., Watanabe, K., Amino, N., and Hase, T. (2000). Loop-mediated isothermal amplification of DNA. Nucleic Acids Res. 28: e63

13. Walker, G. T., Fraiser, M. S., Schram, J. L., Little, M. C., Nadeau, J. G., and Malinowski, D. P. (1992). Strand displacement amplification - an isothermal, in vitro DNA amplification technique. Nucleic Acids Res. 20: 1691-1696.

Chapter 14

Expression of Recombinant Proteins

Outline

14.1: Expression of Foreign Proteins in *E. coli*

Expression refers to the synthesis of protein from a certain gene. The process of expression includes transcription of mRNA from the gene and translation of the mRNA to protein. Although each gene has its own promoter, the promoter can be changed, and a eukaryotic gene can be expressed in bacteria such as *E. coli* if its expression is driven by a bacterial promoter. Since *E. coli* has no splicing system to process mRNA, the cDNA, not genomic DNA, of the eukaryotic gene to be expressed is cloned and used for expression in *E. coli*. The ability to express a foreign protein, commonly referred to as recombinant protein, in *E. coli* is a major breakthrough in medicine because it allows many therapeutic proteins such as insulin, growth hormones, and interferon to be more easily produced. However, expression of a foreign protein in *E. coli* may not always be straightforward. The expressed protein may be toxic to *E. coli* leading to its death and poor yields of the protein. Another possible outcome is that the expressed protein is degraded by *E. coli* proteases or enclosed in vacuoles to become inclusion bodies. Once the protein is packed in inclusion bodies, it usually becomes insoluble and inactive. The possibility also exists that the expressed protein is not harmful to *E. coli* and therefore is not degraded or packed in inclusion bodies.

14.2: Inducible Promoters

To prevent the complications associated with toxic gene products in *E. coli*, the gene is allowed to be expressed only when a sufficient cell density is reached. At this point, an adequate amount of the protein is made even though the host *E. coli* cells are killed shortly after the gene is turn on. To be able to control gene expression, an inducible promoter is used to drive the gene. Many different inducible promoters such as *lac, trp, tac,* lambda P_L, and T7 promoters can be used. The *lac* promoter is normally suppressed by the LacI repressor. When IPTG (isopropyl β-D-thiogalactopyranoside) is added to the growth medium of an *E. coli* culture, LacI repressor is inactivated, and the *lac* promoter starts to function. The *trp* promoter controls the expression of genes responsible for the synthesis of tryptophane. It is normally suppressed by the *trp* repressor which can be inactivated by indol-3-propionic acid or indole-acrylic acid. The *tac* promoter is a hybrid of *lac* and *trp* promoters. Its -35 sequence TTGACA is from the *trp* promoter, and the -10 sequence TATAAT is from the *lac* promoter. Since these are more typical promoter sequences, the *tac* promoter is stronger than both *lac* and *trp* promoters. *tac* promoter can also be induced by IPTG. The lambda P_L promoter is regulated by the CI repressor. The *cI857* gene encodes a temperature sensitive CI repressor. If *cI857* is used to regulate the P_L promoter, the gene driven by the P_L promoter can be induced by simply heating the culture at 42°C for several minutes. Currently, the T7 promoter is the most commonly used promoter for expression of foreign genes in *E. coli*. Since T7 RNA polymerase is required to transcribe a gene driven by the T7 promoter, controlling the production of T7 RNA polymerase will control the expression of the gene. A common approach to control T7 RNA polymerase production is to use *lac* promoter to drive the gene so that it can be induced by IPTG.

14.3: Fusion Proteins

To prevent recombinant proteins from being degraded or enclosed in inclusion bodies, the protein is fused to an *E. coli* protein so that the fusion protein may be considered as native by *E. coli* and not degraded. Commonly used protein tags include dihydrofolate reductase (DHFR), glutathione S transferase (GSH), maltose binding protein (MBP), and thioredoxin. Many different expression vectors, such as pQE40, pQE41, and pQE42 (Fig. 14-1), can be used to fuse the gene to be expressed to the gene encoding a certain tag protein. If the DNA fragment containing the gene to be expressed is inserted into one of the multiple cloning sites of pQE40, pQE41, and pQE42, a recombinant protein fused to DHFR is produced.

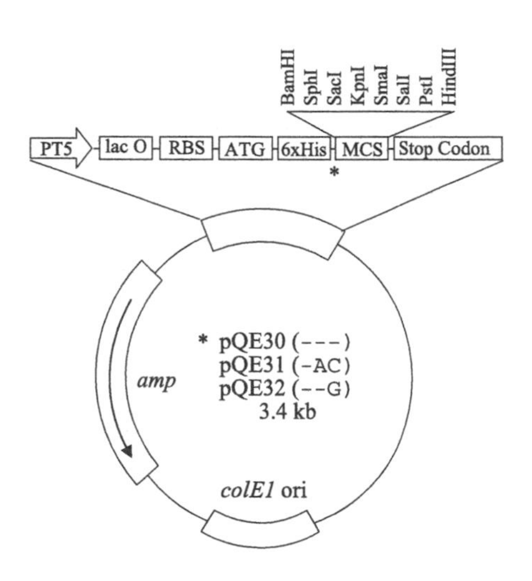

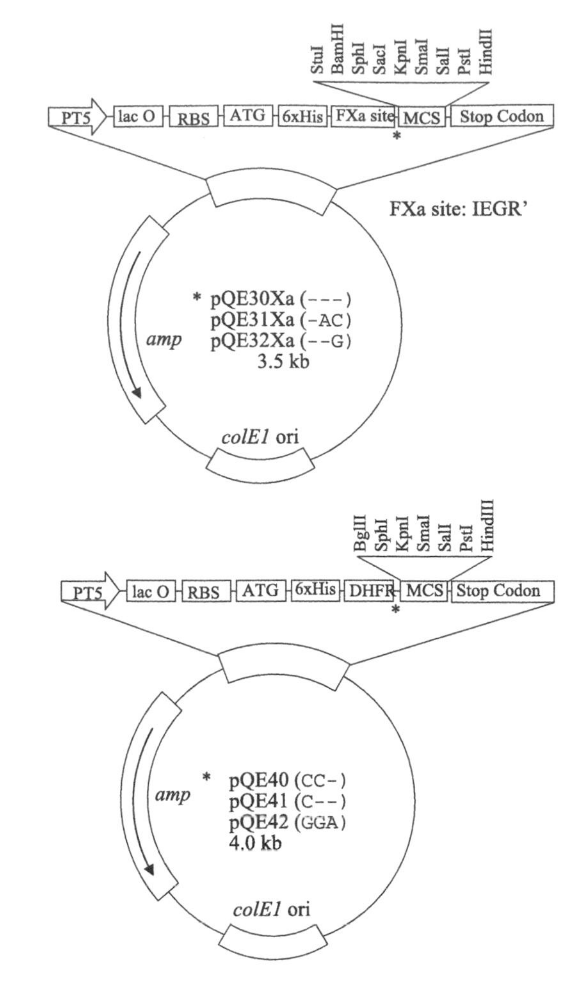


```
                                   BglII   SphI    KpnI  SmaI  SalI   PstI
pQE40  6xHisGGATCCDHFRGGTTCC-AGATCTGCATGCGGTACCCCGGGTCGACCTGCAGCCAAGCTT
pQE41  6xHisGGATCCDHFRGGTTC--AGATCTGCATGCGGTACCCCGGGTCGACCTGCAGCCAAGCTT
pQE42  6xHisGGATCCDHFRGGTTGGAAGATCTGCATGCGGTACCCCGGGTCGACCTGCAGCCAAGCTT
```

Fig. 14-1: pQE Series of Expression Vectors. Asterisk indicates nucleotide differences among each set of the vectors. The extra one or two bases are intended to make the fusion in frame for DNA fragments missing one or two bases of a codon.

14.4: Isolation and Purification of Recombinant Proteins

Another reason for creating fusion proteins is to simplify the isolation and purification process of recombinant proteins. Since an *E. coli* cell contains approximately 3500 different proteins, it is not an easy task to isolate the expressed protein from a mixture containing so many proteins. If the protein is fused to GST, a chromatographic column packed with glutathione-agarose beads can be used to isolate the protein. When an *E. coli* lysate is passed through the column, the GST-tagged protein will bind to glutathione and be retained in the column. Other *E. coli* proteins are then removed by washing the column with a certain buffer, and the GST-tagged recombinant protein is eluted from the column with a buffer containing glutathione.

If the protein is fused to MBP, an antibody affinity column can be used. Anti-MBP antibody is first conjugated to Sepharose CL-4B. The complex is then packed in a column. The MBP-tagged protein is retained in the column when a cell lysate containing the protein is applied to the column. After removing unbound proteins, the MBP-tagged protein is eluted with reagents that can dissociate the protein from the antibody. Similar approaches can be used to purify recombinant proteins fused to DHFR or thioredoxin. The method of intein-mediated purification with an affinity chitin-binding tag (IMPACT) is another method for isolation of recombinant proteins. “Intein” is analogous to intron. Some proteins, such as the RecA protein of *Mycobacterium tuberculosis* and the vacuolar ATPase of yeast, also undergo splicing similar to most eukaryotic mRNAs. The region that is removed is called intein. Almost all inteins identified so far have self-cleavage activity. In the IMPACT system, the protein is fused to an intein which contains a chitin-binding domain. If a cell lysate containing the intein-tagged recombinant protein is passed through a column packed with chitin, the protein will bind to chitin. When the column is flushed with a reducing agent such as dithiothreitol (DTT), β-mercaptoethanol, or cysteine, the intein will self cleave, releasing the recombinant protein from the chitin column.

The most commonly used tag to simplify isolation of recombinant proteins is 6x His. In this method, six histidine residues are fused to either the N-terminal or C-terminal end of the protein. Since a protein with a run of histidines can bind to nickel (Ni), a column packed with nickel-nitrilotriacetate (Ni-NTA) can be used to isolate the protein. The cell lysate containing the 6x His-tagged protein is passed through the column. The unbound proteins are washed away, and the bound protein is then eluted from the column with imidazole which competes with the protein for nickel binding and dissociate the protein from Ni-NTA. All the vectors shown in Fig. 14-1 contain the sequence encoding 6x His. If a DNA fragment containing the gene of interest is inserted into one of the multiple cloning sites, the protein produced will have the 6x His Tag.

14.5: Removal of Tags

Although a tag can facilitate isolation of the recombinant protein, it may affect the function of the protein. Therefore, it may be necessary to remove the tag after the protein has been isolated. To make this process possible, a cleavage signal is inserted between the tag and the protein to be expressed. The vectors pQE30Xa, pQE31Xa, and pQE32Xa (Fig. 14-1) have the sequence encoding such signal. When a DNA fragment containing the gene to be expressed is inserted into one of the multiple cloning sites of these vectors, the recombinant protein will have the clotting factor FXa cleavage signal located between the 6x His tag and the protein. Therefore, the tag can be removed by digesting the protein with FXa which cleaves at the carboxy side of the Arg residue in this signal peptide. Many other proteases can be used including thrombin (activated Factor II) which recognizes LVPR↓G, enterokinase that recognizes DDDK↓M (Asp-Asp-Asp-Lys-Met), and PreScission protease which is from human rhinovirus 3C and recognizes LEVLFQ↓GP (Leu-Glu-Val-Leu-Phe-Gln-Gly-Pro); where arrows indicate the cleavage sites. The intein described above can also be used to remove the tag.

14.6: In-frame Fusion

When linking a tag to the protein to be expressed, the fusion must be in frame, meaning that the translational frame of both the tag and the protein must be maintained. For example, if the sequence of the last two amino acids of the DHFR tag is Gly-Ser and that of the first 3 amino acids of the protein is Met-Asp-Ser, the junction sequence of the recombinant protein must be -Gly-Ser-Met-Asp-Ser- (Fig. 14-2). The corresponding nucleotide sequence of this peptide is -GGT TCC ATG GAT AGC-. Among these nucleotides, GGTTCC are from the DHFR gene. If the last C of these nucleotides is lost during preparation of the DNA fragment for cloning, the sequence of the fusion junction would be -GGT TCA TGG ATA GC- which encodes -Gly-Ser-Trp-Ile-, not the desired -Gly-Ser-Met-Asp-Ser-. This is called "out-of-frame." If the last two C residues of the bases GGTTCC are lost, the fused sequence would be GGT TAT GGA TAG C which encodes -Gly-Tyr-Gly. In this case, a truncated protein is made because TAG is a stop codon.

```
Gly Ser | Met Asp Ser
GGT TCC | ATG GAT AGC

Gly Ser Trp Ile
GGT TCA TGG ATA GC
Gly Tyr Gly ***
GGT TAT GGA TAG C
```

Fig. 14-2: Examples of In-frame and Out-of-frame fusion.

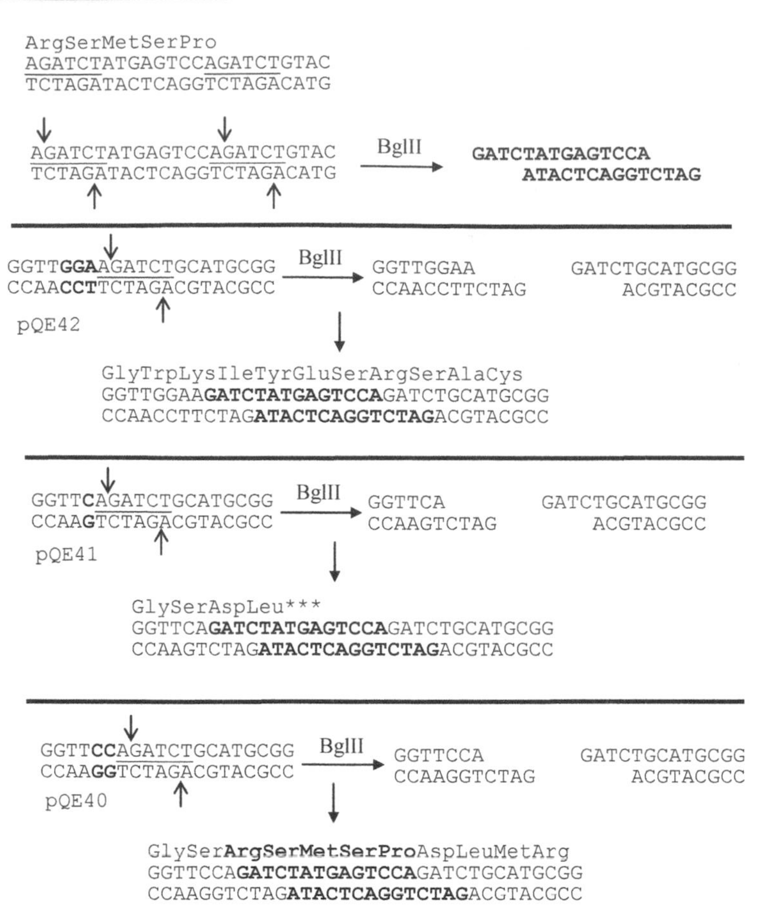

Fig. 14-3: In-frame Fusion with One of the pQE40 Series Vectors.

14.7: Construction of Expression Plasmids

The most critical step in expression of recombinant proteins is construction of the expression plasmid. In addition to promoter, Shine-Dalgarno (SD) sequence also plays a very important role for expression in *E. coli*. The SD sequence is located between the -10 region of the promoter and the translation initiation codon ATG. It facilitates the binding of ribosome to mRNA through the interaction with the 16S rRNA as described in Chapter 2. The distance between the SD sequence and the initiation codon varies with promoters and ranges from 5 – 15 bp. This distance is critical for expression. The SD-ATG distance for the

T7 promoter is 7 bp. To use the T7 promoter to drive gene expression, this distance must be maintained at 7 bp in order to achieve optimal expression. Fig. 14-4 is an example of how an expression plasmid may be constructed using pET8C as the vector.

To clone a gene into pET8C for expression, pET8C is first digested with NcoI and BamHI. The DNA fragment containing the coding region of the gene to be expressed is then ligated between NcoI and BamHI sites, with the 5' end joining to the NcoI site. Therefore, the 5' end of the DNA fragment must be compatible with the NcoI site and its 3' end must be compatible with the BamHI site. Since NcoI digestion results in a 5' overhang sequence of CATG, the 5' end of the fragment must have the same overhang in order to join to the vector. This can be achieved by generating a specific sequence at the 5' end of the fragment by PCR so that when it is cut with a certain restriction enzyme, the CATG overhang is generated.

In the example shown in Fig. 14-4, the sequence of the 5' end of the gene is ---TTTGATGAATCAGCAA---, in which ATG is the initiation codon. The region to be inserted into pET8C begins at this ATG codon. Therefore, a DNA fragment with the sequence ATGAATCAGCAA--- at its 5' end must be generated. The easiest way to generate such a fragment is by PCR. To do so, the 5' primer must have the sequence ATGAATCAGCAA---. Since the CATG overhang must be created, the primer sequence must be CATGAATCAGCAA---. If a T is added in front of this primer sequence, the sequence becomes TCATGAATCAGCAA---, in which TCATGA is the recognition site of BspHI. A PCR product with this sequence will result in the required CATG overhang if it is digested with BspHI. This DNA fragment can then be ligated to the NcoI site of the vector.

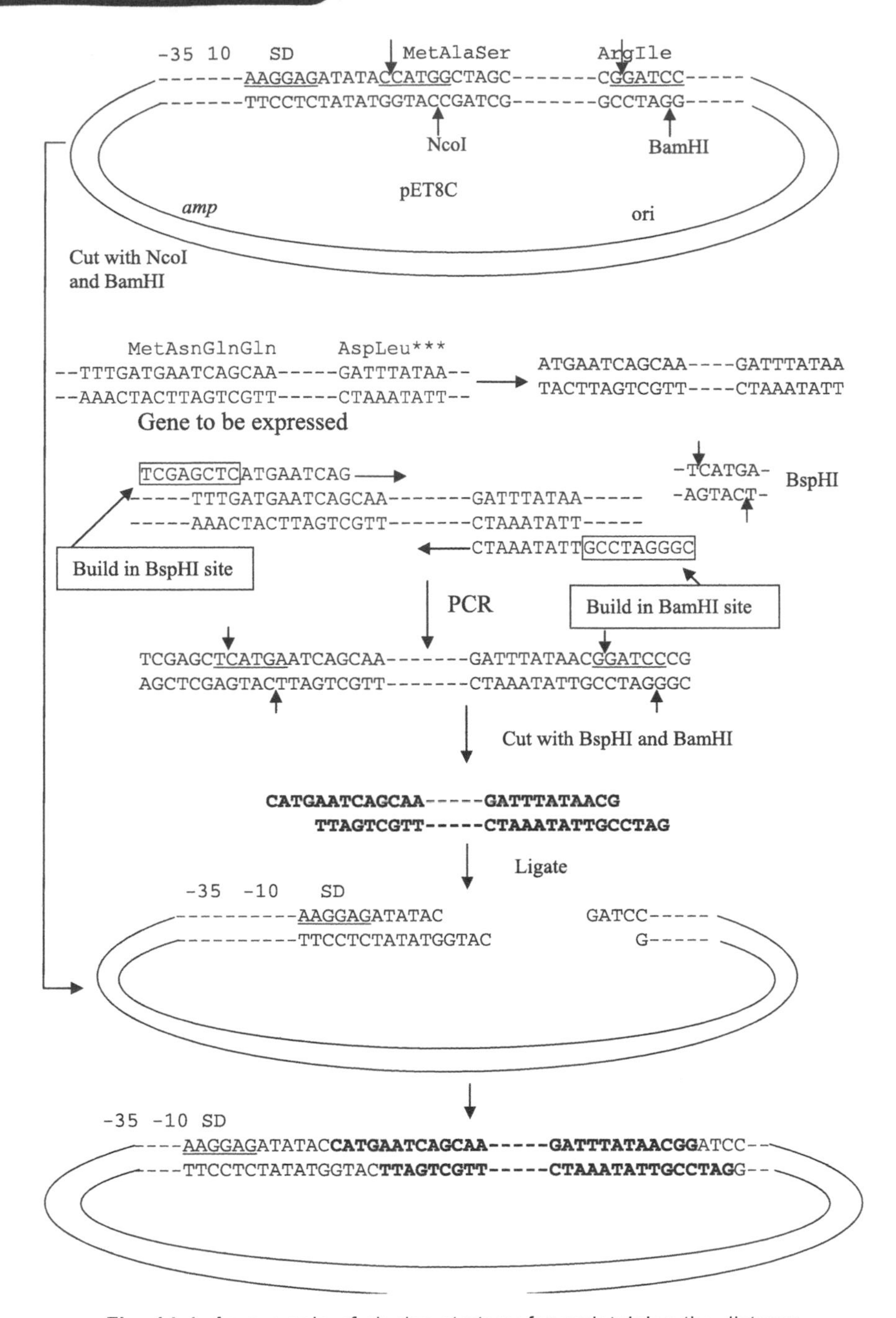

Fig. 14-4: An example of cloning strategy for maintaining the distance between the SD sequence and the initiation codon ATG.

Although restriction enzymes recognize a certain number of bases, e.g., BspHI recognizes a 6 bp cleavage site, it requires more than those 6 bp to bind to DNA. To ensure that BspHI will bind to the PCR product and cut, additional bases TCGAGC are added to the 5' end of the fragment. Therefore, the sequence of the 5' primer becomes TCGAGC<u>TCATGA</u>ATCAGCAA, and the resulting PCR product will have the same sequence at its 5' end. The same approach is used to generate a BamHI site at the 3' end of the PCR product. Therefore, an oligonucleotide with the sequence CGGGATCCGTTATAAATC is used as the 3' primer. The resulting PCR product is digested with BspHI and BamHI and then ligated between the NcoI and BamHI sites of pET8C. In this manner, the distance between the SD sequence and ATG codon of the T7 promoter is not changed and is still 7 bp.

14.8: Detection of Expressed Proteins

Once the expression plasmid is constructed, it is introduced into *E. coli* to express the gene. The first step to determine whether the protein is expressed is to electrophorese the cell lysate of the *E. coli* containing the expression plasmid. The result is compared to that of the electrophoresis of the cell lysate of *E. coli* containing the vector, pET8C in the example shown in Fig. 14-4. If an additional protein band is observed, this band is likely the expressed protein. However, the additional band may not be visible if the size of the expressed protein is the same as one of the *E. coli* proteins. Western blotting is then performed using an antibody against the protein to detect the expressed protein. The protein bands in the gel are transferred to a membrane such as nitrocellulose or polyvinylidene fluoride (PVDF) membrane. The membrane is then reacted with a primary antibody against the protein followed by a secondary antibody against the primary antibody. This secondary antibody is conjugated with an enzyme such as horseradish peroxidase. When H_2O_2 and a compound such as diaminobenzidine (DAB) are added, peroxidase converts H_2O_2 to H_2O and O_2^- which oxidizes DAB to produce a brown color band if the band is the expressed protein.

If an antibody is not available to detect the expressed protein, mini-cells can be used. Mini-cells are produced when *E. coli* divides to produce daughter cells during proliferation. If unequal division occurs, cells that are smaller than normal are produced. These cells are too small to accept *E. coli* genome and therefore are called mini-cells. These cells, however, are big enough to contain the expression plasmid. Since mini-cells also contain all the proteins required for transcription and translation, the gene on the expression plasmid will be expressed. If ^{35}S-methionineis is added to the culture medium of isolated mini-cells, the newly synthesized proteins will become radioactive. Since the average half-life of *E. coli* mRNAs is only 2 minutes, almost all *E. coli* mRNAs are degraded during the mini-cell isolation process, and thus all newly synthesized proteins are encoded by the expression plasmid. Since expression plasmid encodes only 2-3 genes, only 2-3 radioactive bands are visualized. One of them should be the recombinant protein.

Maxi-cells can also be used to detect the expressed protein. If a *recA-*, *recBCD+* *E. coli* is irradiated with UV, its genomic DNA is completely degraded by the RecBCD exonuclease. This is because the cell is defective in the RecA protein and cannot repair DNA breaks caused by UV irradiation. Because the possibility that all plasmids in the cells are damaged by UV is small, there will still be intact plasmids in the cell. Therefore, cells with no genomic DNA are generated. These cells can still carry out transcription and translation. If ^{35}S-methionine is added to the culture medium of these cells, the newly synthesized proteins that are encoded by the expression plasmid are radioactive, similar to mini-cells.

If the T7 promoter is used to drive gene expression, rifampin can be used to inhibit the translation of host *E. coli* genes so that proteins encoded by the expression plasmid can be visualized. In this system, the expression plasmid is introduced into *E. coli* cells that can produce the T7 RNA polymerase used to express the recombinant protein. Since

T7 RNA polymerase is not sensitive to rifampin, the transcription carried out by *E. coli* RNA polymerase but not that by T7 RNA polymerase is inhibited. The production of T7 RNA polymerase is usually driven by the *lac* promoter which can be induced by IPTG (0.1 mM). Similar to the methods for mini-cells and maxi-cells, ^{35}S-methionine is added to the culture containing rifampin and IPTG to visualize newly synthesized proteins that are encoded by the expression plasmid.

14.9: Phage Display

Any protein which does not require posttranslational modifications in order to become functional can be expressed in *E. coli*. Such proteins may include hormones, interferon, and antibodies. Production of antibodies is commonly done using the phage display system in which the M13 bacteriophage is used as the expression vector, and the antibody molecules produced are displayed on the surface of M13 phages (Fig. 14-5). A typical antibody is composed of two heavy (H) and two light (L) chains. Each chain has both constant and variable regions. The variable region in the heavy chain is called V_H and that in the light chain is called V_L. The constant region is required for complement fixation during an antibody reaction. V_H and V_L together make up the antigen-binding site. A recombinant protein consisting of V_H and V_L is called single chain variable fragment (ScFv). Although it does not fix complement, it is a functional antibody for antigen binding. The phage display system is used to produce the ScFV antibody. Two different vectors, M13KE and pHage3.2, are commonly used. M13KE is an M13 phage, and pHage3.2 is a phagemid containing gene III of M13 (Fig. 14-5).

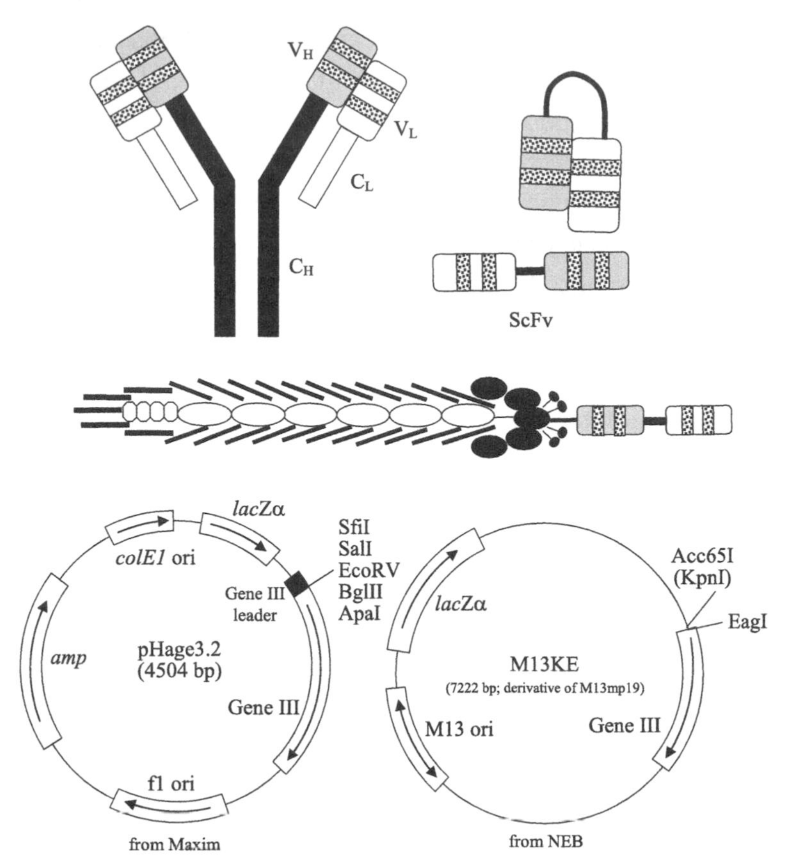

Fig. 14-5: Structure of antibody and phage display vectors for production of ScFv antibody.

If a DNA fragment encoding ScFv is inserted into Acc65I or EagI site of M13KE, a fusion protein of ScFv and M13 gene III protein is produced. Since the gene III protein is a head protein of M13, the recombinant protein thus produced is incorporated into M13 coat in the head region. The recombinant M13 phage particles are then placed in a plate coated with the antigen. If the ScFv antibody can recognize the antigen, the recombinant M13 phage will bind to the antigen and adhere to the plate (Fig. 14-6). These M13 phages are collected and further purified by additional rounds of screening. The resulting pure M13 recombinant clone is then grown to massively produce the antibody. If pHage3.2 is used as the vector, the DNA fragment containing ScFv gene is inserted into SfiI, SalI, EcoRV, BglII, or ApaI site to fuse the gene to M13 gene III. The recombinant plasmid is then introduced into *E. coli* cells containing an M13 helper phage such as M13KO7. The phage particles produced are then screened as described above.

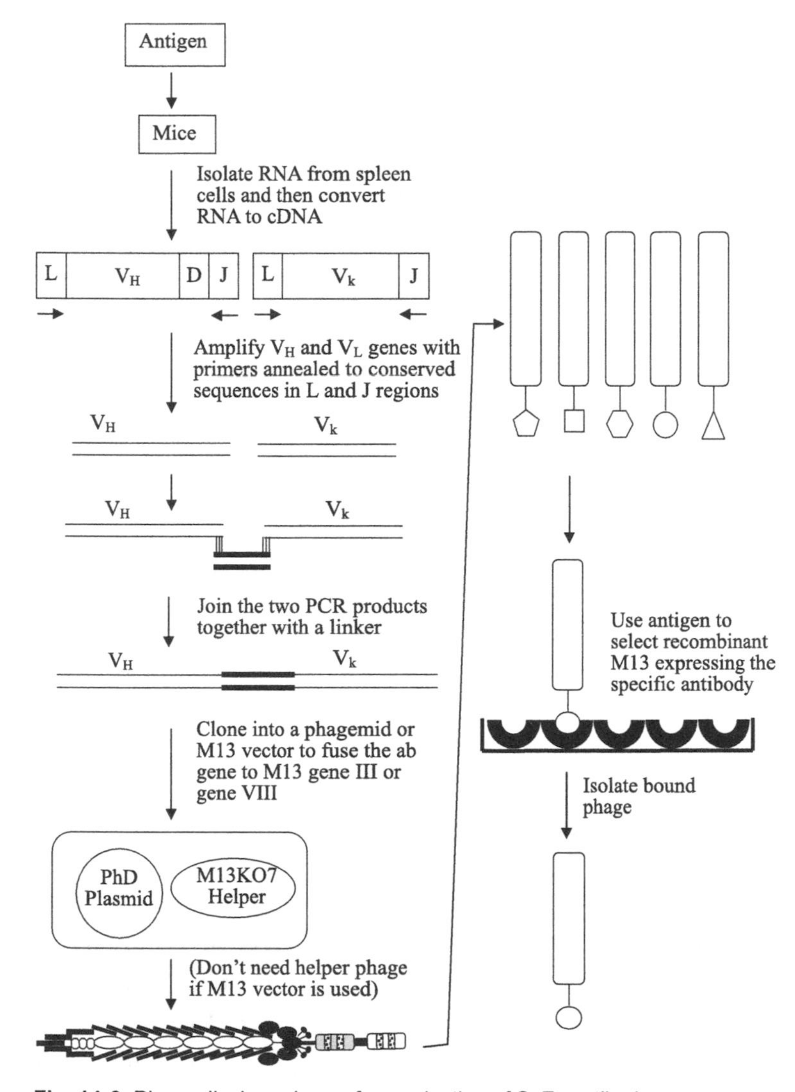

Fig. 14-6: Phage display scheme for production of ScFv antibody.

Antibodies are immunoglobulins. Their heavy and light chains are encoded by two different genes. Each gene is composed of several regions. There are two light chain genes: lambda and kappa. Each light chain gene is composed of a variable (V) region, a joint (J) region, and a constant C region. The heavy chain gene has one additional region, the D region. Each region has several genes. The lambda light chain gene has 30 variable (V) genes, 4 J genes, and 4 constant genes ($C_{\lambda1}$, $C_{\lambda2}$, $C_{\lambda3}$, and $C_{\lambda4}$). The kappa light chain gene has approximately 300 V genes, 5 J genes, and one constant gene. The heavy chain gene has at least 1000 V genes, 15 D genes, 5 J genes, and 9 constant genes including C_{μ}, C_{δ}, $C_{\gamma3}$, $C_{\gamma1}$, $C_{\gamma2}$, $C_{\gamma4}$, C_{ε}, $C_{\alpha1}$, $C_{\alpha2}$ (Fig. 14-7). Each V gene of both heavy and light chains has its own promoter.

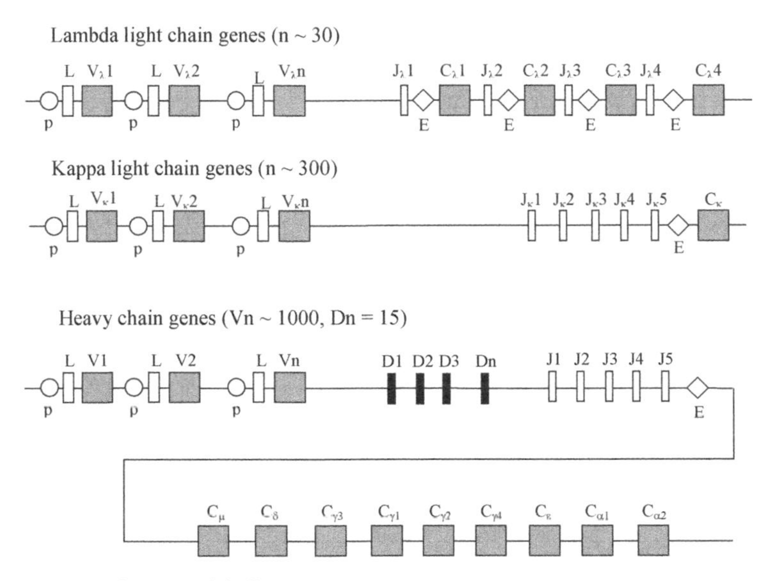

Fig. 14-7: Immunoglobulin genes.

When a lymphocyte is activated to produce antibody against a certain antigen, the immunoglobulin genes, both light and heavy chain genes, will undergo rearrangement. For the lambda light chain gene, one of the V genes is joined to one of the J genes. The rearranged gene is then transcribed, and the pre-mRNA produced is spliced to produce a functional antibody mRNA which is composed of one copy each of V_L, J, and C regions (Fig. 14-8).

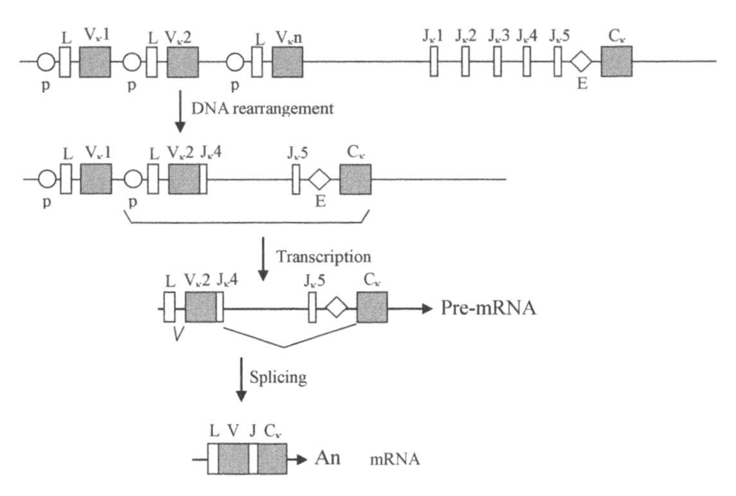

Fig. 14-8: Kappa chain gene rearrangement.

Rearrangement of the heavy chain gene is similar to that of the light chain gene, except that the heavy chain gene has an additional D region. In the first step of rearrangement, a D gene is joined to a J gene. One of the V genes is then joined to this DJ recombinant, resulting in a VDJ joint. The rearranged gene is then transcribed to produce a pre-mRNA, containing one each of the V, D, and J genes and 9 constant genes. The pre-mRNA is then spliced by joining the VDJ region to one of the 9 constant regions to produce an mRNA for IgG, IgA, IgM, IgE, or IgD (Fig. 14-9).

Although the sequence of the V region is variable, all V regions have a leader peptide containing conserved sequences. The 3' end of the J region is also conserved. Therefore, it is possible to use one set of primers that anneal to the L and J regions to amplify all the immunoglobulin mRNAs in lymphocytes. This can be done for both the heavy chain and the light chain genes. The V_H and V_L (V_λ or V_κ) PCR products are then joined together to form the ScFv gene (Fig. 14-6). The DNA fragment is then cloned into the M13KE or pHage3.2 vector to produce recombinant M13 phages with the desired antibody molecules inserted in its protein coat.

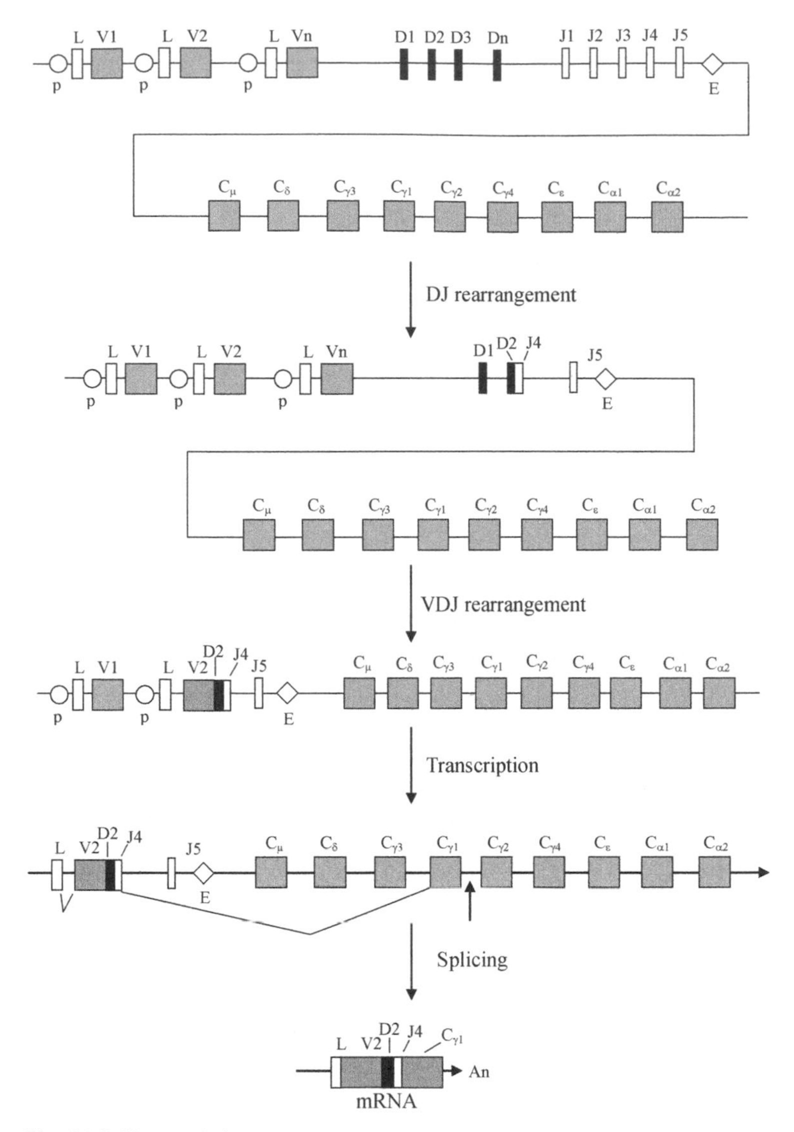

Fig. 14-9: Heavy chain gene rearrangement.

14.10: Eukaryotic Expression Systems

If a protein requires posttranslational modifications to become functional, eukaryotic expression systems are used because *E. coli* does not have proper posttranslational modification enzymes. Posttranslational modification of proteins may include glycosylation, phosphorylation, oligomerization, proteolytic processing, acetylation, methylation, and folding. Commonly used host cells for eukaryotic expression systems include yeasts, insect cells, and mammalian cells. Regardless of the system used, the first step is to construct the expression plasmid using *E. coli* as host cells. The expression plasmid constructed is then introduced into eukaryotic cells to express the protein. Therefore, a plasmid vector capable of replicating in both *E. coli* and in eukaryotic cells is needed. Such a vector is called shuttle vector.

Yeasts that can be used as host cells for expression include *Sacchromyces cerevisiae*, *Schizosacchromyces pombe*, and *Pichia pastoris*. pYES/NT and pYES/CT are examples of yeast expression vectors (Fig. 14-10). These two plasmids have the 2 μ replication origin in addition to *colE1* and f1 replication origins. The 2 μ replication origin comes from a yeast plasmid which is 2 microns in length, and accordingly is called 2 μ plasmid. This replication origin allows the plasmid to replicate in yeast in high copy numbers (50 – 100 copies), thus a higher level of expression of the gene. When a DNA fragment containing the gene to be expressed is inserted into one of the multiple cloning sites of pYES/NT, the gene will be driven by the yeast GAL1 promoter, and the protein produced is tagged with 6x His, Xpress epitope, and V5 epitope at its N-terminal end. Therefore, the protein can be isolated and purified by using the Ni-NTA column or antibodies against the Xpress epitope or the V5 epitope. The Xpress epitope is part of the bacteriophage T7 gene 10 protein; its sequence is -Asp-Leu-Tyr-Asp-Asp-Asp-Asp-Lys-. The V5 epitope comes from the C-terminal end of the P and V proteins of Simian Virus 5. Its sequence is -Gly-Lys-Pro-Ile-Pro-Asn-Pro-Leu-Leu-Gly-Leu-Asp-Ser-Thr-. If pYES2/CT is used as the vector, the protein will be tagged at its C-terminus with the V5 epitope and 6x His. If high expression level of the recombinant protein is harmful to yeast cells, low copy number plasmid vectors are used. Both pYC2/NT or pYC2/CT vectors contain the yeast centromere sequence (CEN6) and the autonomously replicating sequence (ARSH4) (Fig. 14-10). These sequences allow the plasmid to replicate in low copy number, thus expressing the protein at a lower level.

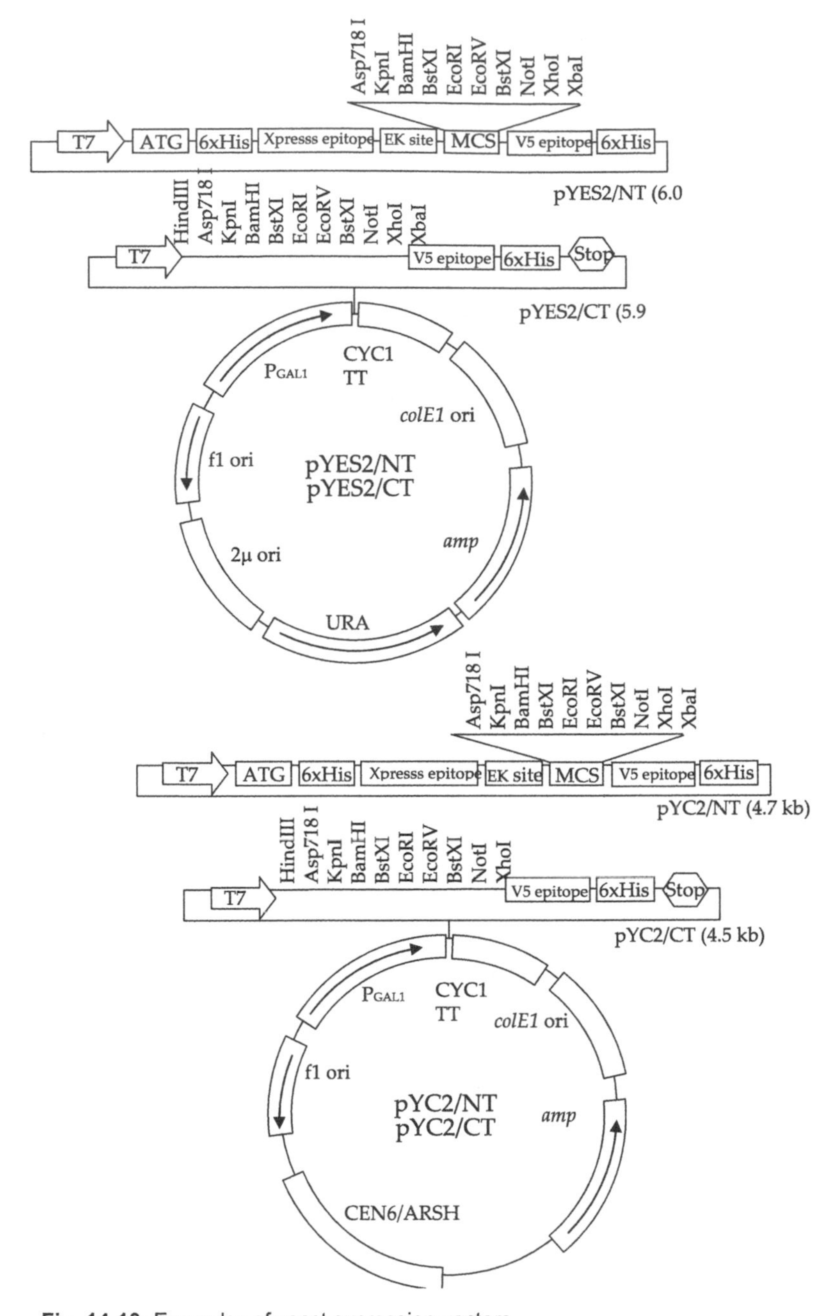

Fig. 14-10: Examples of yeast expression vectors.

Pichia pastoris is another type of yeast. It is very easy to grow and produces high levels of proteins at approximately 12 mg/ml of culture. It can use methanol as its carbon source for growth and therefore is inexpensive to culture. *Pichia pastoris* is currently the most commonly used yeast expression system, and pPICZ and pPICZα are examples of vectors that can be used (Fig. 14-11). The alcohol oxidase (AOX1) gene promoter is used to express the gene in this system. Alcohol oxidase is the enzyme responsible for the first step of methanol metabolism. The AOX protein is 30% of the total protein produced in *Pichia pastoris* as the AOX1 promoter is a very strong promoter.

When a DNA fragment containing the gene to be expressed is inserted into one of the multiple cloning sites of these plasmids, the recombinant protein will be tagged with the C-myc epitope and 6x His. Both plasmids contain the zeocin resistance gene driven by both P_{TEF1} and P_{EM7} promoters. P_{TEF1} is a yeast promoter and will allow the zeocin gene to be expressed in yeast. In *E. coli*, P_{EM7} is used to drive the expression of the zeocin gene. Both plasmids contain the *colE1* replication origin but do not have any yeast replication origin. Therefore, the recombinant plasmid will integrate into yeast chromosome by homologous recombination at the AOX1 gene. pPICZα contains sequence encoding the leader peptide of the yeast α-factor which will allow the expressed protein to be secreted outside the cell, making it easier to purify the recombinant protein. The AOX1 promoter is induced by methanol. pGAPZ and pGAPα are derivatives of pPICZ and pPICZα (Fig. 14-11) in which the glyceraldehyde 3-phosphate dehydrogenase (GAPDH) gene promoter PGAP is used to drive the gene for expression. PGAP is a constitutive promoter; therefore, the gene will be always expressed.

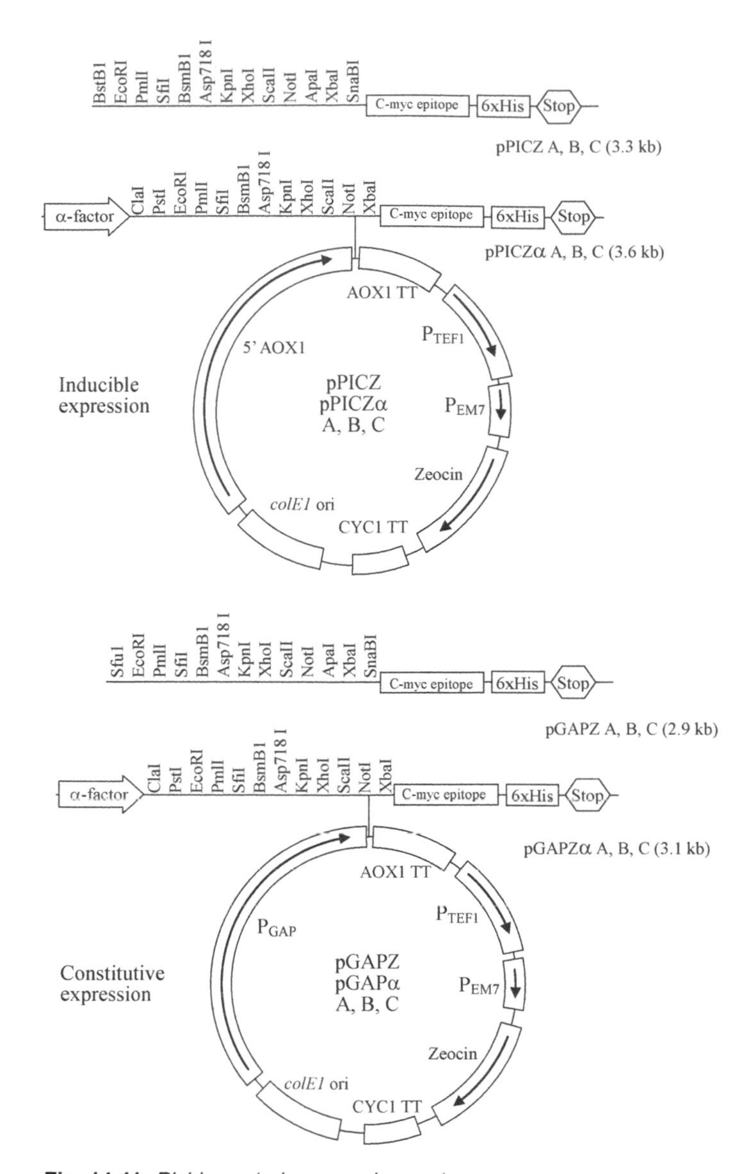

Fig. 14-11: *Pichia pastoris* expression vectors

Although yeast cells have posttranslational modification systems, the modifications may not be the same as those in mammalian cells. Therefore, proteins expressed in yeasts may not be functional. If this is the case, the baculovirus expression system may be used. Baculovirus (autographa californica multicapsid nucleopolyhedrovirus) is a DNA virus and infects insect cells. It may replicate by budding and release viral particles into hemolymph to infect other cells. The virus may also form occlusion bodies that are coated with polyhedrin proteins. Occlusion bodies can survive in adverse environments to infect other insects. Since a large amount of the polyhedrin protein is required for making occlusion bodies, the polyhedrin gene promoter is a very strong promoter. The polyhedrin protein is not required for virus replication in vitro. Therefore, its promoter can be used to drive the expression of foreign genes. This expression system can produce protein at the level of 1 mg/ml of culture, representing 20% of the total protein of baculovirus infected cells. A commonly used cell line is the Sf9 cell line derived from the moth, fall armyworm *Spodoptera frugiperda*.

To express a gene in this system, the DNA fragment containing the gene to be expressed is inserted into the genome of baculovirus. Normally, to insert a DNA fragment into a vector, the vector is first cut with a certain restriction enzyme, and the DNA fragment is then ligated with the linearized vector. However, the baculovirus genome is approximately 100 kb. Any restriction enzyme will cut it into many pieces. Since the possibility of ligating all the pieces back to their original order is very small, a different approach is used to create recombinant baculovirus for expression. The gene to be expressed is first cloned into a plasmid vector behind the polyhedrin gene promoter so that the gene is flanked by two DNA fragments that are also present in the baculovirus genome. When the resulting recombinant plasmid is introduced into Sf9 cells infected with baculovirus, homologous recombination will occur between the plasmid and the baculovirus genome, resulting in a recombinant baculovirus containing the gene to be expressed driven by the polyhedron gene promoter.

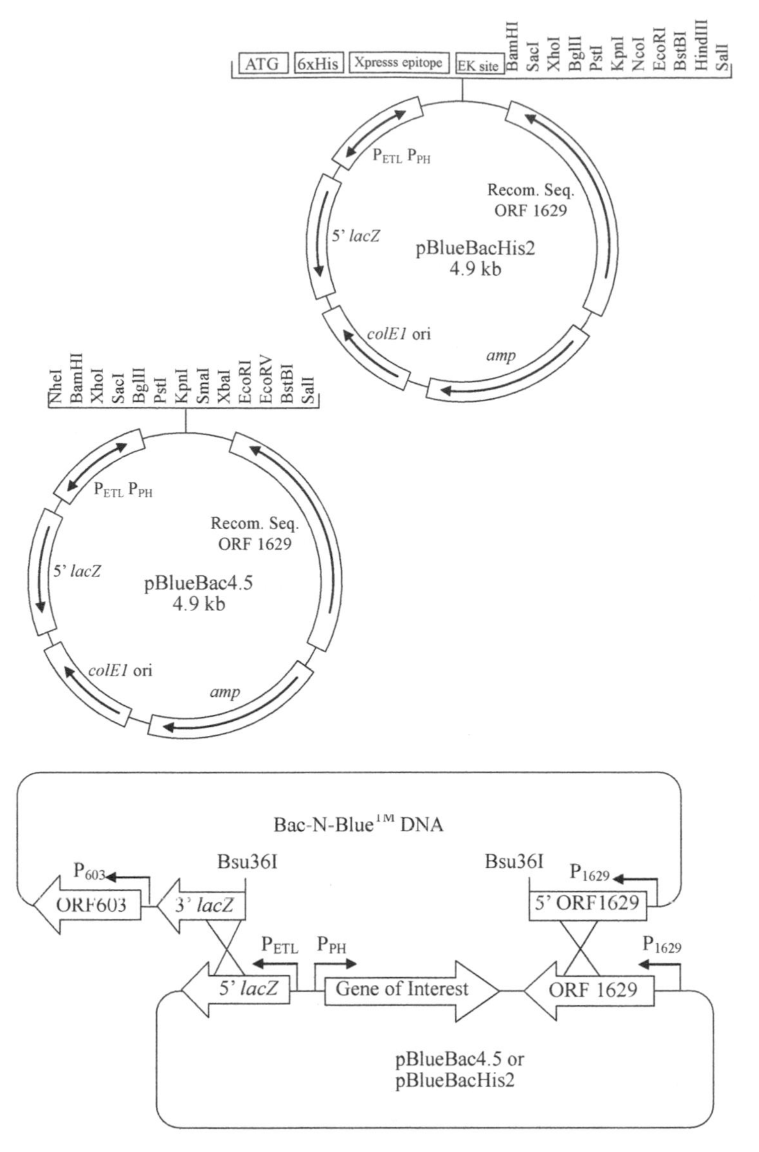

Fig. 14-12: Baculovirus expression system.

An example of this approach is shown in Fig. 14 -12. The gene to be expressed is first cloned into pBlueBacHis2 or pBlueBac4.5. Both plasmids have the *colE1* replication origin and thus can replicate in *E. coli*. It also contains the 5' portion of the *lacZ* gene driven by the P_{ETL} promoter; where ETL stands for early to late. It is a baculovirus gene and is active from early to late stages of the viral life cycle. P_{PH} is the polyhedrin gene promoter. Insertion of a DNA fragment containing the gene to be expressed into one of the multiple cloning sites (MCS) of the vectors will place the gene under the control of P_{PH}. These two vectors also contain ORF 1629 of baculovirus. Insertion of the DNA fragment into the MCS of pBluebacHis2 will tag the protein at its N-terminus with 6x His, Xpress epitope, and the enterokinase cleavage signal. This will simplify the isolation and purification process of the recombinant protein. The recombinant plasmid constructed is then introduced into Sf9 cells containing a baculovirus such as Bac-N-Blue™ available from Invitrogen, Inc. It contains the 3' portion of the *lacZ* gene and will undergo homologous recombination with the introduced expression plasmid at the *lacZ* gene and ORF 1629. This recombination event will result in integration of the gene to be expressed into the baculovirus genome and creation of a functional *lacZ* gene. In the presence of X-gal, cells containing the correct recombinants will turn blue. The recombinant baculovirus is then used to infect Sf9 cells to produce the protein.

If the protein produced with the baculovirus expression system is still not functional, mammalian expression system is used. Its use is very similar to that of the yeast expression system. To express a gene in mammalian cells, selection of an appropriate promoter is critical because some genes are expressed only in certain type of cells. For example, the insulin gene is expressed only in β cells in the pancreas, and the crystalline gene is expressed only in lens cells. If the insulin gene promoter is used to drive the gene, this gene can only be expressed in β cells. The CMV IE gene promoter is active in many types of cells and therefore is commonly used for general expression. The plasmid pcDNA3.1 (Fig. 14-13) is a very popular vector if the CMV promoter is to be used to drive the gene. It has the *colE1* replication origin and therefore can replicate in *E. coli*. When a DNA fragment containing the gene to be expressed is inserted into one of the multiple cloning sites of the plasmid, the recombinant protein will be tagged with 6x His, Xpress epitope, and the enterokinase cleavage signal at the

N-terminus. Since most eukaryotic mRNAs have a poly-A tail, pcDNA3.1 has the polyadenylation signal from the bovine growth hormone gene. In addition, it contains the SV40 replication origin and will replicate in cells that express the SV40 T-antigen. Since the SV40 early gene promoter is located in the SV40 promoter, it will drive the expression of the neomycin resistance gene, making the cells containing the plasmid resistant to antibiotic G418. The SV40 polyadenylation signal will allow a poly-A tail to be added to the neomycin mRNA.

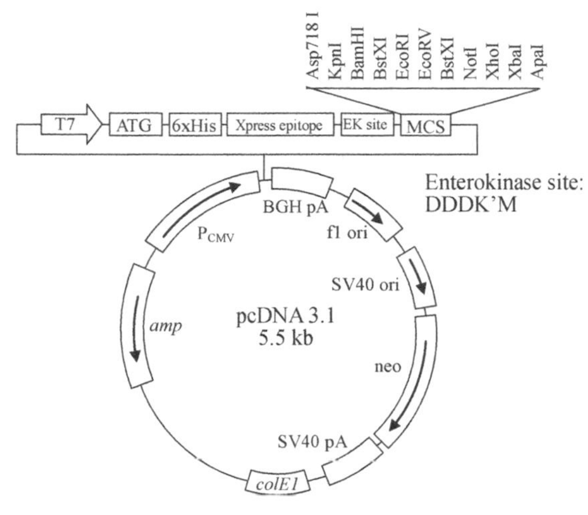

Fig. 14-13: pcDNA 3.1 for expression of recombinant proteins in mammalian cells.

Some eukaryotic mRNAs must undergo splicing in order to become functional. Therefore, some eukaryotic expression vectors have an intron that will allow the pre-mRNA produced to undergo splicing. Examples of such vectors are pIRES-EGFP and pIRESneo shown in Fig. 14-14. pIRES-EGFP contains all the elements required for expression in mammalian cells. The multiple cloning sites are located between the CMV IE promoter and an intron. An internal ribosome entry site (IRES) is located downstream of the intron followed by the gene encoding the enhanced green fluorescence gene (EGFP) and the polyadenylation signal. The IRES allows the EGFP gene to be expressed through internal initiation during translation. Without an IRES, translation will terminate at the stop codon of the gene to be expressed, and the EGFP gene will not be expressed. The EGFP gene will allow detection of cells containing the recombinant plasmid as these cells will be green when visualized under a fluorescence microscope. pIRESneo will allow co-expression of the gene of interest and the neomycin resistance gene making the cells containing the recombinant plasmid resistant to antibiotic G418.

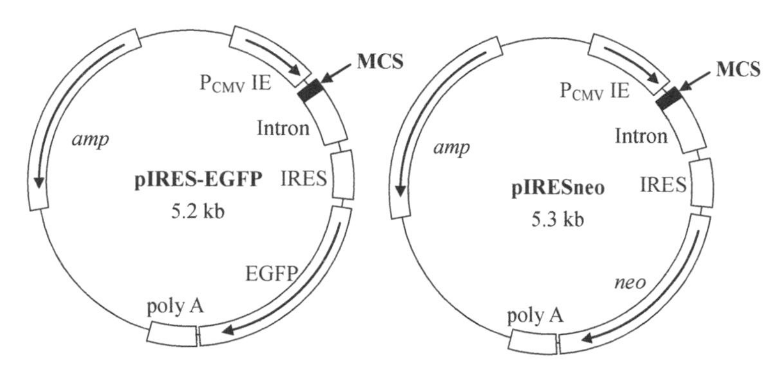

Fig. 14-14: Mammalian cell expression vectors with intron and IRES.

14.11: Regulation of Expression in Mammalian Cells

The eukaryotic expression system is commonly used to study the function of a gene. For such applications, the expression of the gene is made inducible. Therefore, a function gained after induction of the gene is mediated by the gene. The most commonly used regulatory system is the Tet system derived form the *E. coli* tetracycline (Tet) operon. This operon controls the production of the tetracycline efflux protein (also called the TetA protein). The TetA protein pumps tetracycline out of *E. coli* cells so that they are not killed by tetracycline. The TetA protein is produced only when *E. coli* cells are exposed to tetracycline because the Tet operon produces another protein call Tet repressor or the TetR protein. TetR binds to the Tet operator and suppresses the expression of the TetA gene. When *E. coli* is exposed to tetracycline, tetracycline binds to TetR rendering it unable to bind to the Tet operator. Therefore, the TetA gene is expressed, and the tetracycline efflux protein is produced to pump tetracycline out of the cell.

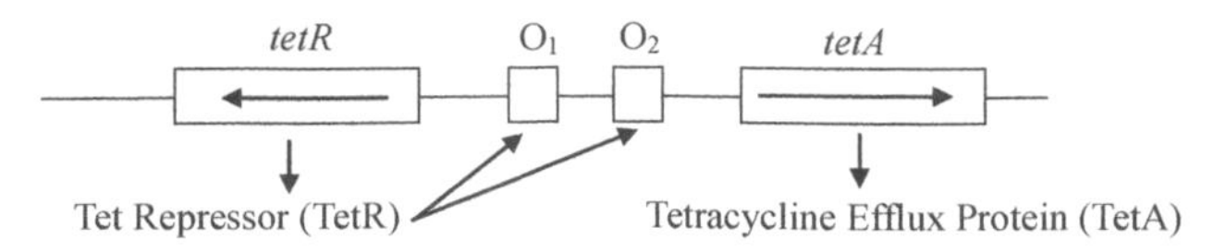

- O_1 regulates both *tetR* and *tetA* genes
- O_2 regulates only *tetA* gene
- Tetracycline binds to and inactivates TetR

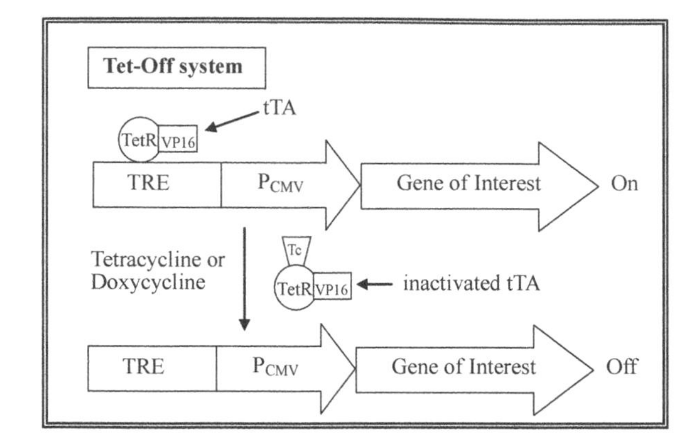

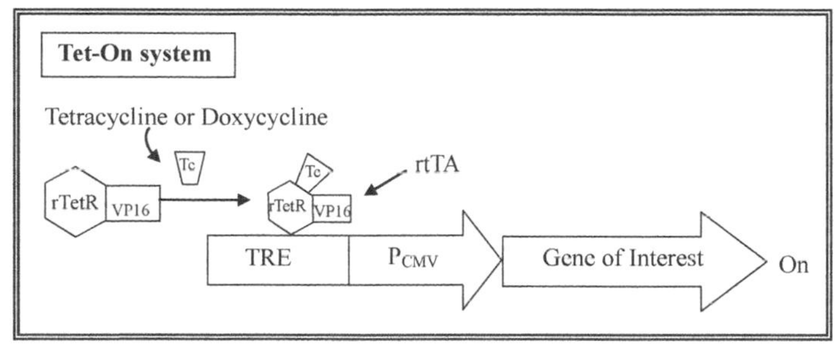

Fig. 14-15: Tet-on and Tet-off regulatory systems.

The Tet regulatory system uses the *tet* operator and TetR to regulate gene expression in mammalian cells. Two different Tet regulatory systems, Tet-off and Tet-on, exist. In the Tet-off system, the expression of the gene is turned off in the presence of tetracycline or its derivative doxycycline. In contrast, the Tet-on system turns on the expression of the gene in the presence of tetracycline or doxycycline (Fig. 14-15). Both systems use the CMV IE promoter (P_{CMV}) to drive the gene. P_{CMV} is linked to the Tet-responsive element (TRE) which is the *tet* operator and becomes active only when a transcription activator (tTA) binds to TRE. The *tet* transcription activator (tTA) is a fusion protein of TetR and VP16 of Herpes simplex virus (HSV). This fusion protein activates P_{CMV} when it binds to TRE. However, in the presence of tetracycline or doxycycline, tetracycline or doxycycline binds to tTA making it unable to bid to TRE to activate P_{CMV}.

In the Tet-on system, the *tet* transcription activator is modified so that it binds to TRE to activate pCMV only when it is bound by tetracycline or doxycycline. This modification changes four amino acids of the TetR protein, including Glu71Lys, Asp95Asn, Leu101Ser, and Gly102Asp. The modified TeR is called reverse TetR (rTetR), and the rTetR-VP16 fusion is called reverse *tet* transcription activator (rtTA) which binds to TRE only when it is bound by tetracycline or doxycycline. Therefore, the Tet-on system is turned on by tetracycline or doxycycline. Doxycycline is more commonly used because it is more active than tetracycline.

14.12: Introduction of Recombinant Plasmid into Eukaryotic Cells

To produce any recombinant protein, the expression plasmid must be introduced into cells. The process in which foreign DNA is introduced into eukaryotic cells is called transfection. The word transfection is derived from infection in which a virus adsorbs to a cell, injects its DNA into the cell, and replicates and produces viral particles in the cell. Since introduction of naked viral DNA into cells can also produce viral particles, the process is therefore called transfection.

Transfection can be achieved by several different methods. Microinjection injects the DNA into cells by using a very fine capillary. This method is used only when very few cells need to be transfected. Chemical transfection precipitates DNA to fine particles that are then taken up by cells via endocytosis. Calcium phosphate and DEAE dextran are the two most commonly used chemicals for this purpose. Another method is lipofection in which DNA is enclosed by lipids to form a liposome. When the liposome is fused with the cell membrane, DNA is introduced into the cell. Transfection can also be done by electroporation. DNA is mixed with cells, and the cells are shocked with electricity so that the membrane becomes porous allowing the DNA to get in. Electroporation usually kills 80 - 90% of the cells. However, most of the remaining cells are transfected. Currently, lipofection is the most commonly used method.

Since none of the transfection methods is 100% efficient, there must be a way to select cells that have taken up the expression plasmid. The most common approach is to use vectors containing a dominant selection marker. When cells take up the vector, they can be selected. Several such markers can be used. The neomycin-resistance gene (*neo*) encodes aminoglycoside phosphotransferase which can inactivate G418. G418 is an aminoglycoside antibiotic isolated from *Micromonospora rhodorangea*. It is very similar to gentamycin B1. The gene encoding the hygromycin B phosphotransferase can also be used as a selection marker as cells containing this gene are resistant to hygromycin B. The *sh ble* gene from *Streptoalloteichus hindustanus* confers cell resistance to zeocin which is a glycopeptide antibiotic. The *sh ble* gene is more commonly referred to as the zeocin-resistance gene. Blasticidin S isolated from *Streptomyces griseochromogenes* is a peptidyl nucleoside antibiotic and can also be used to select cells containing the plasmid with the *bsr* gene from *Bacillus cereus*.

The green fluorescence protein (GFP) is also a commonly used marker to detect transfected cells. If the GFP gene is placed on the same plasmid that contains the gene to be expressed, cells that fluoresce green are the ones containing the gene to be expressed. The GFP gene may be expressed by a different promoter or by the same promoter as the one driving the gene to be expressed. In the latter case, an IRES is placed between the two genes as described above (Fig. 14-13).

14.13: Gateway Expression System

To express a gene in any eukaryotic system, the gene is first cloned into a plasmid using *E. coli* as the host for the plasmid. The expression plasmid is then isolated from *E. coli* and introduced into cells in a certain eukaryotic expression system such as *Pichia pastoris*. If the protein produced by this system is not functional, the gene is recloned into a different expression vector such as that for expression in the baculovirus system or mammalian cells. Therefore, many subclonings may need to be done to find the best expression system. To simplify the process, the Gateway expression system is developed. The principle of the Gateway expression system is based on the integration and excision of lambda phage genome. When lambda phage genome is integrated into *E. coli* chromosome, it is always inserted into the attachment (*att*) site, called *attB*, in the *E. coli* genome. The lambda phage also has an *att* site, called *attP*. This is the place where a circular lambda genome is cut and inserted into *E. coli* genome. Therefore, this integration is a site-specific recombination event. Both *attB* and *attP* are composed of 23 bp; 15 of which are identical and are called the core sequence, designated as O region. The sequences flanking the O region of the lambda *att* site are designated as P and P', and those flanking the O region of the *E. coli att* site are designated as B and B'. Therefore, the lambda *att* site is POP' and that of the *E. coli* is BOB'. When the lambda genome is integrated into *E. coli* chromosome, two *att* sites are created. The one located on the left end of the lambda genome is designated as *attL* (BOP') and the other one is designated *attR* (POB') (Fig. 14-16).

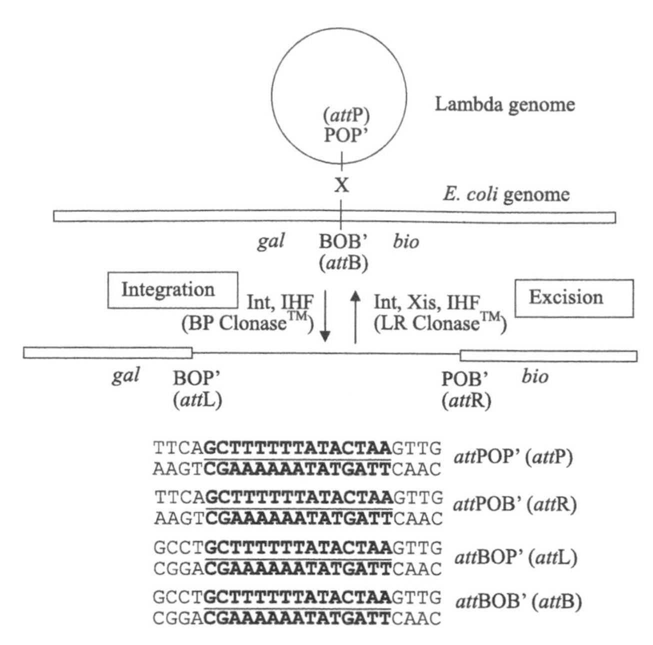

Fig. 14-16: The att sites of lambda phage and E. coli and their role in integration and excision of lambda genome.

The enzyme responsible for lambda genome integration is the lambda integrase (Int) in conjunction with the integration host factor (IHF) of *E. coli*. When lambda phage switches its life cycle from lysogenic to lytic, the integrated lambda genome is excised from the *E. coli* genome by the lambda excisionase (Xis), also in conjunction with IHF. All of these enzymes recognize the *att* site, and the *att* sites are the only cis elements required for this integration-excision process. Therefore, both integration and excision events can take place in a test tube. If a DNA fragment containing a gene flanked by two *att* sites is mixed with another DNA fragment containing a different gene also flanked by two *att* sites, these two genes will exchange places if Int, Xis, and IHF are present (Fig. 14 -17).

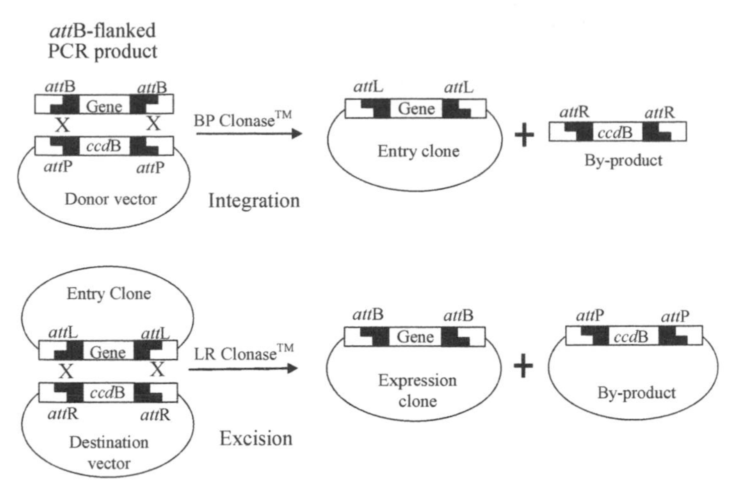

Fig. 14-17: Principles of the Gateway Cloning System.

Invitrogen, Inc. provides two enzyme mixtures called BP Clonase and LR Clonase. The BP Clonase contains Int and IHF, and the LR Clonase contains Xis, Int, and IHF. BP Clonase will carry out exchange of a DNA fragment containing the gene flanked by two *att* sites with the other one also flanked by two *att* sites located on a plasmid. This process is equivalent to the integration event of lambda genome. If the two different genes flanked by *att* sites are located on two different plasmids, LR Clonase is used. This process is equivalent to the excision event of lambda genome. A DNA fragment flanked by two *att* site can be produced by PCR using primers with the *att* sequence built in at the 5' end of each primer. Another way to place a DNA fragment between two *att* sites is to insert a PCR product into the cloning site of a vector such as pGR8/GW/TOPO (Fig. 14-18). This will result in a recombinant plasmid containing the gene of interest flanked by two *att* sites. If this gene is to be expressed in a certain expression system such as the mammalian system, the recombinant plasmid is mixed with another plasmid such as pT-Rex-DEST30. In the presence of LR Clonase, the gene to be expressed located on pGR8/GW/TOPO is transferred automatically to pT-Rex-DEST30 under the control of the CMV promoter.

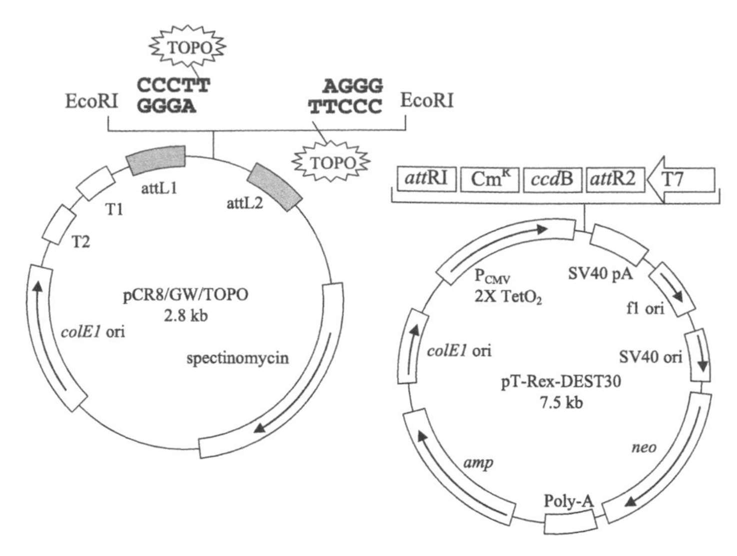

Fig. 14-18: Examples of vectors of the Gateway cloning system.

Summary

To produce a foreign protein in *E. coli*, the gene to be expressed is cloned into a plasmid. The recombinant plasmid is then introduced into *E. coli* so that the gene is transcribed, and the mRNA is translated. This whole process is called expression. Several outcomes may result when a foreign protein is expressed in *E. coli*. The expressed protein may be harmful to *E. coli* or may be degraded or packaged by *E. coli* to become inclusion bodies. Once a protein is enclosed in inclusion bodies, it usually becomes insoluble and inactive. To prevent the expressed protein from killing the *E. coli* host, an inducible promoter is used to drive the gene. Many inducible promoters are available, including the *lac, trp, tac*, Lambda P_L, and T7 promoters. *lac* and *tac* promoters can be induced with IPTG. The *trp* promoter is induced by indole-3-propionic acid or indole-acrylic acid. The lambda P_L promoter can be controlled by the *cI857* gene. Since the CI857 protein is temperature sensitive and is inactivated at 42°C, the P_L promoter can be induced by heating. T7 promoter must have T7 RNA polymerase in order to function. Therefore, it can be regulated by controlling the production of T7 RNA polymerase.

In order to prevent the expressed protein from being degraded or packaged into inclusion bodies, the protein to be expressed is fused to an *E. coli* protein such as dihydrofolate reductase (DHFR), glutathione S transferase (GST), maltose-binding protein (MBP), or thioredoxin. *E. coli* may consider the fusion protein its own and may not degrade it. The use of fusion proteins can also simplify isolation and purification of the expressed protein. For example, anti-DHFR antibodies can be used to isolate the expressed protein using affinity column chromatography. If the protein is fused to GST, glutathione-conjugated agarose beads can be used to isolate the protein. The most commonly used method is to fuse six histidine residues (6x His) to the N- or C-terminus of the protein. The protein is then isolated using a column packed with Ni-NTA.

After the protein has been isolated, the *E. coli* protein fused to the protein of interest may need to be removed. To do this, the cleavage signal of a certain protease is inserted between the tag and the protein to be expressed. This is achieved by inserting the gene to be expressed into a plasmid containing sequences encoding the tag and the cleavage signal. Commonly used proteases include thrombin, enterokinase, clotting factor Xa, and PreScissionase. Their cleavage signals are LVPRG, DDDKM, IEGR, and LEVLFQGP, respectively. If the protein is fused to an intein, the tag can be removed by activating the self-cleavage function of intein with dithiothreitol (DTT), β-mercaptoethanol, or cysteine.

The most important step in expression is selection of an appropriate promoter. In addition to the -35 and -10 sequences, the Shine-Dalgarno (SD) sequence also plays an important role for gene expression in *E. coli*. The SD sequence is located between the -10 sequence and the translation initiation codon ATG. The distance between these two elements is critical for expression and varies with different promoters. For example, this distance in the T7 promoter is 7 bp. If the T7 promoter is to be used to drive the gene for expression, this distance must be kept at 7 bp in order to achieve optimal expression.

Once the expression plasmid is constructed, it is introduced into *E. coli* to produce the protein. The expressed protein can be detected using several different methods. A regular polyacrylamide gel electrophoresis of the cell lysate of *E. coli* containing the expression plasmid is always performed first to detect bands representing the expressed protein. A Western blot analysis using antibody against the protein is then performed to confirm the identity of the expressed protein. If the T7 promoter is used to drive the gene, rifampin can be used to inhibit the production of the *E. coli* host protein as *E. coli* RNA polymerase is sensitive to rifampin, but T7 RNA polymerase is resistant to rifampin.

Many therapeutic proteins can be produced in *E. coli* including insulin, growth hormone, interferons, and antibodies. Antibodies are usually produced using the phage display system. The gene encoding the antibody to be produced is cloned into an M13 vector so that the antibody is fused to the gene III protein of M13. This fusion protein is then incorporated into the protein coat and displayed on the surface of M13. The recombinant M13 is then isolated and grown in large scale to produce the antibody. To isolate the antibody gene for expression, mice are first injected with the antigen to induce antibody production. mRNAs are then isolated from lymphocytes and subjected to RT-PCR with primers that anneal to the leader peptide and the J region of the variable region of the heavy chain and the light chain genes. The PCR products produced are joined together and inserted into the vector for expression.

Some proteins require posttranslational modifications in order to become functional. The modifications may include glycosylation, phosphorylation, acylation, methylation, proteolytic cleavage, and folding. Since *E. coli* does not have posttranslational modification systems, it cannot be used to express these proteins. Eukaryotic expression systems that can be used include yeast, baculovirus, and mammalian systems. The expression plasmid is first constructed using *E. coli* as host cells and then introduced into appropriate cells for expression.

Yeast cells that can be used include *Sacchromyces cerevisiae*, *Schizosacchromyces pombe*, and *Pichia pastoris*. Among them, *Pichia pastoris* is most commonly used. The promoter used in this system is usually the AOX1 promoter which can be induced with methanol. Methanol is also used as the carbon source to grow *Pichia pastoris*. In the baculovirus system, the polyhedrin gene promoter is used to express the gene. The gene to be expressed is first cloned behind the polyhedrin gene promoter on a plasmid. The recombinant plasmid is then introduced into insect cells, such as the Sf9 cells from the fall armyworm, that are infected with baculovirus. Since the DNA sequences flanking the gene to be expressed on the recombinant plasmid are also present in baculovirus, homologous recombination between these sequences will take place resulting in a recombinant baculovirus containing the gene to be expressed driven by the polyhedrin gene promoter. The recombinant proteins produced in Sf9 cells are isolated and used. To express a gene in mammalian cells, the CMV promoter is most commonly used. The expression plasmid is first constructed and then introduced into mammalian cells for expression.

Plasmid DNA can be introduced into yeast cells by transformation in a way similar to that for *E. coli*. Lipofection is the most commonly used method to introduce DNA into other eukaryotic cells. The DNA is mixed with lipid reagents to form a liposome which is then fused with cells to be transfected. To be able to select transfected cells, expression vectors usually contain dominant selection markers so that the transfected cells will become resistant to antibiotics such as G418, hygromycin, zeocin, or blasticidin S. If a mammalian cell system is used to study the function of a gene, a regulatory system is required to control the expression of the gene. The most commonly used system is the Tet-on or Tet-off system. In the Tet-on system, the expression is regulated by the HSV VP16 and Tet repressor fusion protein. This transcription activator can only bind to the Tet-responsive element to activate the gene in the presence of tetracycline or doxycycline. In the

Tet-off system, tetracycline or doxycycline binds to the transactivator preventing it from binding to the Tet-responsive element to initiate expression.

Since many different expression systems are available for different purposes, it may be necessary for a gene to be recloned into a different system. To simplify this process, Invitrogen, Inc. developed the Gateway expression system based on the integration and excision of lambda genome. In the presence of integrase, the lambda genome is integrated into *E. coli* genome. The integrated lambda genome is excised by the excisionase when lambda switches its life cycle from lysogenic to lytic. Both integrase and excisionase recognize a special sequence call *att*. A DNA fragment containing a gene flanked by two *att* sites located on a plasmid can be replaced by another DNA fragment containing the gene to be expressed also flanked by two *att* sites in the presence of integrase in a test tube. Similarly, two different plasmids containing different genes each flanked by two *att* sites can exchange locations in the presence of excisionase. Therefore, a gene to be expressed can first be cloned into a plasmid so that it is flanked by two *att* sites. This recombinant plasmid is called the entry clone. To express the gene in a eukaryotic expression system such as in the mammalian system, the entry clone is mixed with a destination vector containing an appropriate promoter such as the CMV promoter which drives a certain gene flanked by two *att* sites. In the presence of excisionase and integrase, these two genes will exchange positions, thereby placing the gene to be expressed under the control of the CMV promoter on the destination vector. With this system, it is not necessary to reclone the gene into a different system.

Sample Questions

1. Which of the following are important for gene expression in prokaryotic cells: a) AAUAAA, b) -35 region, c) -10 region, d) SD sequence.
2. Which of the following is critical for optimal gene expression in prokaryotic cells: a) the distance between -10 and -35 regions, b) the distance between -10 and SD

sequences, c) the distance between -35 and SD sequences, d) the distance between SD and initiation codon ATG.

3. A foreign protein is usually fused to an *E. coli* protein during expression in order to: a) minimize degradation by *E. coli* protease, b) prevent formation of inclusion bodies, c) simplify purification process, d) ensure posttranslational modifications.
4. If posttranslational modifications are required for the function of a protein, which of the following expression systems can be used: a) *Bacillus subtilus*, b) mammalian cells, c) *Pichia pastoris*, d) Baculovirus.
5. Kozak sequence a) enhances binding of mRNA to 40 S ribosome subunit during translation, b) is functionally equivalent to the SD sequence of prokaryotic genes, c) has the following consensus sequence CCG[A/G]CCATGG, d) enhances replication.
6. In the Tet-on expression system, the gene is turned on in the presence of: a) tetracycline, b) doxycycline, c) ampicillin, d) zeocin.
7. In the Tet-off expression system, the gene is turned off in the presence of: a) tetracycline, b) doxycycline, c) ampicillin, d) zeocin.
8. If the T7 promoter is used to express a foreign gene in *E. coli*, which of the following can be used to suppress production of *E. coli* proteins: a) ampicillin, b) tetracycline, c) rifampin, d) chloramphenicol.

Suggested Readings

1. Adler, H. L., Fisher, W. D., Cohen, A., and Hardigree, A. A. (1967). Miniature *E. coli* cells deficient in DNA. Proc. Natl. Acad. Sci. USA 57: 321-326.
2. Adler, H. I., Terry, C. E., and Hardigree, A. A. (1968). Giant cells of *Escherichia* coli. J. Bacteriol. 95: 139-142.
3. Baron, U., Gossen, M., and Bujard, H. (1997). Tetracycline-controlled transcription in eukaryotes: novel transactivators with graded transactivation potential. Nucleic Acids Res. 25: 2723-2729.
4. Campbell, E. A., Korzheva, N., Mustaev, A., Murakami, K., Nair, S., Goldfarb, A., and Darst, S. A. (2001). Structural mechanism for rifampicin inhibition of bacterial RNA polymerase. Cell 104: 901-912.
5. Clark-Curtis, J. E. and Curtiss, R. (1983). Analysis of recombinant DNA using *Escherichia* coli minicells. Meth. Enzymol. 101: 347362.
6. Gellert, M. (1992). Molecular analysis of VDJ recombination. Annu. Rev. Genet. 26: 425-446.
7. Gossen, M. and Bujard, H. (1992). Tight control of gene expression in mammalian cells by tetracycline-responsive promoters. Proc. Natl. Acad. Sci. USA 89: 5547-5551.
8. Gossen, M., Freundlieb, S., Bender, G., Muller, G., Hillen, W., and Bujard, H. (1995). Transcriptional activation by tetracycline in mammalian cells. Science New Series 268: 1766-1769.
9. Hartley, J. L., Temple, G. F., and Brashc, M. A. (2000). DNA cloning using in vitro site-specific recombination. Genome Research 10: 1788-1795.
10. Honjo, T., Kinoshita, K., and Muramatsu, M. (2002). Molecular mechanism of class switch recombination: linkage with somatic hypermutation. Annu. Rev. Immunol. 20: 165-196.

11. Ito, H., Fukuda, Y., Murata, K., and Kimura, A. (1983) Transformation of intact yeast cells treated with alkali cations. J. Bacteriol. 153: 163–168.

12. Khachatourians, G. G., Clark, D. J., Adler, H. I., and Hardigree, A. A. (1973). Cell growth and division in *Escherichia* coli: a common genetic control involved in cell division and minicell formation. J. Bacteriol. 116: 226-229.

13. Landy, A. (1989). Dynamic, structural, and regulatory aspects of lambda site-specific recombination. Annu. Rev. Biochem. 58: 913-949.

14. Marsh, P. (1986). ptac-85, an *E. coli* vector for expression of non-fusion proteins. Nucleic Acids Res. 14: 3603.

15. Meier, I., Wray, L. V., and Hillen, W. (1988). Differential regulation of the Tn*10*-encoded tetracycline resistance genes *tetA* and *tetR* by the tandem *tet* operators O_1 and O_2. EMBO J. 7: 567-572.

16. Paulus, H. (2000). Protein splicing and related forms of protein autoprocessing. Annu. Rev. Biochem. 69: 447-496.

17. Schatz, D. G., Oettinger, M. A., and Schlissel, M. S. (1992). VDJ recombination: molecular biology and regulation. Annu. Rev. Immunol. 10: 359-383.

18. Shockett, P., Difilippantonio, M., Hellman, N., and Schatz, D. G. (1995). A modified tetracycline-regulated system provides autoregulatory inducible gene expression in cultured cells and transgenic mice. Proc. Natl. Acad. Sci. USA 92: 6522-6526.

19. Smith, G. P. (1985). Filamentous fusion phage: novel expression vectors that display cloned antigens on the virion surface. Science 228: 1315-1317.

20. Tabor, S. and Richardson, C. C. (1985). A bacteriophage T7 RNA polymerase/ promoter system for controlled exclusive expression of specific genes. Proc. Natl. Acad. Sci. USA 82: 1074-1078.

21. Wigler, M., Silverstein, S., Lee, L. S., Pellicer, A., Cheng, Y. C., and Axel, R. (1977). Transfer of purified herpes virus thymidine kinase gene to cultured mouse cells. Cell 11: 223-232.

Chapter 15

Mutagenesis, Transgenic Animals and RNA Interference

Outline

15.1: Purpose of Mutagenesis

Mutagenesis is the process in which the nucleotide sequence of a gene is changed. It is usually performed to determine the function of a gene. For example, if mutation of a gene in a cell causes the cell to lose its ability to produce β-galactosidase, this gene must be involved in β-galactosidase production. Mutagenesis can be done randomly by treating cells with physical or chemical agents such as UV or mitomycin C that damage DNA. Cells are then screened for the desired mutation. This can be a very labor intensive process. The discovery of restriction enzymes enabled mutagenesis to be performed at specific sites, greatly simplifying the process. To perform site-directed mutagenesis, DNA is first cut with a certain restriction enzyme, and mutations are then created at the cleavage site.

15.2: Site-directed Mutagenesis

Mutations that can be created at a specific site include frame shift, deletion, insertion, and single-base changes. To produce a frame-shift mutation, the DNA fragment containing the gene is cleaved with a certain restriction enzyme such as BamHI, which creates 5' overhangs at the resulting ends (Fig. 15-1). The overhangs are then filled in with dNTPs using an appropriate DNA polymerase such as the Klenow fragment, and the filled ends are ligated together. This process inserts four extra bases (GATC) to the DNA at the BamHI site and changes the translational frame of the gene, leading to production of a truncated or nonfunctional protein. If two BamHI sites are present, the gene can be mutated by deleting the region located between the two BamHI sites. This can be done by cutting the DNA with BamHI and ligating the two BamHI sites together. If the gene contains a single BamHI site, it can be mutated by inserting an unrelated DNA fragment into the BamHI site.

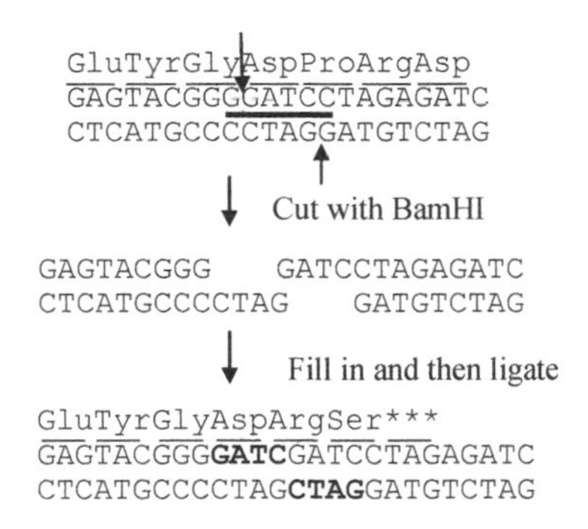

Fig. 15-1: An example of frame shift mutation.

15.3: Single-base Mutation

An example of an application of single-base mutagenesis is shown in Fig. 15-2. Two open reading frames designated ORF1 and ORF2 are found on the DNA fragment shown. These two ORFs overlap but are in opposite orientations. ORF1 is encoded by the upper strand, and ORF2 is encoded by the lower strand of the DNA. To determine the function of ORF1, a mutation in ORF1 needs to be made. Since ORF1 and ORF2 overlap, all the methods mentioned above will mutate both ORF1 and ORF2. The only way to change ORF1 without affecting the function of ORF2 is through single-base mutagenesis. In this example, the sequence of ORF1 is 5'-CGT-TGG-CAG-CTC-TCC-3' which codes for Arg-Trp-Glu-Leu-Ser. Its complement (ORF2) 5'-G-GAG-AGC-TGC-CAA-CG-3' encodes Glu-Ser-Cys-Glu. If the 10th nucleotide of ORF2 is changed from C to T, its third codon will be changed from TGC to TGT. Since both TGC and TGA code for the same amino acid, cysteine, this change does not alter the coding potential of ORF2. However, this C to T change converts the TGG codon of ORF1 to TGA which is a stop codon; therefore, protein synthesis derived from ORF1 will terminate at the newly created stop codon, resulting in the production of a truncated protein.

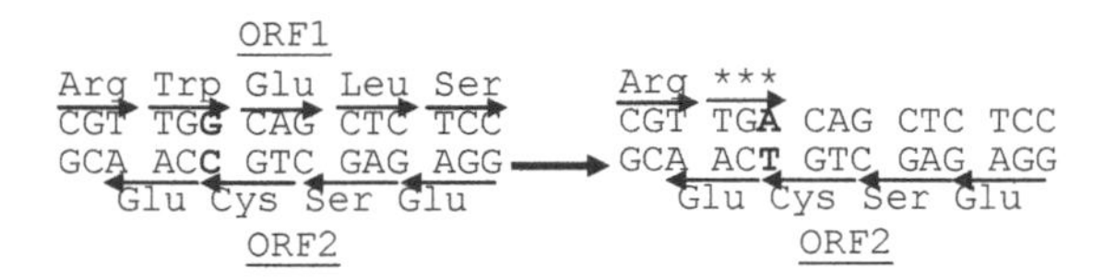

Fig. 15-2: Application of single-base mutation.

The easiest way to mutate a single base at a specific site is through PCR as shown in Fig. 15-3 in which the second T in the sequence ---GATCGTAGAC ---, is to be changed to G, represented by #. The basic strategy is to build the desired mutation in PCR primers that are then used to amplify the DNA containing the gene to be mutated into two pieces. When these two PCR products are joined together, the newly generated DNA will have the desired mutation. The mutation is built into the sequence of the 5' primer to be used to amplify the 3' portion of the gene and the 3' primer to be used to amplify the 5' portion of the gene. These primers are called mutagenic primers. Each primer is approximately 30 bases long with the mutated base located in the middle, flanked by approximately 15 bases on each side. In this example, the sequence of the relevant portion of the 5' primer to be used to amplify the 3' portion of the gene is --- GATCGGAGAC ---, and that of the 3' primer to be used to amplify the 5' portion of the gene is --- CTAGCCTCTG ---. When these primers are paired with other primers to amplify the gene, the two PCR products generated will have identical sequences at the 3' end of the 5' portion of the gene and the 5' end of the 3' portion of the gene. When these two PCR products are denatured, the upper strand of the 5' portion of the gene will anneal to the lower strand of the 3' portion of the gene. If the two single-stranded regions of the hybrid are filled in, a DNA fragment containing the desired mutation is created (Fig. 15-3).

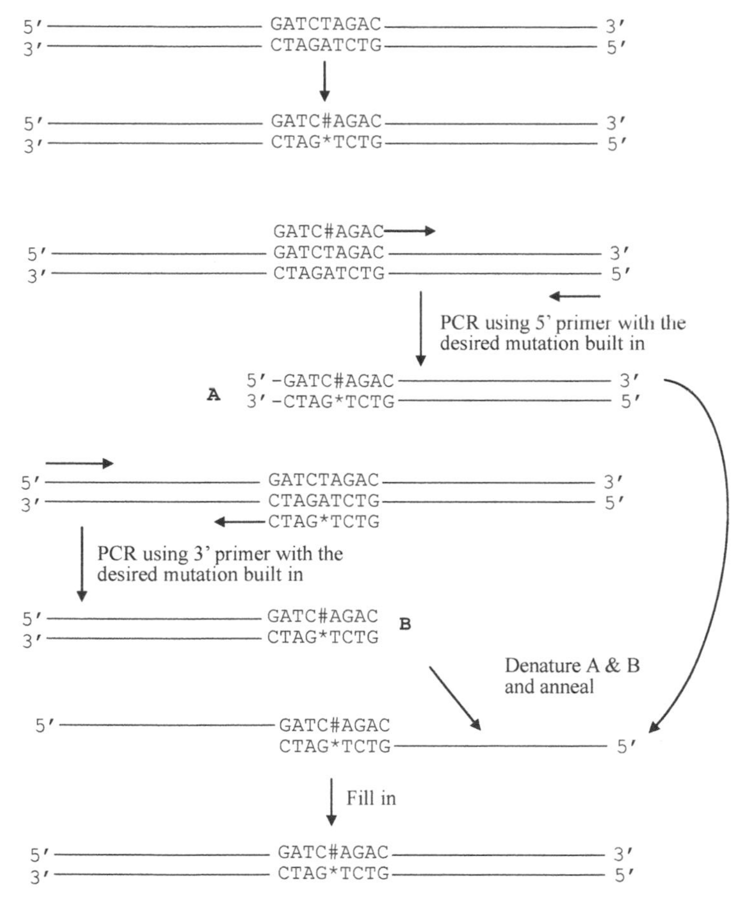

Fig. 15-3: Single-base mutation by PCR.

15.4: Insertion and Deletion by PCR

PCR can also be used to mutate a gene by insertion or deletion. This strategy is identical to that of single-base mutagenesis. The extra bases to be inserted are built in the 5' primer used to amplify the 3' portion of the gene and the 3' primer used to amplify the 5' portion of the gene. As shown in Fig. 15-4, the sequence GTAGC is to be

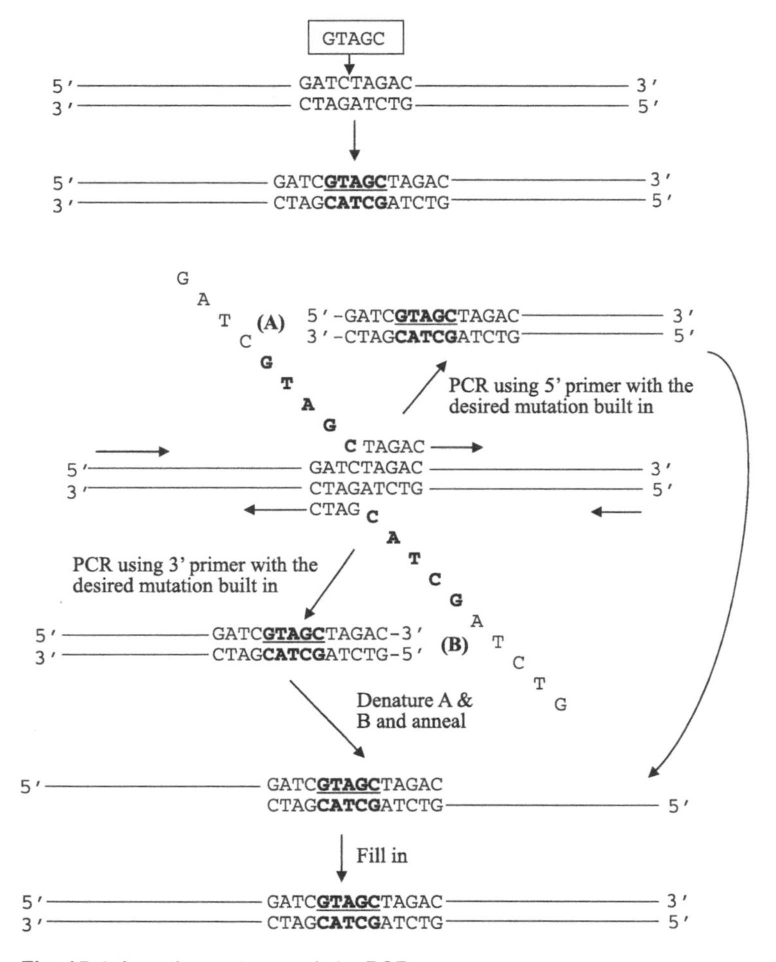

Fig. 15-4: Insertion mutagenesis by PCR.

inserted between C and T of the sequence --- GATCTAGAC ---. In this case, the sequence of the 5' primer for amplification of the 3' portion of the gene is ---GATCGTAGCTAGAC ---; where the boxed bases are those to be inserted. The number of bases flanking the boxed bases should be approximately 20 on each side. The sequence of the 3' primer is --- GTCTAGCTACGATC--- with the boxed bases also flanked by approximately 20 bases of the target sequences on each side. When these primers are paired with their respective primers, two PCR products are produced. Since the sequence of the 3' end of the 5' portion of the PCR product and that of the 5' end of the 3' portion of the PCR product are identical, the upper strand of the 5' portion of the gene and the lower strand of the 3' portion of the gene will anneal to each other when the two PCR products are denatured and mixed together. If the two single-stranded regions are filled in, a DNA fragment with the inserted sequence is generated.

To delete a portion of a gene using PCR, the sequence of the 5' primer used to amplify the 3' portion of the gene is made to be identical to the sequence flanking the region to be deleted on the upper strand. The sequence of the 3' primer used to amplify the 5' portion of the gene is identical to the sequence flanking the region to be deleted on the lower strand (Fig. 14-5). All other steps in this process are the same as those described above.

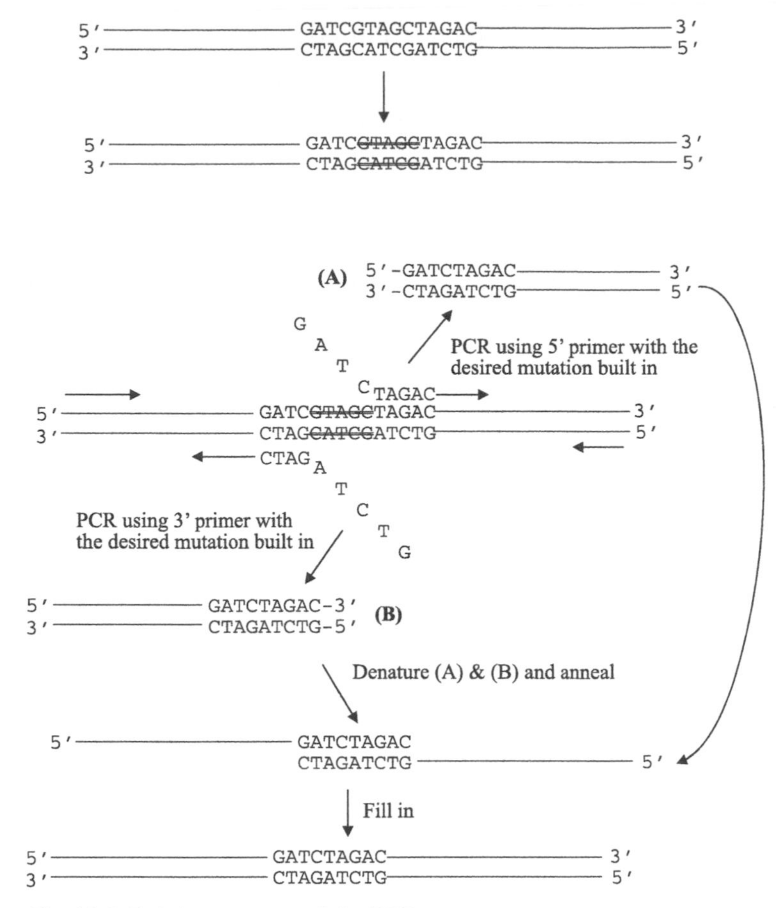

Fig. 15-5: Deletion mutagenesis by PCR.

15.5: Advanced Methods for PCR Mutagenesis

The PCR methods described above amplify the gene to be mutated in two pieces using mutagenic primers. The two PCR products are subsequently joined. This is necessary because most DNA polymerases such as Taq and Pfu DNA polymerases used for PCR can only amplify DNA fragments less than 5 kb. In recent years, there have been new enzymes discovered or old enzymes modified that can amplify DNA fragments as big as 20 kb. Because of these improvements, methods for PCR mutagenesis have also changed. Newer methods enable an entire plasmid carrying the gene to be amplified with mutagenic primers. One example of such is the GeneTailor method developed by Invitrogen, Inc. This method can amplify DNA fragments up to 8 kb using Pfx DNA polymerase isolated from *Pyrococcus kodakaraensis*. Similar to Pfu DNA polumerase, Pfx DNA polymerase has proofreading capabilities. It is also much more efficient than Pfu DNA polymerase in DNA synthesis. The high fidelity Platinum Taq DNA polymerase can also be used. It is actually a mixture of Taq DNA polymerase and *Pyrococcus* GB-D thermostable DNA polymerase. In this system, the errors created by Taq DNA polymerase are corrected by the *Pyrococcus* GB-D thermostable DNA polymerase.

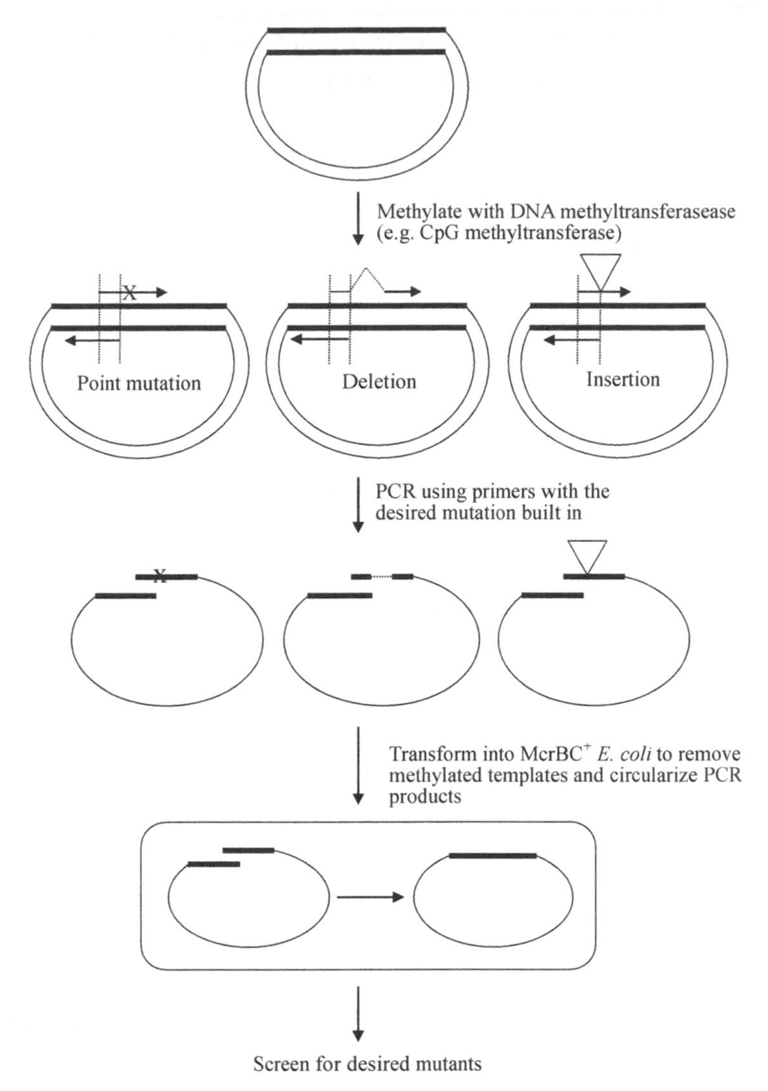

Fig. 15-6: GeneTailor PCR mutagenesis method.

In the GeneTailor method (Fig. 15-6), the plasmid DNA containing the gene to be mutated is first methylated with a cytosine methyltransferase such as CpG methyltransferase to methylate the majority of C residues on the plasmid. Two primers called "Forward" and "Reverse" primers with the desired mutation built in are used. The Forward primer is composed of three parts (Fig. 15-7). Its 5' portion consists of 15-20 bases that are required for the primer to anneal to the template. Adjacent to this portion is the sequence of the desired mutation. The mutation can be a single-base change, insertion, or deletion. To create a single-base change, the sequence in this region is that of desired mutation. To create an insertion, the sequence in this region is the sequence of bases to be inserted; up to 21 bases can be inserted in this method. To create a deletion, no extra bases are placed in this region. This region is followed by 15 bases that can anneal to the template.

The Reverse primer has two parts. Its 5' portion consists of 15-20 bases that are complementary to the 5' portion of the Forward primer, and the 3' portion consists of approximately 15 bases required for the primer to anneal to the template. The Forward primer anneals to the lower strand of the template, and the Reverse primer anneals to the upper strand of the template. The PCR product thus generated is the entire plasmid with the desired mutation and the additional 15-20 bases at each end of the PCR product. When this DNA is introduced into *E. coli* cells, the two ends will be joined to form a covalently closed circular DNA. Since the PCR mixture contains both the newly synthesized PCR products and the original plasmid template which contains the wild type gene, a method is needed to remove the original template. This is achieved by introducing the mixture into *E. coli* cells that are McrBC+, such as DH5α. In this type of cells, the original template will be degraded by the McrBC restriction enzyme because it is methylated. Therefore, only the PCR products with the desired mutation will remain and replicate.

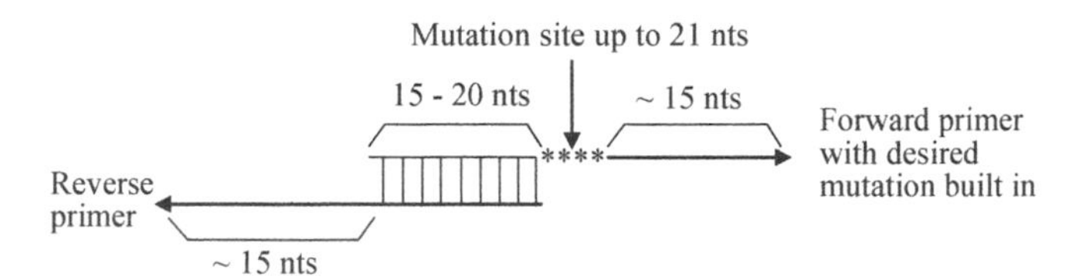

Fig. 15-7: Primers used for GeneTailor PCR mutagenesis.

The QuickChange mutagenesis kit from Strategene, Inc. works similarly to the GeneTailor kit from Invitrogen, Inc. One difference is that the two primers used, 25 – 45 bases long, are completely complementary to each other. Therefore, the ends of the PCR products have 25 – 45 base pairs that are identical. The other difference is that the template DNA is made in Dam+ *E. coli*. Therefore, the A residue in every GATC sequence is methylated. After PCR, the PCR mixture is digested with DpnI which recognizes GATC and cuts the DNA only when the A residue of the sequence is methylated. Thus, only the PCR products containing the desired mutation remain. The DNA polymerase used in this system is PfuUltraII which is a modified Pfu DNA polymerase with a higher template binding ability and is capable of amplifying DNA fragments as large as 20 kb as described in Chapter 13.

15.6: Homologous Recombination

All the methods described above only allow mutations to be created on DNA molecules. To mutate a gene within a cell, homologous recombination is used. Homologous recombination is the process by which DNA molecules with similar sequences within a cell exchange places. Homologous recombination can take place in both prokaryotic and eukaryotic cells. An example of homologous recombination is chromatid exchange between maternal and paternal chromosomes or sister chromatid exchange on the same chromosome. There are 23 pairs of chromosomes in human cells. Each pair consists of one paternal and one maternal chromosome. Although the nucleotide sequences of these two chromosomes are not identical, they are very similar and therefore can exchange portions of the chromosomes. The process of homologous recombination is shown in Fig. 15-8. A nick is created on one strand of each of the two homologous DNA molecules represented by thick and thin lines, respectively. The broken end of the thin-lined DNA then invades the thick-lined DNA, and that of the thick-lined DNA invades the thin-lined DNA. If the DNA is cut at the cross junction, a portion of the thin-lined DNA is integrated into the thick-lined DNA molecule, and a portion of the thick-lined DNA is integrated into the thin-lined DNA molecule. Since the sequences of the thin- and thick-lined DNAs are not identical, the resulting recombinant DNA contains a portion with mismatched bases. DNA mismatch repairs will then take place. In the thin-lined DNA molecule, if the integrated thick-lined DNA is used as the template to repair the other strand, a double-stranded thin-lined DNA molecule containing a portion of the thick-lined DNA is generated. The same process also occurs for the thick-lined DNA molecule containing a portion of the thin-lined DNA. Therefore, the sequence of the recombinant DNA produced is distinct from both the original thin- and thick-lined DNA molecules. The same set of chromosomes may undergo homologous recombination at different places in different cells. This is why siblings, except identical twins, have different personalities and do not look the same even though they inherit the same chromosomes from the same parents.

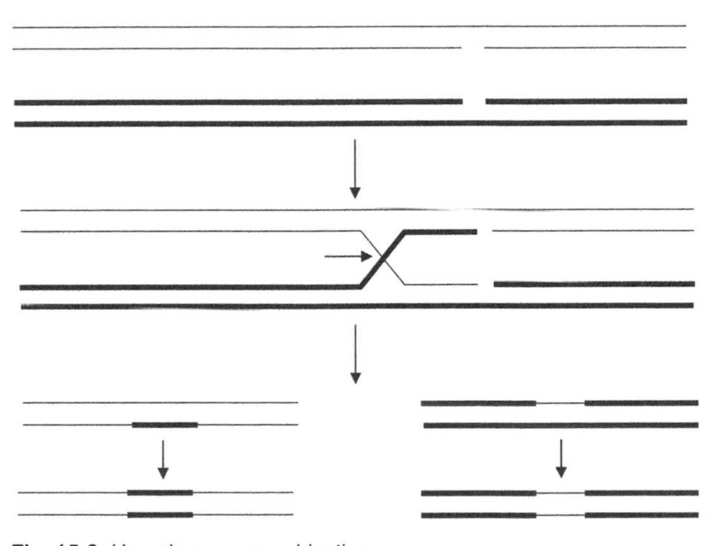

Fig. 15-8: Homologous recombination.

15.7: Insertion Mutagenesis by Homologous Recombination

To mutate a gene within a cell by insertion mutagenesis, the gene to be mutated is first cloned into a plasmid, and an unrelated DNA fragment is then inserted into the gene located on the plasmid (Fig. 15-9). This recombinant plasmid is called targeting plasmid since it is used to target the gene to be mutated. When the targeting plasmid is introduced into cells, homologous recombination will take place between the sequences on the plasmid and those on the chromosome. As the result, the wild type gene on the chromosome is replaced by the gene contained in the insert on the plasmid. Thus, the gene on the chromosome is mutated.

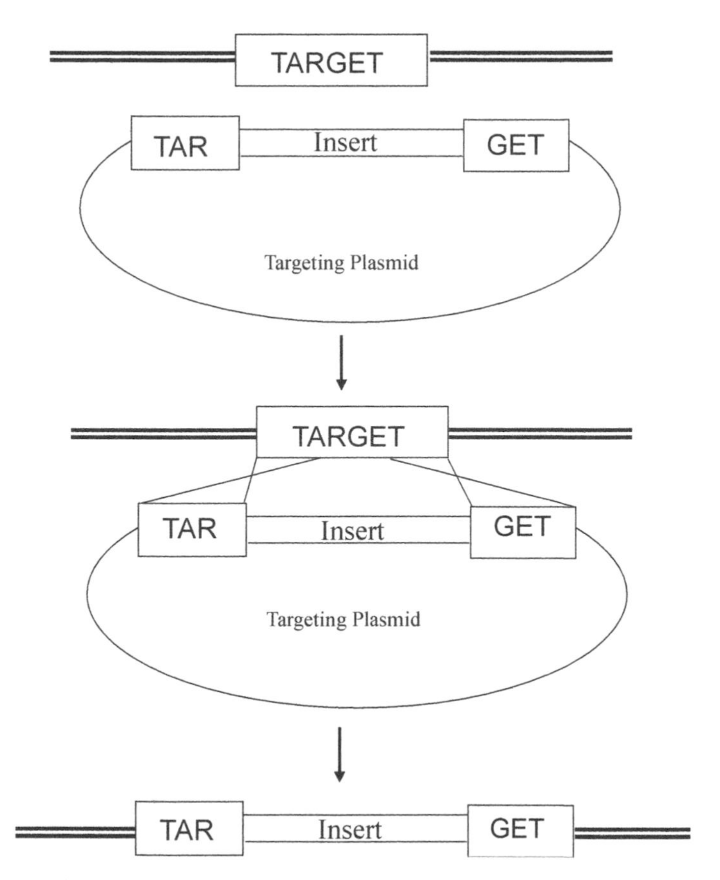

Fig. 15-9: Insertion mutagenesis by homologous recombination.

To mutate a gene within a cell, the targeting plasmid must be introduced into the cell. Since not all cells will take up the targeting plasmid, a method that allows selection of cells containing the targeting plasmid is needed. To achieve this, a DNA fragment containing the neomycin-resistance gene is used to perform the insertion mutagenesis of the gene on the plasmid (Fig. 15-10). Therefore, cells that have taken up the targeting plasmid will become resistant to G418 and can be selected by growing the cells in culture medium containing G418 which will kill those that did not take up the targeting plasmid.

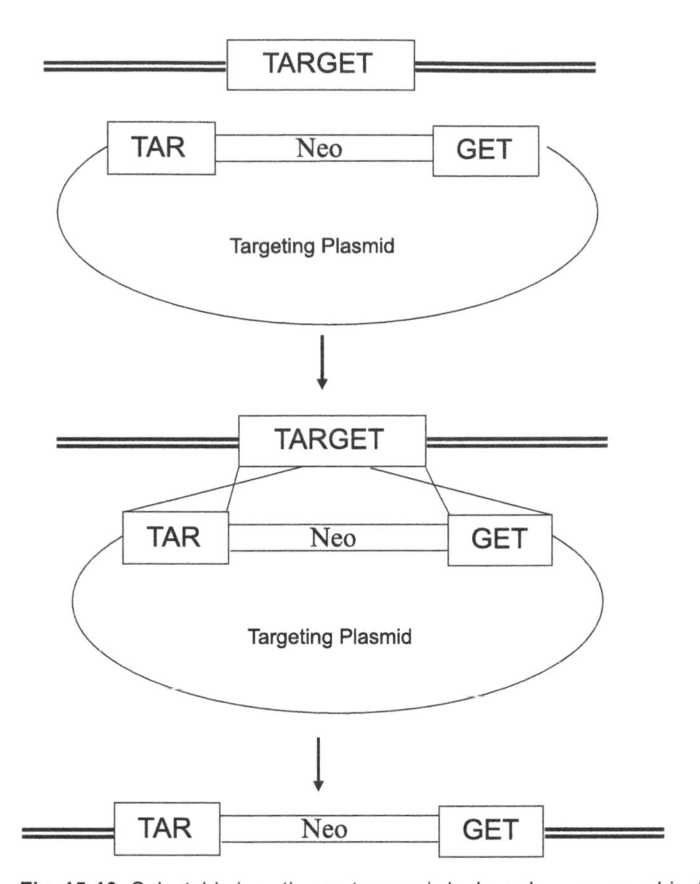

Fig. 15-10: Selectable insertion mutagenesis by homologous recombination.

15.8: Single and Double Crossing Over

The homologous recombination process described above takes place between the two sequences flanking the inserted DNA on the targeting plasmid and the homologous sequences on the chromosome. This is called double crossing over. However, homologous recombination can also occur at only one side, and thus is called single crossing over. When single crossing over occurs, the entire targeting plasmid is integrated into the chromosome (Fig. 15-11). This integration

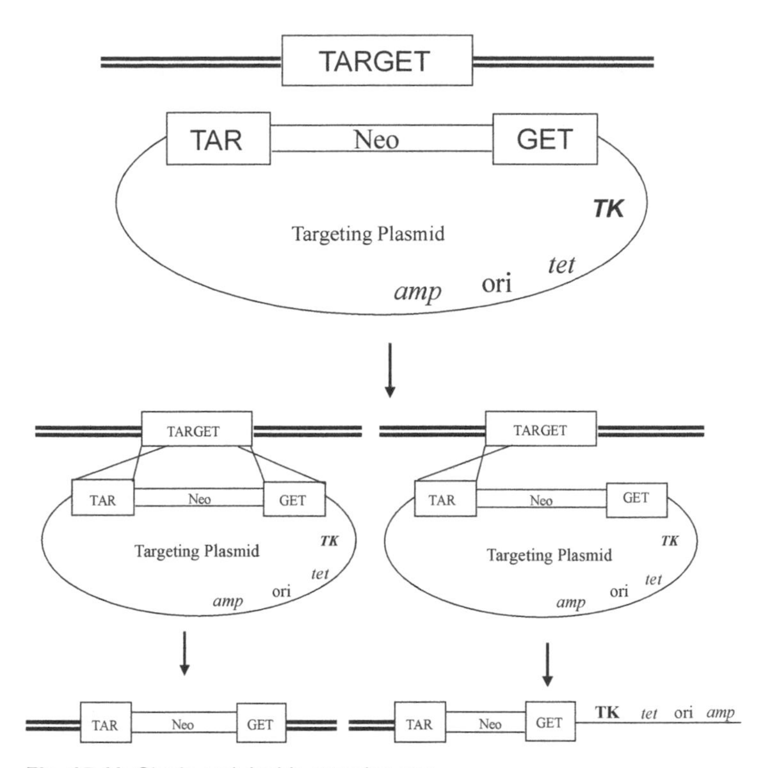

Fig. 15-11: Single and double crossing over.

may affect the function of genes located around the insertion site or even the entire chromosome. Therefore, double crossing over is desired during homologous recombination. Since it is not possible to force homologous recombination to occur by double crossing over, the process of double crossing over recombination is selected for by using a vector containing the thymidine kinase (TK) gene from herpes simplex virus type 1 (HSV-1) (Fig. 15-11). Thymidine kinase can phosphorylate ganciclovir causing it to become a toxic compound. If the homologous recombination takes place by single crossing over, the entire targeting plasmid including the TK gene is integrated into the chromosome. Therefore, the cells will express thymidine kinase and can be killed by ganciclovir. In contrast, the TK gene will not be integrated into the chromosome if the homologous recombination takes place by double crossing over, and the cells will be resistant to ganciclovir.

15.9: Deletion Mutagenesis by Homologous Recombination

Homologous recombination can also be used to mutate a gene by deletion. The procedure is similar to insertion mutagenesis described above. A DNA fragment containing the neomycin-resistance gene is first inserted between two *loxP* (locus of crossing over of P1 bacteriophage) sites on a vector containing the TK gene (Fig. 15-12). A DNA fragment of approximately 2 kb containing the 5' portion of the gene to be mutated is then inserted at the left hand side of the left *loxP* site. Another DNA fragment also approximately 2 kb containing the 3' portion of the gene to be mutated is inserted at the right hand side of the right *loxP* site. The targeting plasmid thus constructed is introduced into cells that can produce the Cre (cyclization recombinase) protein of P1 bacteriophage. The Tet-on system is usually used to control the expression of Cre in these cells. Cells that have taken up the targeting plasmid are selected by growing the transfected cells in culture medium containing G418. G418 resistant cells are then cultured in the presence of ganciclovir to select those in which the homologous recombination has taken place by double crossing over. This homologous recombination will replace the wild type gene on the chromosome with the neomycin-resistance gene flanked by two *loxP* sites on the targeting plasmid. The integrated 1.2-kb neomycin gene is then removed by adding doxycycline to the culture medium of cells to turn on the production of the Cre protein which mediates a site-specific recombination between the two *loxP* sites, thus deleting the region between these two *loxP* sites. Therefore, the majority of the target gene is deleted and replaced by one *loxP* site which is only 34 bp long.

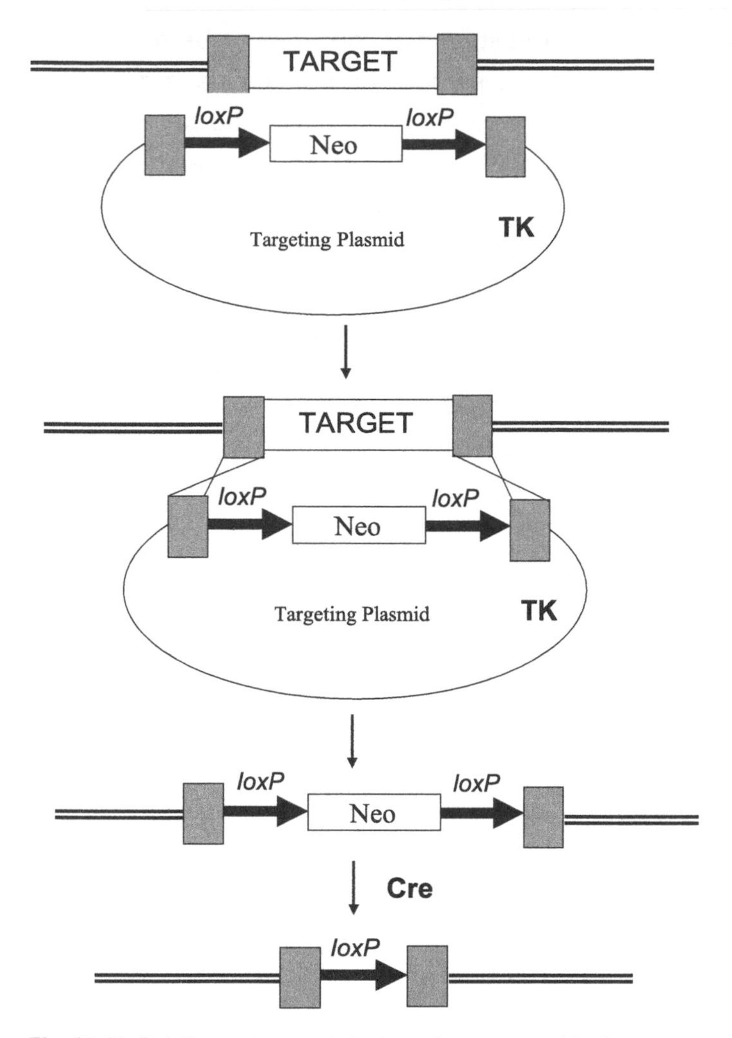

Fig. 15-12: Deletion mutagenesis by homologous recombination.

15.10: Gene Replacement by Homologous Recombination

Replacement of a defective (bad) gene in a cell with a good gene can also be achieved through homologous recombination. A DNA fragment containing the good gene is first inserted in front of the neomycin resistance gene flanked by two *loxP* sites on a vector containing the TK gene. A DNA fragment of approximately 2 kb containing the region upstream of the gene to be replaced is then inserted at the 5' side of the good gene on the plasmid. Another DNA fragment also approximately 2 kb containing the region downstream of the gene to be replaced is inserted at the right hand side of the right *loxP* site. The targeting plasmid is then introduced into cells in which the production of the Cre protein is controlled by the Tet-on system. The transfected cells are grown in culture medium containing G418 and ganciclovir. G418 and ganciclovir resistant cells are then cultured in a medium containing doxycycline to turn on the production of Cre to remove the neomycin resistance gene. The resulting product is a defective gene replaced by a good gene linked to a copy of the *loxP* site (Fig. 15-13).

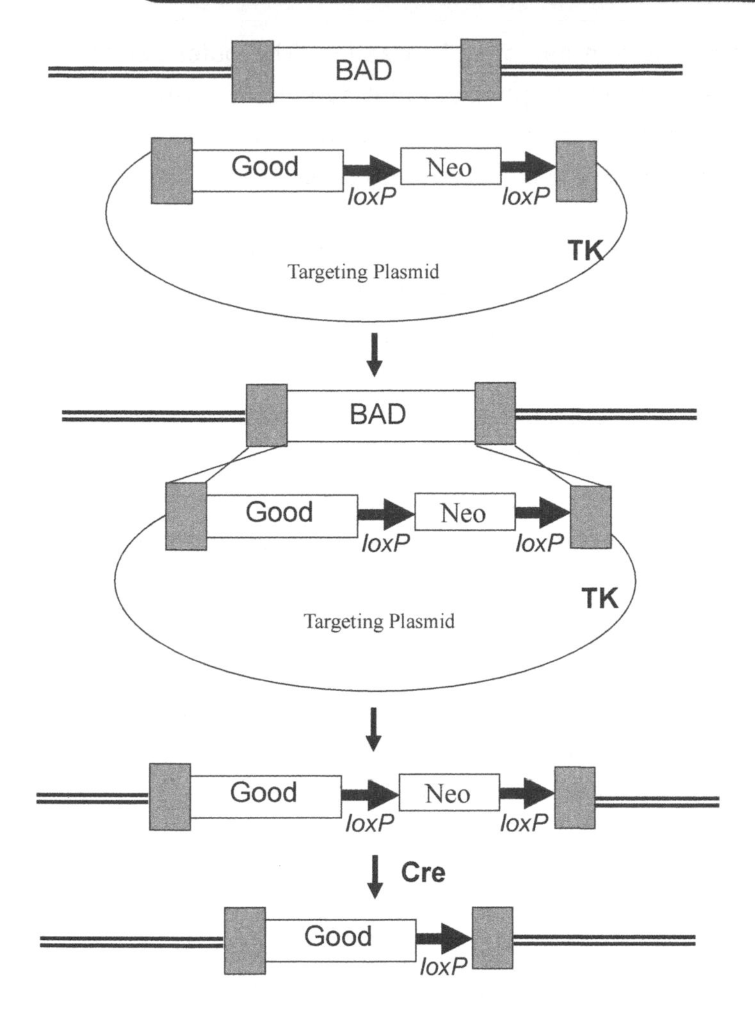

Fig. 15-13: Gene replacement by homologous recombination.

15.11: Transgenic Animals

Transgenic animals are animals with a transgene deliberately introduced into their genome. They can be used to study the function of a certain gene or to produce a certain protein for therapeutic or investigative purposes. Creation of transgenic animals mostly starts with manipulating the transgene in embryonic stem (ES) cells that are evolved from fertilized eggs. A fertilized mouse egg starts to divide within 12 to 15 hours, and a ball-like structure called a blastocyst is formed by the fifth day after fertilization. The outer cell layer of the blastocyst is called trophoblast that eventually forms the placenta. Inside the blastocyst is a fluid-filled cavity with a cell mass called embryoblast which is the source of ES cells.

ES cells can be grown in culture without loosing their potential to produce all the cells of a mature animal. The first step in the creation of a transgenic animal is to introduce a targeting plasmid into ES cells to insert (knock in) or delete (knock out) a gene in the genome of the cell by homologous recombination. Transfected ES cells are then grown in a culture medium containing G418 which selects for cells that have taken up the targeting plasmid and then in culture medium containing ganciclovir to select for homologous recombination via double crossing over. The selected cells are grown and analyzed to ensure that they contain the desired transgene in the correct genomic location. These cells are then microinjected into a blastocyst isolated from a pregnant mouse. The blastocyst containing the injected ES cells is then implanted into the uterus of a pseudopregnant mouse (surrogate mother) that is created by mating a female mouse with a vasectomized male mouse. This mating stimulates the required hormonal changes that make the uterus receptive for implantation.

If the implantation is successful, some of the pups will have the desired transgene. This can be confirmed by clipping a small piece of tissue from the tail of the pup and examining its DNA. At this stage, mice with the desired mutation are usually heterozygous. These mice are then mated with other heterozygous mice to obtain offspring with homozygous transgene. To allow easier identification of transgenic mice, different strains of mice are used. For example, the ES cells are from a mouse with black fur, and the blastocyst is from a mouse with brown fur. The embryo derived from the blastocyst is then implanted into a white-furred surrogate mother mouse. Chimeric pups produced with patchy black fur in the brown-fur background would have the transgene.

If an essential gene is knocked out, the transgenic animal will not survive. To overcome this problem, conditional knockout is performed. The most commonly used technique for conditional knockout is the Cre/*loxP* system described above. The transgene is placed between two *loxP* sites that are in a direct repeat orientation. The animal containing this *loxP*-flanked transgene is mated with another transgenic animal containing the Cre gene. This Cre gene is controlled by an inducible system such as the CreERT2 system in which the Cre gene is fused to a mutated estrogen receptor (ER) gene. The CreERT2 protein cannot bind the endogenous estrogen and thus remains in the cytoplasm. However, the CreERT2 protein can bind tamoxifen which is an estrogen competitor. When the CreERT2 protein is bound by tamoxifen, it is translocated into the nucleus where the Cre portion of the CreERT2 fusion protein cleaves the transgene located between the two *loxP* sites. Therefore, the transgene is knocked out only when the transgenic animals are treated with tamoxifen.

In addition to using ES cells, transgenic animals can also be produced by injecting the targeting plasmid into a pronucleus of a fertilized egg before the two pronuclei from sperm and egg are fused to form the diploid zygote nucleus. The targeting plasmid is usually injected into the male pronucleus because it is bigger. The injected egg is allowed to form a zygote and divide to a two-cell embryo which is then implanted into the uterus of a surrogate mother as described above.

Transgenic animals can be used to produce certain therapeutic proteins such as alpha-1-antitrypsin and antithrombin. They can also be used as model animals to study a certain disease. A good example of this is the use of transgenic mice to study polio virus infections. Normal mice cannot be infected with the polio virus because they lack the necessary receptors. With the transgenic technology, a human polio virus receptor gene can be knocked in, creating transgenic mice that are susceptible to polio virus infection. A major application for transgenic animals is for elucidating the function of a gene by knocking out the gene and the determining what function is lost.

15.12: RNA Interference

Creating knockout transgenic animals is a very labor intensive, time-consuming, and expensive process. The discovery that small double-stranded RNA (dsRNA) can be used to partially suppress (knock down) gene expression has revolutionized the gene knockout technology. Gene expression affected by RNA is called RNA interference (RNAi) which was first described by Napoli et al. (1990). They introduced a pigment-producing gene into petunia cells and found that the expression of both the introduced gene and the homologous endogenous gene was suppressed. RNA interference was subsequently observed in almost all eukaryotes. In 1998, Fire et al. reported that injection of dsRNA corresponding to a portion of the target mRNA into gonads of the nematode *Caenorhabditis elegans* silenced the target gene. This finding became the foundation for the commonly used technique of gene knock down by means of small interfering RNA (siRNA).

Small interfering RNA is one of several types of small RNAs known to exist. These small RNAs may affect gene expression by inhibiting translation or altering chromatin structure and thus suppressing transcription. They are believed to play a major role in the defense against viral infections in plants. In humans and other animals, they have been shown to be involved in regulating developmental timing, hematopoietic cell differentiation, cell death, cell proliferation, and oncogenesis. Repeat-associated siRNAs (rasiRNAs) and micro RNAs (miRNAs) are two well-studied small RNAs with regulatory functions; rasiRNAs have been shown to inhibit transcription by remodeling the heterochromatin through histone 3 (Lys9) methylation. MicroRNAs (miRNAs) are synthesized in the nucleus of a cell by RNA polymerase II as long (up to 1000 nt) RNA, called primary miRNAs (pri-miRNAs) that are m7G capped and polyadenylated like mRNAs. The pri-miRNAs are then processed by a dsRNA-specific endonuclease, Drosha, in conjunction with a dsRNA-binding protein, called Pasha in *Drosophila* and DGCR8 (DiGeorge syndrome critical region gene 8) in humans, to become hairpin RNAs of 70–100 nt in length, called pre-miRNAs.

Pre-miRNAs are then transported to the cytoplasm via an Exportin-5 dependent mechanism and processed by the enzyme Dicer, which is a member of the RNase III family, in conjunction with a dsRNA-binding protein, Loqs (Loquacious) in *Drosophila* and TRBP (HIV TAR RNA-binding protein) in humans, to become mature, double-stranded miRNAs of 21 -23 nucleotides in length with dinucleotide 3' overhangs. These small double-stranded miRNAs are then converted to single-stranded. The strand complementary to the target mRNA is annealed to the target mRNA forming an RNA-induced silencing complex (RISC) in which the Argonaute protein degrades the target mRNA. The loading of siRNAs to RISC is mediated by the R2D2 protein which was first found in *Drosophila*. The R2D2 protein contains two dsRNA-binding domains in tandem (R2) and forms a heterodimeric complex with one of the Dicer proteins, Dcr-2 (D2). If the siRNAs or miRNAs do not pair perfectly with their targets, they do not cause target mRNA degradation. Instead, they block translation by causing degradation of the polypeptide as it emerges from the ribosome or by stalling the movement of ribosome after initiation. It has been estimated that miRNA genes represent 2-3% of all human genes and regulate as many as 30% of human genes.

Functional miRNA for gene silencing is dsRNA of 21-23 nucleotides in length. This is the basis of the current RNAi technology. To knock down the expression of a gene, a 21-base dsRNA molecule with dinucleotide 3' overhangs containing a sequence corresponding to the target mRNA is chemically synthesized and then introduced into target cells by transfection to silence the target gene. As the control, scrambled siRNAs that have the same nucleotide composition as the gene specific siRNAs but are incapable of targeting intracellular mRNA are used. Another approach for RNA silencing is to clone the gene encoding the specific siRNA into a cloning vector. The siRNA gene to be cloned is chemically synthesized as 19 bases of inverted repeats. The two repeats are separated by a spacer of approximately 10 bases and are linked to a polythymidine tract at the 3' end in order to terminate transcription by RNA polymerase III. When this recombinant vector is introduced into target cells, dsRNAs with short hairpin structures are produced. These dsRNAs are called shRNA (short hairpin RNA). They are then processed in the same manner as pri-mRNAs in the cell to become functional siRNAs. Two commonly used RNA polymerase III promoters in the production of shRNAs are the human U6 snRNA and H1 RNA gene (a component of RNase P) promoters. In addition to studying the function of genes, siRNAs have the potential to function as therapeutic agents. Many siRNA-based drugs targeting various viruses such as hepatitis C, HIV, and respiratory syncytial virus as well as several neurodegenerative diseases are being developed.

Summary

The function of an unknown gene can be determined by mutating the gene, and the resulting loss of function of a cell can be attributed to the mutated gene. The process of mutating a gene is referred to as mutagenesis. To mutate a gene, cells can be treated with certain physical or chemical agents to damage DNA and then screened for the desired mutation. This type of mutagenesis is called random mutagenesis. A more commonly used method is site-directed mutagenesis in which a mutation is created at a specific site. Mutations that are commonly created include frame shift, deletion, insertion, and single-base changes. To create a frame-shift mutation, the DNA containing the gene to be mutated is cut with a certain restriction enzyme such as BamHI which generates 5' overhang ends. When the overhangs are filled, four extra bases are inserted into the BamHI site, and the translational frame of the gene is changed. Insertion of a DNA fragment into the BamHI site located in the gene will also destroy the gene. If the gene contains two BamHI sites, the gene can be cut with BamHI and religated. This process will result in deletion of a portion of the gene. The most commonly used method for single-base mutagenesis is PCR. The sequence of the desired mutation is built into PCR primers to generate PCR products with the desired mutation. PCR can also be used for insertion or deletion mutagenesis. To perform insertion mutagenesis, the sequence to be inserted is built in the PCR primers. For deletion mutagenesis, the sequences of the primers are the sequences flanking the region to be deleted.

Techniques based on the principle of homologous recombination are used to mutate a gene in a cell. For insertion mutagenesis, an unrelated DNA is inserted into the gene which has been cloned into a plasmid. The resulting targeting plasmid is introduced into cells. Since the targeting plasmid contains sequences that are also present in chromosomes in the cell, homologous recombination will occur between these sequences resulting in replacement of the gene on the chromosome with the mutated gene on the targeting plasmid. To select the cells that have taken up the targeting plasmid, the gene on the plasmid is disrupted by inserting a DNA fragment containing the neomycin resistance gene. Cells that do not take up the targeting plasmid are killed when they are cultured in medium containing G418, which is an analogue of gentamycin B1.

The homologous recombination that takes place at both regions flanking the target gene is called double crossing over which is the desired recombination event. However, homologous recombination can also take place at only one region. In this single crossing over event, the entire targeting plasmid is integrated into the chromosome, which may affect the function of genes near the integration site of even the entire chromosome. To select for homologous recombination events that occur by double crossing over, a vector containing the TK gene of HSV-1 is used to construct the targeting plasmid. If single crossing over occurs, the cells containing the integrated plasmid can be eliminated by ganciclovir because these cells will produce thymidine kinase which converts ganciclovir to a toxic compound.

To delete a gene by homologous recombination, a targeting plasmid is constructed such that the neomycin resistance gene is flanked by two *loxP* sites in a direct repeat orientation. A DNA fragment from the 5' portion of the gene is placed upstream of the left *loxP* site, and another DNA fragment from the 3' portion of the gene is placed downstream of the right *loxP* site. When the targeting plasmid is introduced into cells, the neomycin resistance gene flanked by the two *loxP* sites will replace the target gene on the chromosome as the result of homologous recombination. If the cells can produce the Cre protein, the integrated neomycin resistance gene will be deleted as the result of site-specific recombination between the two *loxP* sites. Therefore, the target gene on the chromosome is deleted and replaced by a 34-bp *loxP* sequence.

To replace a defective gene in cells with a good one, the good gene is first cloned into a targeting plasmid in front of the neomycin resistance gene which is located between two *loxP* sites. A DNA fragment from the region upstream of the 5' end of the gene to be replaced is then inserted at the 5' side of the good gene on the plasmid, and another DNA fragment from the region downstream of the gene to be replaced is inserted at the 3' side of the right *loxP* site. When this targeting plasmid is introduced into cells, recombination will take place between the two homologous sequences on the targeting plasmid and those on the chromosome, resulting in replacement of the bad gene with the good one. The neomycin resistance gene which is integrated together with the good gene into the chromosome is then removed by inducing the production of the Cre protein in the cell.

Homologous recombination is also the basic technique used to create transgenic animals. A targeting plasmid is first introduced into embryonic stem (ES) cells or a pronucleus of a fertilized egg. ES cells containing the desired transgene are then injected into a blastocyst which is then implanted in the uterus of a pseudopregnant mouse. When pups are born, the ones with the desired heterozygous transgene are mated with other heterozygous transgenic animals to obtain animals homozygous for the transgene. If a fertilized egg is used, the egg containing the targeting plasmid is allowed to divide into a two-cell embryo and then implanted into the uterus of a surrogate mother. Conditional knockout is performed if the gene to be knocked out is essential. The transgene is placed between two *loxP* sites. The resulting transgenic animal is mated with another transgenic animal containing an inducible Cre gene. When the new transgenic animal is induced to produce the Cre protein, the transgene is deleted.

Since transgenic animals are very expensive to create, gene knock down using siRNA (small interfering RNA) is becoming the method of choice for investigating the function of a certain gene. Double-stranded RNAs with sequence corresponding to the gene to be targeted are chemically synthesized. These dsRNAs are typically 21 to 23 nucleotides long with dinucleotide 3' overhangs and are introduced into target cells by transfection. Inside the cell, siRNAs are converted to single-stranded, and the strand complementary to the target mRNA anneals to the target mRNA forming the RNA-induced silencing complex (RISC) which degrades the mRNA. SiRNAs may also be derived from miRNAs (micro RNAs) that are transcribed from micro RNA genes. The length of miRNAs may range from 200 to 1000 nucleotides. Micro RNAs are capped and polyadenylated like mRNAs. They are processed in the nucleus to become short hairpin RNAs (shRNAs) and further processed in the cytoplasm to become siRNAs of 21 to 23 bases long with dinucleotide 3' overhangs. If the sequence of an siRNA is perfectly complementary to the target mRNA, the siRNA usually causes degradation of the target mRNA. In contrast, it causes nascent protein degradation or interferes with protein synthesis if its sequence is not perfectly complementary to that of the target mRNA. Another type of small RNA is called repeat-associated siRNA (rasiRNA) which is transcribed from repetitive sequences in the genome. RasiRNA has been shown to alter chromatin structure and inhibit transcription of genes.

Sample Questions

1. The basic technique for gene knockout and gene replacement is: a) PCR, b) homologous recombination, c) transposition, d) replication.
2. The *loxP* sequence is a) the packaging signal of P1 phage, b) recognized by P1 Cre protein, c) the packaging signal of lambda phage, d) commonly used in gene knockout.
3. The thymidine kinase of herpes simplex virus phosphorylates a) ganciclovir making it toxic to cells, b) dNTP, c) DNA, d) RNA.
4. Site-directed mutagenesis may be achieved by a) making mutations on DNA randomly with chemical or physical agents, b) making frameshift mutation at a specific site, c) deletion of a portion of a gene, d) insertion of a DNA fragment into a gene.

Suggested Readings

1. Bronson, S. K. and Smithies, O. (1994). Altering mice by homologous recombination using embryonic stem cells. J. Biol. Chem. 269: 27155-27158.
2. Capecchi, M. (1989). Altering the genome by homologous recombination. Science 244: 1288-1292.
3. Elbashir, S. M., Harborth, J., Lendeckel, W., Yalcin, A., Weber, K., Tuschl, T. (2001). Duplexes of 21-nucleotide RNAs mediate RNA interference in cultured mammalian cells. Nature 411:494–498.
4. Fire, A., Xu, S., Montgomery, M. K., Kostas, S. A., Driver, S. E., Mello, C. C. (1998). Potent and specific genetic interference by double-stranded RNA in *Caenorhabditis elegans*. Nature 391: 806–811.
5. Gossen, M. and Bujard, H. (2002). Studying gene function in eukaryotes by conditional gene inactivation. Annu. Rev. Genet. 36: 153-173.
6. Gu, H., Zou, Y.-R. and Rajewsky, K. (1993). Independent control of immunoglobulin switch recombination at individual switch regions evidenced through Cre-*loxP*-mediated gene targeting. Cell 73: 1155-1164.
7. Guo, F., Gopaul, D. N., and Van Duyne, G. D. (1997). Structure of Cre recombinase complexed with DNA in a site-specific recombination synapse. Nature 389: 40-46.
8. Kowalczykowski, S. C. and Eggleston, A. K. (1994). Homologous pairing and DNA strand-exchange proteins. Annu. Rev. Biochem. 63: 991-1043.
9. Lantinga-van leeuwen, I. S., Leonhard, W. N., van de Wal, A., Breuning, M. H., Verbeek, S., de heer, E., and Peters, D. J. M. (2006). Transgenic mice expressing tamoxifen-inducible Cre for somatic gene modification in renal epithelial cells. Genesis 44: 225-232.
10. Napoli, C., Lemieux, C., and Jorgensen, R. (1990). Introduction of a chimeric chalcone synthase gene into petunia results in reversible co-suppression of homologous genes in trans. Plant Cell 2:279–289.
11. Niemann, N., Kues, W., and Carnwash, J. W. (2005). Transgenic farm animals: present and future. Rev. Sci. Tech. Off. Int. Epiz. 24: 285-298, 2005.
12. O'Gorman, S., Fox, D. T., and Wahl, G. M. (1991). Recombinase-mediated gene activation and site-specific integration in mammalian cells. Science 251: 1351-1355.
13. Sontheimer, E. J. (2005). Assembly and function of RNA silencing complexes. Nat. Rev. Mol. Cell. Biol. 6: 127–138.

Chapter 16

Viral Vectors

Outline

16.1: Elements Required for Viral Vectors

In gene therapy, a gene called transgene is introduced into target cells to correct a certain defect. Although several different methods can be used to introduce transgenes into cells including microinjection, chemical transfection, electroporation, and lipofection, none of these methods can achieve the efficiency required for gene therapy. Since viruses can get into cells by infection with an efficiency near 100%, they are used to carry transgenes into cells for gene therapy. Viruses that are more commonly used as vectors for gene therapy include retrovirus, lentivirus, adenovirus, adeno-associated virus (AAV), and herpes simplex virus type 1 (HSV-1). The method by which a virus is used to introduce foreign DNA into cells is called transduction.

Because viruses can kill infected cells when they replicate, their replication function must be disabled in order to become a useful vector for gene therapy. The basic strategy is to clone the transgene into a plasmid containing essential viral elements. Genes that are required for viral replication and packaging are contained in a different plasmid.

When these two plasmids are introduced into appropriate packaging cells, recombinant viral particles are produced. These viral particles can infect and inject packaged DNA into target cells but cannot replicate. The introduced transgene is then transcribed, and the mRNA produced is translated, producing a specific protein.

16.2: Retroviral Vectors

Retroviruses can cause tumors. Since the genome of a retrovirus is RNA, it is also called RNA tumor virus. A retroviral particle contains two plus-stranded RNA molecules of approximately 10 kb each (Fig. 16-1). These two RNA molecules are the genome of the virus. The viral

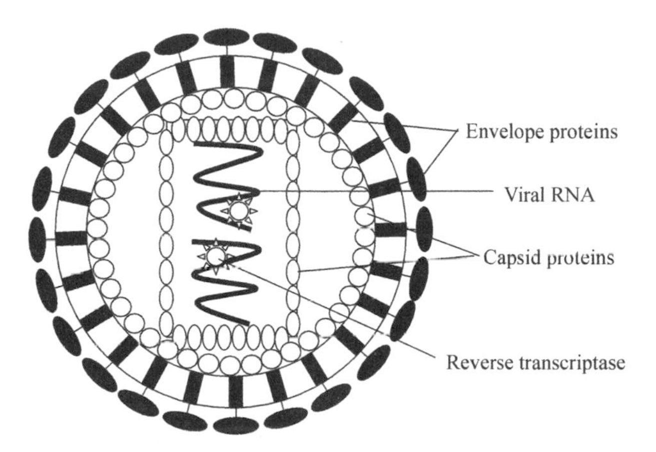

Fig. 16-1: Structure of retrovirus.

particle also contains a reverse transcriptase which converts RNA to DNA when the virus gets inside cells. The DNA is then integrated into the chromosome of infected cells and transcribed to produce more viral genomic RNAs. These RNAs are packaged by viral proteins to become a capsid which is then enclosed by a membrane and other viral proteins to form infectious viral particles. This is how retroviruses proliferate. Some of the genomic RNAs produced in infected cells are processed to become various mRNAs that are then translated to produce different viral proteins required for proliferation and packaging of viral particles.

The most characteristic feature of a retroviral genome is the presence of a long terminal repeat (LTR) of approximately 600 bases at each end of the genomic RNA. LTR contains the promoter, enhancer, and a tRNA-binding site where viral replication initiates. The tRNA-binding site is also called primer binding site (PBS) because retroviruses use tRNA as the primer for replication. Different viruses use different tRNAs to initiate transcription of the viral genome. The human immunodeficiency virus (HIV) uses lysine tRNA as the primer. The tryptophane tRNA is used by avian myoblastosis virus (AMV), and the proline tRNA is used by Moloney murine leukemia virus (MMLV) as the primer. The viral genome also contains a very important sequence called Ψ (psi) which is the packaging signal of the virus. A viral RNA must have Ψ in order to be packaged to become a viral particle. On the other hand, any RNA containing the Ψ sequence of a certain virus may be packaged by proteins of that virus. The Ψ sequence is located near the 3' end of the left LTR. The retroviral genome contains three major genes including *gag*, *pol*, and *env* genes. The *gag* gene encodes the capsid protein such as p17 of HIV. The *pol* gene is responsible for the production of the reverse transcriptase. The *env* gene encodes proteins, such as gp41 and gp120 of HIV, that are inserted in the viral membrane. A typical retroviral vector contains only the two LTRs and the Ψ signal (Fig. 16-2). A transgene is usually inserted between Ψ and the right LTR. The resulting recombinant plasmid is a transgene vector.

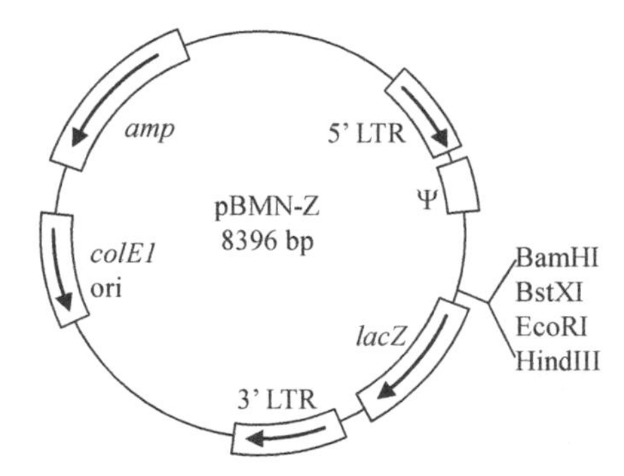

Fig. 16-2: An example of a retroviral vector.

Commonly used retroviral vectors are derived from murine

leukemia virus (MLV). These vectors can carry a gene of approximately 7.5 kb. The recombinant transgene vector is introduced into cells that can produce Gag, Pol, and Env proteins (Fig. 16-3). These cells are

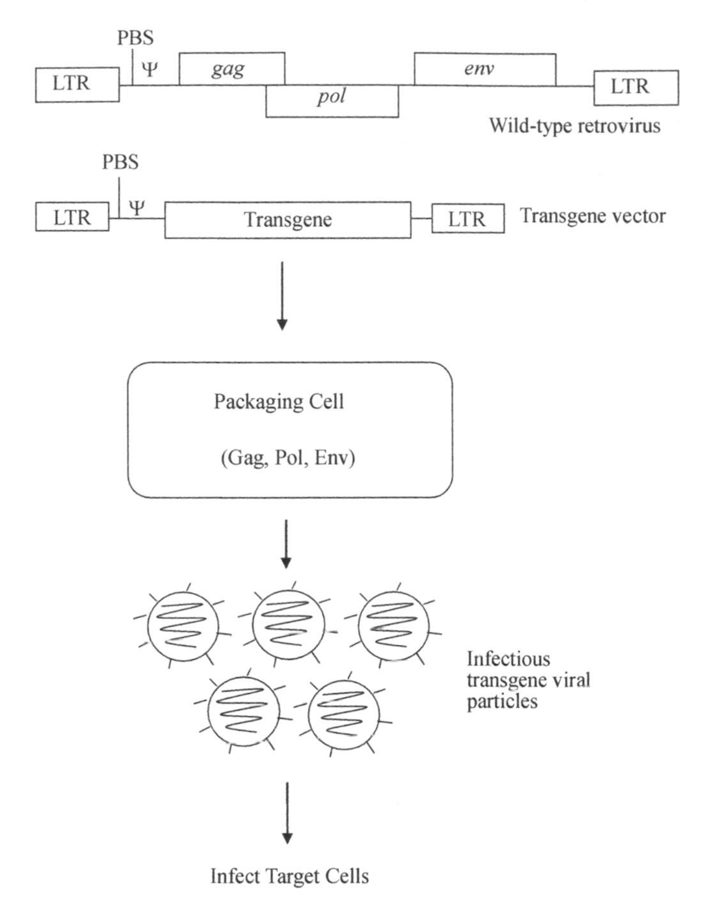

Fig. 16-3: Production of retroviral transgene particles.

referred to as packaging cells. Examples of packaging cell lines include PG13, PA317, GP+E86, and GP+env Am12 derived from the mouse NIH3T3 cell line. The Ampho-Phoenix and Eco-Phoenix cell lines derived from the human HEK293 cell line are also commonly used. In packaging cells, the RNA polymerase II will transcribe the region between the two LTRs located on the transgene vector. Since the RNAs produced contain the Ψ signal, they are packaged by the Gag protein to form a capsid. When the capsid buds out of the cell, it is enclosed by the cell membrane that buds out together with the capsid. Since the Env proteins are inserted in the cell membrane, the viral particle formed has the retroviral Env proteins. The Env proteins are responsible for infection; therefore, the viral particles produced are infectious even though the packaged RNA is not a typical retroviral genome. When these viral particles infect target cells, the packaged RNA of each viral particle is injected into the cell. The reverse transcriptase that is also packaged in the viral particle will convert the RNA to DNA and integrate the DNA into the chromosome of the target cell. The RNA polymerase of the target cell then transcribes the transgene to produce the therapeutic protein. This is the basic principle of gene therapy using retroviral vectors.

The advantage of using a retroviral vector for gene therapy is its high efficiency of infection. Almost all target cells can be infected if sufficient viral particles are present. It can also infect many types of cells. Since the transgene is integrated into the chromosome of target cells, permanent expression of the gene can be achieved. The major disadvantage of retroviral vectors is that the viral particles only infect dividing cells. The retroviral vector was the vector used in the first human gene therapy trial. During this trial, a functional adenosine deaminase gene was introduced into bone marrow cells of a patient with severe combined immunodeficiency (SCID). The marrow cells containing the transgene were then injected back into the patient.

A potential problem with the use of retroviral vectors is the possibility of producing replication competent retroviruses (RCR). This may be due to homologous recombination between the transgene vector and endogenous viruses, resulting in RCR. Another potential problem is activation of proto-oncogenes located near the integration site of the transgene. This activation is usually caused by the enhancer located in the LTR of the transgene vector. Because of these potential problems, lentivirus vectors were developed. Lentiviruses are not known to activate proto-oncogenes and can infect non-dividing cells.

16.3: Lentiviral vectors

Commonly seen lentiviruses include human immunodeficiency virus (HIV), simian immunodeficiency virus (SIV), and feline immunodeficiency virus (FIV). Two types of HIV's exist: HIV-1 and HIV-2. The lentiviral vectors used at present are derived from HIV-1. HIV normally only infects $CD4^+$ cells. Therefore, its host specificity must be changed in order to make it a useful vector. This problem is solved by packaging HIV vectors with glycoproteins from other viruses such as the vesicular stomatitis virus (VSV). If the vector is packaged by the glycoprotein of VSV, the packaged viral vector will be able to infect other types of cells. The process in which a virus is packaged by the glycoproteins of other viruses is called pseudotyping. VSV mainly infects domestic animals such as cows, pigs, and horses. Sheep, monkeys, deer as well as humans may also be infected. The infection causes oral and nasal blisters. It is very painful when the blisters rupture, and infected animals die from starvation.

To pseudotype HIV carrying the transgene with VSV glycoproteins, the HIV transgene vector is introduced into cells that can produce VSV glycoproteins and other HIV proteins required for packaging. Since the VSV glycoproteins are inserted in the membrane of the packaging cells, the HIV-transgene capsid will be enclosed by the membrane containing VSV glycoproteins when it buds out the packaging cells. Because VSV can infect many types of cells, the pseudotyped transgene HIV particles have a much broader host range than the native HIV. The VSV glycoproteins may cause cell fusion and activate complement which destroys the HIV transgene viral particles. Therefore, glycoproteins of other viruses such as Ross River virus (RRV) and Semliki Forrest virus (SFV) have been tried. Both RRV and SFV are alphaviruses and can infect many different animals and many types of cells. Mosquitoes are their vectors for transmission. RRV can cause fever, rash, and arthritis. SFV causes encephalitis.

In addition to changing the host range of HIV, the U3 portion of the

3' LTR is also deleted in the HIV vector (Fig. 16-4). An HIV without this region cannot replicate in infected cells and is called self-inactivating

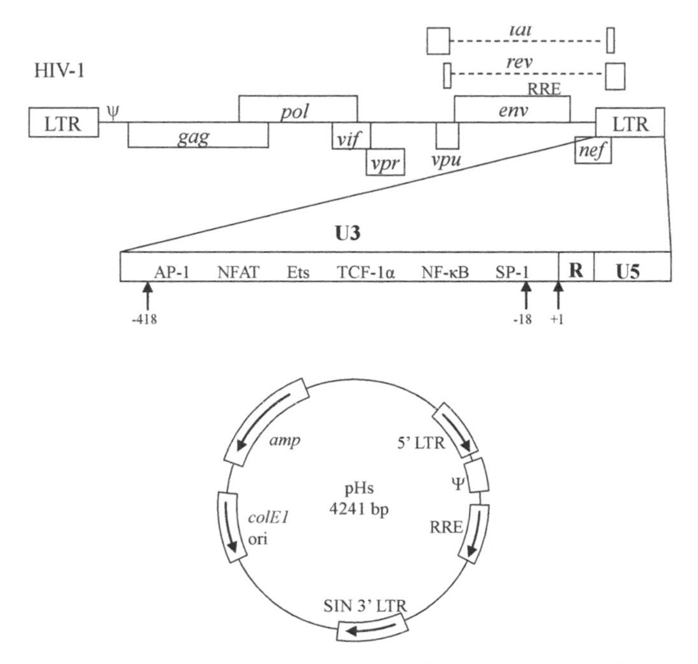

Fig. 16-4: Genomic structure of HIV-1. LTR is composed of U3, R, and U5 regions. U3 contains the HIV major promoter represented by TATA and binding sites of transcription factors AP-1, NF-AT, Ets, TCF-1α, NF-κB, and SP-1. SIN 3' LTR is the LTR in which the region between nucleotide -18 and -418 of the U3 region is deleted. pHs is one example of an HIV-1 vector. RRE stands for Rev responsive element, and Ψ is the packaging signal of HIV.

HIV (SIN HIV). When SIN HIV is used as the vector, no replication competent retrovirus is produced. Since the U3 region contains the enhancer, SIN HIV does not activate proto-oncogenes located near the integration site. The RNA genome of HIV-1 is 9.6 kb. Its LTR is 640 bases long including the 450 nt U3, 100 nt R, and 90 nt U5 regions. HIV-1 has a total of 9 genes. Among them, *gag*, *pol*, and *env* are structural genes; rev (regulator of viral gene expression) and tat (transcription activator) are regulatory genes; and *vif* (viral infectivity factor), *vpr* (viral protein R), and *vpu* (viral protein U) are accessory genes. Two Tat proteins are present. One of them is encoded by exon 1 of the tat pre-mRNA, and the other is encoded by both exon 1 and 2 of the mRNA. Tat activates the promoter located in the LTR responsible for transcription of the genomic RNA. The transcription starts from the first base of the R region of the 5'LTR and ends at the last base of the R region of the 3'LTR. Since a polyadenylation signal is located at this position, the genomic RNA produced is polyadenylated. The genomic RNA can be packaged into viral particles or processed to approximately 30 different mRNAs.

HIV-1 mRNAs can be classified into three types. The first type is the unspliced genomic RNA responsible for production of Gag and Gag-Pol fusion proteins. The second type is incompletely spliced RNA encoding Env, Vif, Vpu, Vpr, and the one-exon Tat proteins. This type of HIV mRNA is considered incompletely spliced because it still contains splicing signals. The third type is the completely spliced RNA that codes for Rev, Nef, and the two-exon Tat proteins. Usually, only completely spliced mRNAs are exported from the nucleus to the cytoplasm for translation. However, some of the HIV proteins are encoded by unspliced and incompletely spliced mRNAs. Therefore, there must be a mechanism by which these mRNAs are exported to the cytoplasm. This is mediated by the 13-kDa Rev protein. The Rev protein binds to the 240-nt Rev responsive element (RRE) and transports mRNAs that have the RRE signal to the cytoplasm for translation.

The 27-kDa Nef protein can inhibit the production of CD4 proteins in HIV infected cells. Since CD4 protein molecules are inserted in the membrane, its decrease in quantity allows more Env proteins to be inserted in the membrane, thus increasing the production of more viral particles. When the RNA genome is converted to DNA, the DNA is integrated into the chromosome of infected cells. The 14-kDa Vpr protein enhances this integration process. The Vpu protein is 16 kDa. Its function is similar to that of Nef and can inhibit the production of the CD4 protein. It also enhances the release of viral particles from infected cells. The 23-kDa Vif protein enables HIV to replicate in lymphocytes or macrophages.

One example of SIN HIV vector is shown in Fig. 16-4. It is a derivative of pBR322 containing 5'LTR and the SIN 3'LTR. The packaging signal Ψ is located near the 5' LTR, and the RRE is located between Ψ and the SIN 3'LTR. When a transgene is inserted into this vector between RRE and SIN 3' LTR, a transgene vector is constructed. The recombinant transgene vector is then introduced into packaging cells such as HEK293T cells to replicate and be packaged. The packaging cells must be able to produce the Gag, Pol, and Rev proteins of HIV. This is usually done by co-introducing pMDLg which encodes the Gag and Pol proteins and pRSV/Rev which encodes the Rev protein of HIV. In addition, a plasmid encoding a certain glycoprotein is needed. If the vector is to be pseudotyped with VSV glycoprotein, the plasmid pCI-MD.G can be used. pCI-RRV or pCI-SFV is used if the vector is to be pseudotyped with RRV or SFV glycoprotein, respectively. The viral particles produced in the packaging cells are then used to infect target cells for gene therapy.

16.4: Adenovirus Vectors

Adenoviruses are non-enveloped viruses with double-stranded DNA genome packaged in a 20-facet icosahedral capsid (Fig. 16-5). Each facet is composed of 12 hexon homotrimers. The hexon protein (protein II) is the major protein of the capsid. Two other abundant proteins of the capsid are the penton base (protein III) and penton fiber (protein IV) proteins. Penton fibers are responsible for the infectivity of adenoviruses. More than 50 serotypes of human adenoviruses exist. Adenoviral vectors are mainly derived from serotypes Ad2 and Ad5. The genome of adenoviruses ranges from 34 to 48 kb. The ends of the adenovirus genome have approximately 100 bp that are identical in sequence but in an inverted repeat orientation, referred to as the inverted terminal repeat (ITR). Adenoviral genes can be classified as early (E1 – E4) and late genes (L1 – L5). The E1 gene (E1A and E1B) is most critical as inactivation of the E1 gene renders the virus unable to replicate. Therefore, an E1 gene defective adenovirus can be used as a vector for gene therapy. The basic principle is to insert the transgene into an E1 deletion mutant. The recombinant virus is then introduced into cells such as HEK293 cells that can produce the E1 proteins for replication and packaging. The viral particles produced are then used to infect target cells, thus introducing the transgene into the cells.

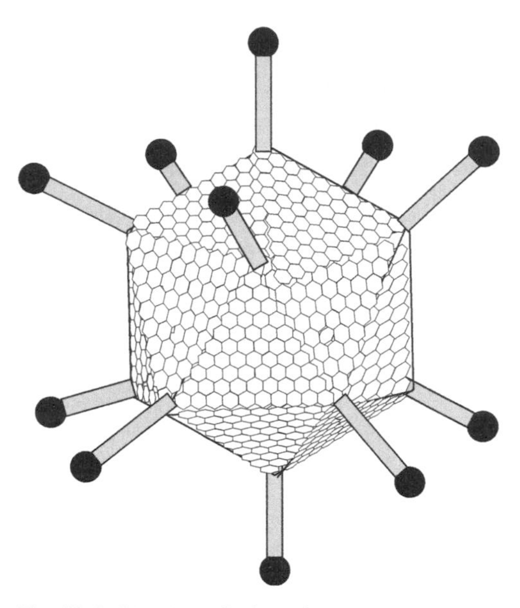

Fig. 16-5: Structure of adenovirus.

The major advantage of using an adenovirus as the vector is that it can infect many types of cells with high efficiency. The disadvantage is that the viral genome does not integrate into the chromosome of infected cells. Therefore, the transgene is expressed only transiently, usually less than 10 weeks. Since more than 85% of people have been exposed to adenoviruses, the vector may trigger adverse inflammatory responses, leading to cell damage or even death.

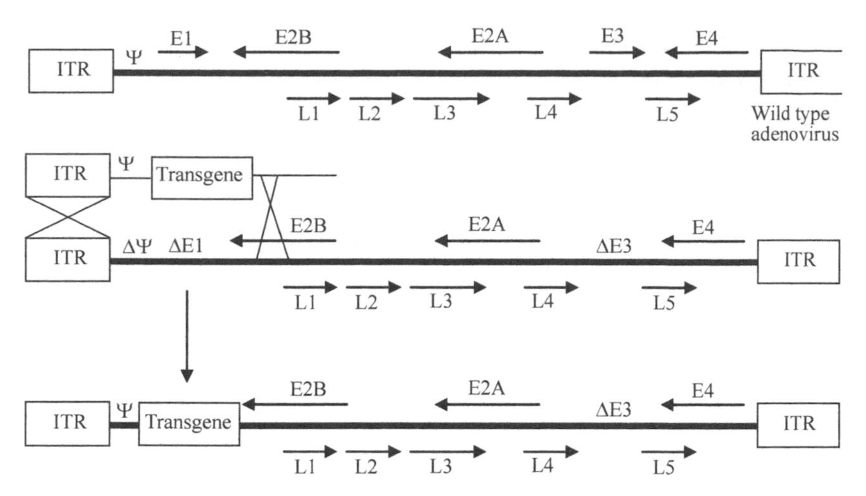

Fig. 16-6: First generation of adenoviral vector.

The E1-deleted adenovirus is the first generation adenoviral vector (Fig. 16-6). To use it, the transgene is first cloned into a plasmid containing one ITR, the adenovirus packaging signal (Ψ), and a portion of the E1 gene such that the transgene is flanked by the packaging signal and a portion of the E1 gene. This recombinant plasmid is then introduced into the cells that can produce the E1 protein and also contain an adenovirus missing the packaging signal (Ψ), E1, and E3 genes. This defective virus together with the host cells can produce all the proteins required to replicate the virus but cannot be packaged to form viral particles because it does not have the packaging signal (Ψ). Since the plasmid contains sequences that are also present in the virus, homologous recombination between these sequences will occur, resulting in a recombinant adenovirus with the packaging signal and its E1 gene replaced by the transgene. This recombinant virus will then replicate and be packaged to become viral particles. The viral particles produced are then used to infect target cells. Since the recombinant virus does not have the E1 gene, it cannot replicate in target cells, but the transgene will be transcribed, and the mRNA produced will be translated. However, this recombinant transgene adenovirus still contains the majority of the viral genes and will produce most of the viral proteins that may induce host inflammatory responses. To address some of these issues, adenoviral vectors have undergone several changes. The second generation adenoviral vectors have most of the early genes including E1, E2, E3, and E4 deleted, but are still not satisfactory. The third generation adenoviral vectors is called gutless adenoviral vector. It contains only the ITRs and the packaging signal (Ψ). Since an adenovirus must be approximately 35 kb in order to be packaged, the vector contains stuffer DNA that does not encode any proteins. The gutless vector cannot replicate without the presence of a helper virus and is also called helper-dependent adenoviral vector. The helper virus is also defective in the E1 gene, and its packaging signal (Ψ) is flanked by two *loxP* sites. In cells such as CRE/293 that can produce E1 and the Cre protein, this virus can replicate and produce all adenoviral proteins but cannot be packaged because the packaging signal (Ψ) is deleted when the Cre protein is produced.

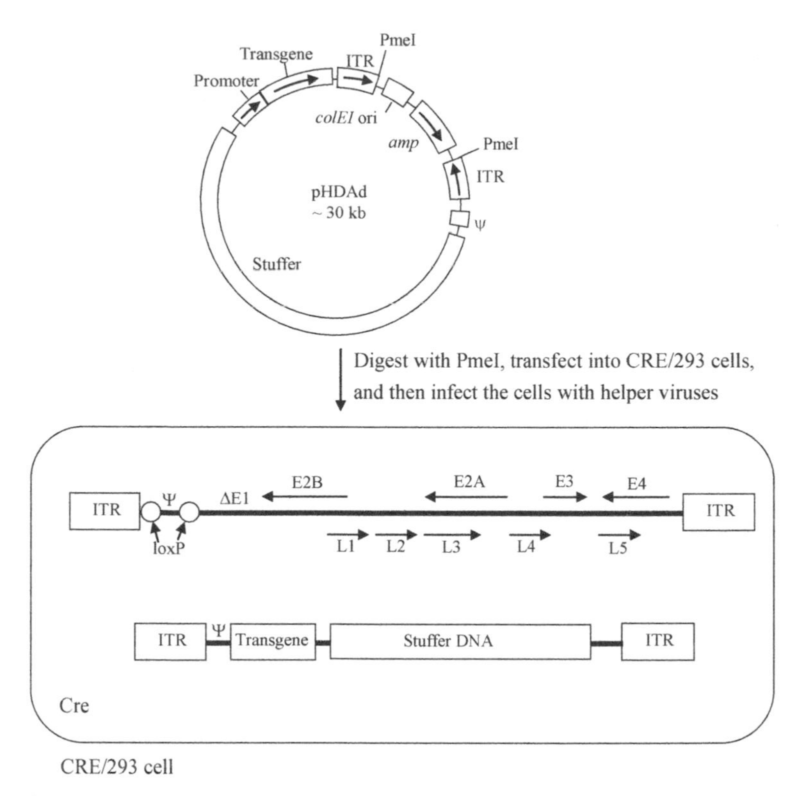

Fig. 16-7: Gutless adenoviral vector.

To use this vector for gene therapy, the transgene is first cloned into a plasmid containing essential elements of the gutless vector. One such plasmid is shown in Fig. 16-7. The resulting recombinant plasmid is then cut with a certain restriction enzyme to remove unwanted regions. The resulting DNA which contains the transgene and the stuffer DNA located between the two ITRs is then introduced into cells containing the helper virus. The adenoviral proteins present in the cells will replicate the DNA containing the transgene and package the replicated DNA to become infectious adenoviral particles. These viral particles are then used to infected target cells.

16.5: Adeno-associated Viral Vectors

The adeno-associated virus (AAV) is a parvovirus (Fig. 16-8). It cannot replicate without adenovirus and therefore is called

Fig. 16-8: Morphology of AAV.

adeno-associated virus. AAV genome is a single-stranded DNA of approximately 5 kb with a 145-bp inverted terminal repeat at both ends. There are 11 types of AAVs, and AAV2 is most commonly used as the vector. AAV contains only two genes: *rep* and *cap*. The *rep* gene encodes Rep78, Rep68, Rep52, and Rep40 proteins. Rep78 and Rep68 are responsible for AAV replication and integration of its genome into the chromosome of host cells. Rep52 and Rep40 are responsible for assembly of viral particles. The *cap* gene encodes coat proteins VP1, VP2, and VP3 in the ratio of 1:1:10. If the *rep* and *cap* genes are substituted with the transgene, a transgene AAV vector is generated. To package the AAV transgene vector, two different plasmids together with the transgene vector are introduced into 293T cells that can produce the E1 protein of adenovirus (Fig. 16-9). One of the plasmid contains the

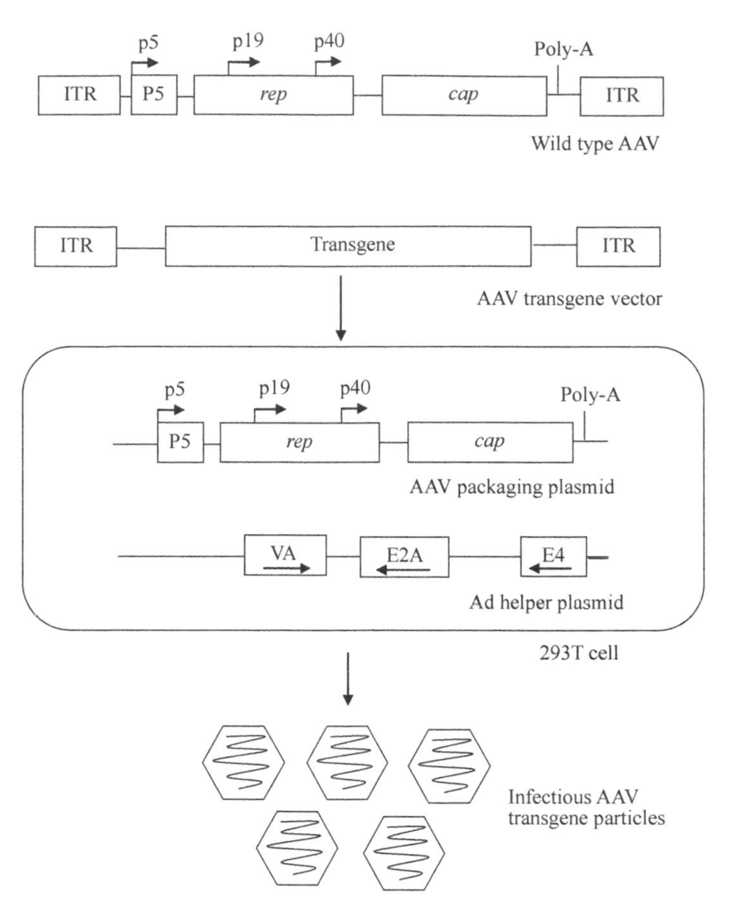

Fig. 16-9: Production of AAV transgene viral particles.

rep and *cap* genes of AAV. The other plasmid contains genes encoding E2A, E4, and VA proteins of adenovirus that are required to replicate and package the AAV transgene vector. The packaged AAV transgene particles are released outside the cell and can be harvested and used to infect target cells.

The major advantage of using AAV as the vector is that it is nonpathogenic and nontoxic. The wild type AAV genome always integrates into a specific site on chromosome 19 when it gets inside cells. However, recombinant AAV may lose this site-specific integration ability. If the transgene is integrated into the chromosome, permanent expression of the transgene may be achieved. The disadvantage of the AAV vector is that it can only carry a transgene less than 4.7 kb.

16.6: Herpes Simplex Virus Vector

Herpes simplex virus looks like a sunny-side-up fried egg under electron microscope. The egg yolk portion is the capsid which is enclosed in a tegument and an envelope. Two major types of HSV exist: HSV-1 and HSV-2. Most HSV vectors are derived from HSV-1. The genome of HSV-1 is a double-stranded DNA of approximately 150 kb. The viral particle also contains the virion-associated protein VP16 which initiates HSV viral gene expression. The HSV genome is composed of two fragments: long and short fragments. Both fragments are flanked by inverted repeats in which the packaging signals (Pac) are located. The Pac signal is the only sequence required for packaging; therefore, any DNA of approximately 150 kb can be packaged by HSV proteins if it contains the Pac signal.

HSV has approximately 80 genes that can be classified into two groups: essential and accessory genes (Fig. 16-10). Genes located in

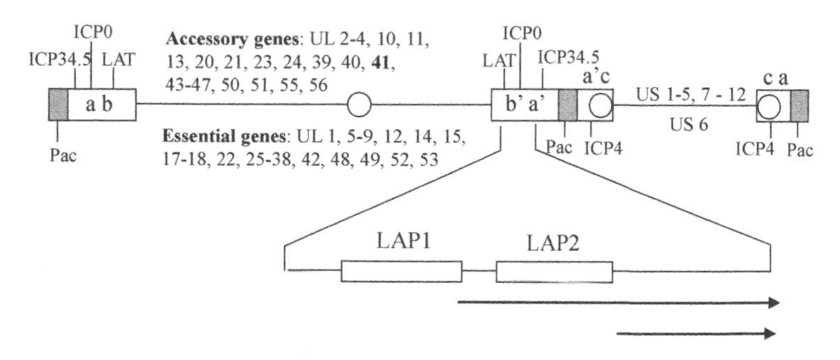

Fig. 16-10: Genomic structure of HSV-1.

the long fragments are designated as UL, and those located in the short fragment are designated US. When HSV becomes latent, it usually resides in nerve cells.

Two types of HSV vectors exist. One is the replication defective vector. This type of vector is very similar to the replication defective adenoviral vector. The virus is modified to become a vector by deleting one or more of the genes that are essential for HSV replication. Essential genes that are commonly deleted include ICP34.5, ICP6, ICP0, ICP4, ICP22, and ICP27; where ICP stands for infected cell protein. The other type of HSV vector is called amplicon vector which is a plasmid containing an HSV replication origin and the packaging signal Pac. When HSV replicates in infected cells, some portions of the genome may be deleted. If the deleted portion contains an HSV replication origin, it will replicate in infected cells. These DNA elements are called amplicons. Since they may interfere with the replication of the HSV genome, they are also referred to as defective interfering (DI) particles.

Since approximately 50% of HSV genes are nonessential, the HSV vector can carry a transgene greater than 30 kb. This is the major reason why HSV is being developed as a vector for gene therapy. Another advantage of using HSV as the vector is that it can infect many types of cells including nondividing cells. Its disadvantage is cytotoxicity. During latency, the virus resides in nerve cells, and only the latency-associated promoter (LAP) is active. Therefore, HSV can be used to target nerve cells for long-term expression of the transgene.

To use HSV vector, the transgene is cloned into an amplicon or a defective viral vector. The recombinant DNA is then introduced into cells containing a BAC plasmid encoding genes that are required for HSV replication and packaging. The proteins produced will package the replicated transgene vectors to become infectious viral particles. These viral particles are then used to infect target cells.

Summary

In gene therapy, the therapeutic gene needs to be introduced into the majority of cells containing the defective gene in order to correct the defect. Although several methods are available for introducing foreign DNA molecules into cells, none of them can achieve the efficiency required for gene therapy. Since viruses can infect cells with almost 100% efficiency, they are developed as vectors for gene therapy. The most common ones include retrovirus, lentivirus, adenovirus, adeno-associated virus, and herpes virus type 1. Viral vectors must fulfill two requirements to be useful for gene therapy: 1) they must not be able to replicate when introduced into target cells, and 2) they must be able to replicate in vitro and be packaged into infectious viral particles.

The genome of retroviruses is single-stranded RNA with a pair of long terminal repeats (LTR). It also contains a packaging signal (Ψ) and sequences encoding the *gag*, *pol*, and *env* genes. A retroviral vector contains only the LTRs and the packaging signal. The transgene is inserted between the two LTRs. When the recombinant plasmid is introduced into cells that can produce Gag, Pol, and Env proteins, the region located between the two LTRs is transcribed, and the RNA produced are packaged to become infectious viral particles. These particles are then used to infect target cells, thereby, introducing the transgene into cells. Retrovirus can infect many types of cells. Since the transgene carried by the retroviral vector is integrated into the chromosome of the cells, permanent expression of the transgene can be achieved. The disadvantage of the retroviral vector is that it can only infect dividing cells. Lentivirus is a type of retrovirus. It can infect nondividing cells and therefore is developed as a vector for gene transfer. The most commonly used lentivirus vector is derived from HIV-1.

HIV-1 mainly infects CD4+ cells. To be able to use it as a vector, its host range must be increased. This is achieved by packaging the HIV transgene vector with glycoproteins from vesicular stomatitis virus,

Ross River virus, or Semiliki Forrest virus. To prevent the HIV vector from producing replication competent retroviruses, the U3 portion of the 3'LTR is deleted. The resulting LTR is called SIN 3'LTR. Since this deletion removes the enhancer in the 3'LTR, this HIV vector does not activate protooncogenes located near the integration site. Some of the HIV proteins are translated from unspliced or incompletely spliced HIV genomic RNA. These RNAs are exported from the nucleus to the cytoplasm by the Rev protein which binds to the Rev responsive element (RRE). Therefore, HIV vectors must contain the RRE. A typical HIV vector is a plasmid containing two LTRs, the packaging signal, and RRE. The transgene is inserted between RRE and the 3' LTR. The HIV transgene vector is then introduced into packaging cells to replicate and be packaged. The most commonly used packaging cell line is 293T. Four different plasmids are simultaneously introduced into 293T cells in order to replicate and package the transgene vector. The first plasmid is the transgene vector. The second one is pMDLg encoding HIV Gag and Pol proteins. The third one is pRSV/Rev containing the *rev* gene. The fourth one is pCI-MD.G, pCI-RRV, or pCI-SFV encoding the glycoprotein of VSV, RRV, and SFV, respectively.

Adenoviruses must have the E1 gene in order to replicate. Theoretically, any adenovirus defective in E1 can be used as a vector. The first generation adenoviral vector is constructed by replacing the E1 gene with the transgene. The recombinant virus is then introduced into HEK 293 cells which contains an integrated E1 gene. In this cell, the recombinant virus will replicate and be packaged to become infectious viral particles. When these viral particles are used to infect target cells, the transgene-containing adenovirus gets into cells but does not replicate because it lacks the E1 gene. This virus, however, still has all the other adenoviral genes, and the proteins produced may induce inflammatory responses since most people have been exposed to adenoviruses. To solve this problem, gutless adenoviral vector is developed. This is a plasmid containing only the ITRs and packaging signal of adenovirus. The transgene is inserted between the packaging signal and one of the ITRs. The recombinant plasmid is then introduced into CRE/293 cells containing a helper adenovirus whose packaging signal is located between two *loxP* sites. When the cells are induced to produce the Cre protein, the packaging signal is deleted. Therefore, this virus can produce adenoviral proteins but cannot be packaged. Since the transgene vector has the packaging signal and ITRs, it will replicate and be packaged to become infectious viral particles that are then used to infect target cells. The gutless adenoviral vector can carry a transgene up to 30 kb in size and can stay in target cells longer than other types of adenoviral vectors.

Adeno-associated virus (AAV) is nonpathogenic and therefore can be used as a vector. It cannot replicate without adenovirus. AAV contains only two genes: *rep* and *cap* genes. When the transgene is inserted between the two ITRs of AAV on a plasmid, an AAV transgene vector is created. The recombinant plasmid is then introduced into cells that can produce the Rep and Cap proteins of AAV and E1A, E1B, VA, E2A, and E4 proteins of adenovirus. In these cells, the AAV transgene vector will replicate and be packaged to become infectious viral particles. AAV can only carry a transgene less than 5 kb.

Herpes simplex virus has a double-stranded DNA genome of approximately 150 kb. Although it contains at least 80 genes, 50% of these genes are nonessential. Therefore, HSV can carry a transgene of at least 30 kb. This is the major reason why HSV is used as a vector. There are two types of HSV vectors. One is replication defective virus in which one or more of the essential genes are deleted. The transgene is inserted, and the recombinant virus is then introduced into cells containing a BAC plasmid which contains all the HSV genes required to replicate and package the transgene vector. The other type of HSV vector is the amplicon vector which is a plasmid containing an HSV replication origin and packaging signal. The transgene is cloned into the plasmid, and the recombinant plasmid is introduced into cells that contain a BAC plasmid capable of producing all HSV proteins required to replicate the vector and package the replicated DNA to become infectious viral particles. These particles are then used to infect target cells thereby introducing the transgene into cells. HSV almost always become latent after acute infection. During latency, HSV resides in nerve cells. The only promoter which is active during latency is the latency-associated promoter (LAP). If the LAP promoter is used to drive the transgene, long-term expression of the transgene can be achieved in nerve cells. This is another major reason why HSV is useful as a vector.

Sample Questions

1. Which of the following viruses can be used for gene transfer: a) retrovirus, b) adenovirus, c) adeno-associated virus, d) herpes virus
2. Retroviruses usually infect: a) nondividing cells, b) dividing cells, c) neuronal cells, d) red blood cells
3. Gutless adenoviral vector is also called a) adenovirus without capsid, b) helper-dependent adenoviral vector, c) adenovirus with no genome, d) adenovirus without the inverted terminal repeats.
4. The major reason why adeno-associated virus is used for gene therapy is a) large cloning potential, b) non-pathogenic, c) the ability to integrate into a specific site, d) easy to package.
5. Reasons why herpes virus is being developed to become a gene therapy vector are a) nontoxic, b) large cloning potential, c) nonpathogenic, d) possibility of achieving long-term expression in nerve cells.

Suggested Readings

1. Alba, R., Bosch, A., and Chillon, M. (2005). Gutless adenovirus: last-generation adenovirus for gene therapy. Gene Ther. 12: S18-S27.
2. Burton, E. A., Wechuck, J. B., Wendell, S. K., Goins, W. F., Fink, D. J., and Glorioso, J. C. (2001). Multiple applications for replication-defective herpes simplex virus vectors. Stem Cells 19: 358-377.
3. Fink, D. J., DeLuca, N. A., Yamada, M., Wolfe, D. P., and Glorioso, J. C. (2000). Design and application of HSV vectors for neuroprotection. Gene Therapy 7: 115-119.
4. Fink, D. J. and Glorioso, J. C. (1997). Engineering herpes simplex virus vectors for gene transfer to neurons. Nat. Medicine 3: 357-359.
5. Fisher, K. J., Choi, H., Burda, J., Chen, S.-J., and Wilson, J. M. (1996). Recombinant adenovirus deleted of all viral genes for gene therapy of cystic fibrosis. Virology 217: 11-22.
6. Glorioso, J. C., DeLuca, N. A., and Fink, D. J. (1995). Development and application of herpes simplex virus vectors for human gene therapy. Annu. Rev. Microbial. 49: 675-710.
7. Hillgenberg, M., Hofmann, C., Stadler, H., and Loser, P. (2006). High-efficiency system

for the construction of adenovirus vectors and its application to the generation of representative adenovirus-based cDNA expression libraries. J. Virol. 80: 5435-5450.

8. Kahl, C. A., Marsh, J., Fyffe, J., Sanders, D. A., and Cornetta, K. (2004). Human immunodeficiency virus Type-1-derived Lentivirus vectors pseudotyped with envelope glycoproteins derived from Ross River virus and Semliki Forest virus. J. Virol. 78: 1421-1430.

9. Kay, M. A., Glorioso, J. C., and Naldini, L. (2001). Viral vectors for gene therapy: the art of turning infectious agents into vehicles of therapeutics. Nat. Medicine 7: 33-40.

10. Mulligan, R. C. (1993). The basic science of gene therapy. Science 260: 926-932.

11. Robbins, P. D. and Ghivizzani, S. C. (1998). Viral vectors for gene therapy. Pharmacol. Ther. 80: 35-47.

12. Smith, A. E. (1995). Viral vectors in gene therapy. Annu. Rev. Microbiol. 49: 807-838.

13. Wang, Q. and Finer, M. H. (1996). Second-generation adenovirus vectors. Nat. Medicine 2: 714-716.

14. Zufferey, R., Dull, T., Mandel, R. J., Bukovsky, A., Quiroz, D., Naldini, L., and Trono, D. (1998). Self-inactivating lentivirus vector for safe and efficient in vivo gene delivery. J. Virol. 72: 9873-9880.

Chapter 17

Special Techniques

Outline

Nuclear run-on assay
Reporter assay
Electrophoresis mobility shift assay
DNase I footprinting
Chromatin immuno precipitation
Two-hybrid system
Three hybrid system
Primer extension for determination of transcription start site
S1 nuclease analysis
RNase A protection assay
Single strand conformation polymorphism
Heteroduplex assay
Denatured gradient gel electrophoresis
Random amplification of polymorphic DNA or arbitrary primed PCR
Amplified fragment length polymorphism
mRNA differential display
DNA microarray
Serial analysis of gene expression
Massively parallel signature sequencing
Sequencing by oligonucleotide ligation and detection
Total gene expression analysis
Representational difference analysis

17.1: Nuclear Run-on Assay

The nuclear run-on assay is used to determine the rate of

transcription of a gene. During transcription, many mRNA molecules are transcribed from the same template. The transcription of the second mRNA molecule may start before the first mRNA is completely transcribed; therefore, many mRNA molecules may attach to the same template during transcription. A gene with a higher transcription rate would have more mRNA molecules associated with the template; thus, the rate of transcription of a certain gene can be determined by measuring the number of mRNA molecules associated with the template.

To determine the number of mRNA molecules associated with a certain template, nuclei of the cells expressing the gene are isolated. When these nuclei are placed in a test tube containing nucleotides (ATP, CTP, GTP, and UTP), transcription that has been initiated will continue, but no new transcription is initiated. Completion of the already initiated transcription reactions in a test tube containing radioactive nucleotides such as α-^{32}P-UTP will produce radioactive mRNA molecules. These radioactive mRNA molecules are then hybridized to DNA spots representing different genes on a membrane. Since many spots can be placed on the same membrane, this method can measure the transcription rates of many different genes simultaneously. The radioactivity of each spot is then normalized to that of a house-keeping gene such as the β-actin gene. A gene with a higher transcription rate would have a higher radioactivity on the hybridized spot. The nuclear run-on assay is also called nuclear run-off assay because it actually allows the transcripts to run off the template in vitro. Since the rate of transcription is proportional to promoter strength, the nuclear run-on assay is not commonly performed since promoter strength can be more easily measured.

17.2: Reporter Assay

The stronger the promoter the greater the amounts of mRNA are transcribed and protein translated. Measuring protein levels, however, may not reflect promoter strength because the steady state level of a protein is the equilibrium between its production and degradation. Furthermore, protein levels cannot be determined if no appropriate antibody is available. In a reporter assay, the promoter to be assayed is used to drive a reporter gene whose activity can be easily measured. One example of a reporter gene is the chloramphenicol acetyltransferase (CAT) gene. CAT can transfer the acetyl group from acetyl-CoA to chloramphenicol to form 1-acetyl-chloramphenicol, 3-acetyl-chloramphenicol, and 1, 3-acetyl-chloramphenicol. If the acetyl-CoA used for the reporter assay is labeled with isotopes such as ^{14}C, the acetylated forms of chloramphenicol are radioactive. When they are separated by thin-layer chromatography, the radioactivity on the spots representing different forms of acetylated chloramphenicol can be determined. The higher the radioactivity, the stronger the strength of the promoter is. The luciferase gene from firefly or renilla can also be used as the reporter gene. Luciferase can convert luciferin to light. Since measuring light is much easier than radioactivity, the luciferase assay is much more commonly used then the CAT assay.

The reporter assay can also be used to locate a certain promoter. If a DNA fragment is known to contain a promoter, but its exact location is not known, the DNA fragment can be cut into several smaller fragments. Each small fragment is then inserted in front of a promoterless reporter gene. If one of the inserted smaller fragments results in expression of the reporter gene, the promoter is located in that fragment. Similar approaches can be used to find a certain enhancer. A DNA fragment is inserted in front of a reporter gene driven by a promoter which is inactive without an enhancer. If insertion of a DNA fragment in front of this promoter causes expression of the reporter gene, the DNA fragment must contain an enhancer.

17.3: Electrophoretic Mobility Shift Assay

In addition to an enhancer, other regulatory elements may also be required for gene expression. Enhancers and regulatory elements are not functional unless they bind transcription factors. The electrophoretic mobility shift assay (EMSA) is a method used to detect binding of a certain protein to DNA. A DNA fragment approximately 200 bp is first labeled at its 5' or 3' end and then incubated with the nuclear extract of a certain type of cells. If a protein in the nuclear extract binds to the DNA fragment, a protein-DNA complex is formed. Since the size of this protein-DNA complex is bigger than the naked DNA fragment, its mobility in electrophoresis is slower. Since binding of a protein to a DNA molecule changes the mobility of the DNA, this method is also called gel-shift or gel-retardation assay.

17.4: DNase I Footprinting

Transcription factors recognize and bind to specific sequences. DNase I footprinting is used to determine the binding sequence of a transcription factor or a DNA-binding protein. The principle of DNase I footprinting is similar to that of DNA sequencing. A DNA fragment is labeled at its 5' end, incubated with the protein, and then digested with DNase I. Since DNase I is a nonspecific endonuclease, it will cut DNA at any base. As in DNA sequencing, different DNA molecules are allowed to be cut at different positions to generate DNA fragments that have the same 5' end but differ from each other by only one base in length. When these DNA fragments are electrophoresed on a gel, a DNA ladder is observed. If a protein binds to the DNA fragment to be analyzed, the place where the protein binds will not be cut by DNase I; therefore, this DNA sample will result in a ladder missing several bands in the gel. If the sequencing reaction of the same DNA fragment is electrophoresed on the same gel, the sequence of the missing bands can be determined. This sequence is the binding sequence of the protein (Fig. 17-1).

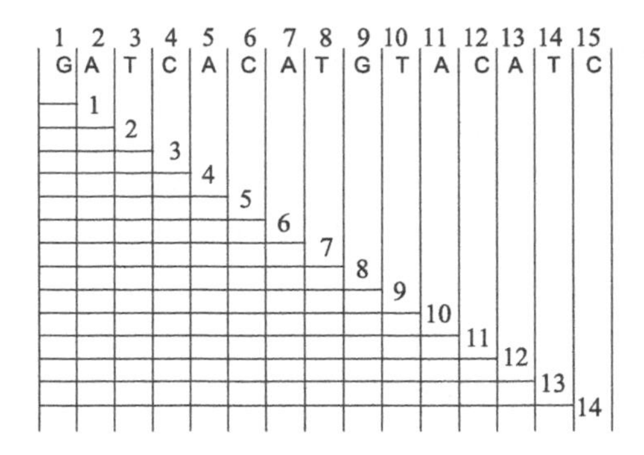

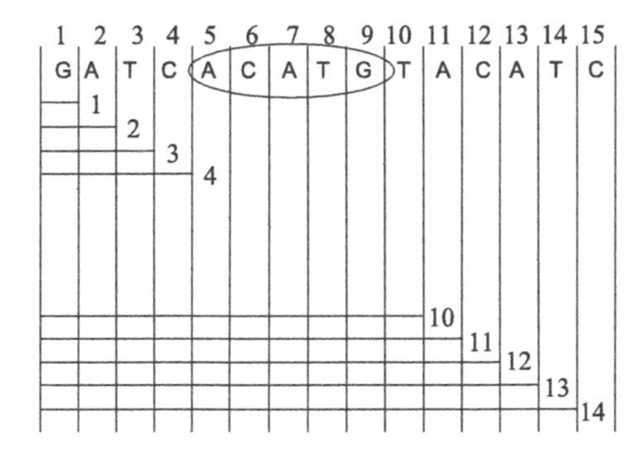

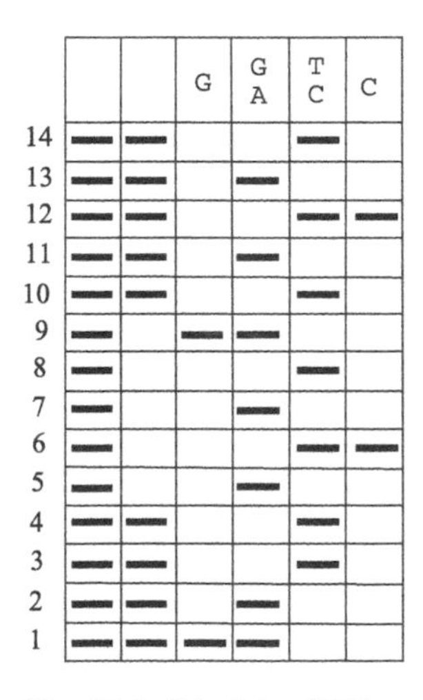

Fig. 17-1: Principle of DNase I footprinting. (A). Digestion of 5' end-labeled DNA fragment with DNase I results in 5' end-labeled DNA fragments with variable 3' ends, differing from each other by one base. (B). A protein that binds to the DNA fragment protects it from being digested by DNase I at where the protein binds. (C) electrophoregram of DNase I digested DNA along with the sequencing reaction of the same DNA fragment. The missing bands between positions 5 and 9 in the second lane indicate the protein binding region.

17.5: Chromatin Immunoprecipitation

Transcription factors bind to DNA to initiate gene expression. Chromatin immunoprecipitation (ChIP) is the method used to identify or isolate a DNA fragment which interacts with a certain transcription factor. Chromatin is a portion of chromosomes; it is a complex of DNA and protein. To perform ChIP, cells are first treated with 1- 5% paraformaldehyde to cross link proteins to chromatin. The cells are then sonicated to break the chromosome into pieces of 0.3 – 2.5 kb. Antibody against the transcription factor is then used to react with the resulting chromatin fragments. The fragment to which the antibody binds is then precipitated by protein A-sepharose CL-4B beads. The DNA fragments precipitated are amplified and sequenced to determine the binding sequence of the transcription factor.

Since a transcription factor may bind to numerous places on the chromosome, numerous DNA fragments may be precipitated. To identify these DNA fragments, a method referred to as ChIP-on-Chip is developed. The precipitated DNA fragments are labeled and then hybridized to a DNA microarry (DNA chip) containing all or the majority of genes in the cell. When the hybridized spots are identified, the genes to which this transcription factor binds are identified.

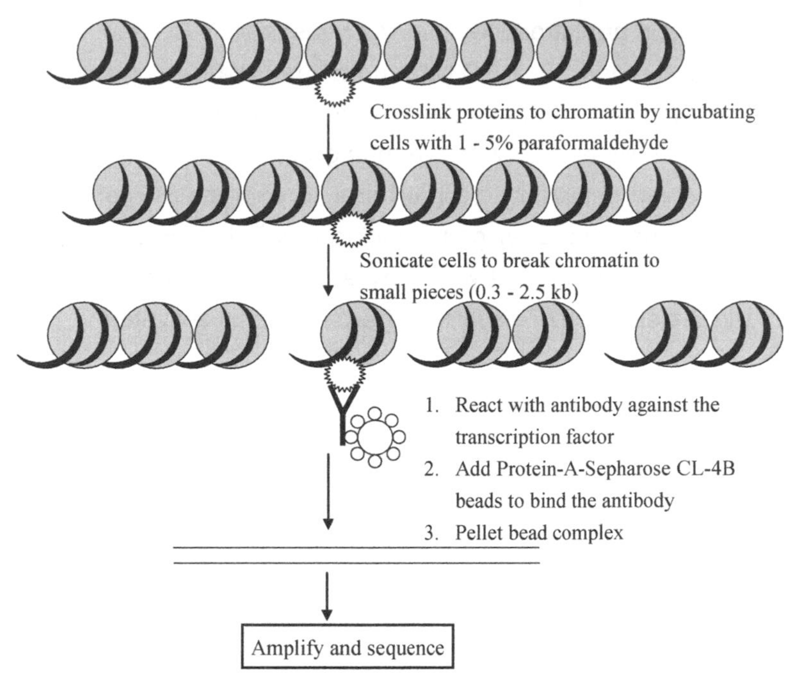

Fig. 17-2: Principle and procedures of ChIP.

17.6: Two-hybrid System

Many proteins must interact with other proteins in order to carry out their functions. The two-hybrid system is a method for detection and identification of interacting proteins based on the structure and property of transcription factors. Most transcription factors have two functions: DNA binding and transcription activation. These two functions are carried out by two different domains of a transcription factor. Each of these two domains is functional even when they are separated. The yeast two-hybrid system uses the yeast transcription factor Gal4 to study protein-protein interactions. In this system, a known protein, protein X, is fused to the binding domain (BD) of Gal4, and its target protein, protein Y, is fused to the activation (AD) domain of Gal4. A reporter gene driven by a Gal4 responsive promoter is integrated into the chromosome of the yeast cell used in this system. When the plasmid encoding X-BD fusion protein and another one encoding Y-AD fusion protein are introduced into the yeast cell, X-BD and Y-AD fusion proteins are produced. X-BD will bind to the Gal4 binding site located in front of the reporter gene. If the Y protein is the target of the X protein, Y-AD fusion will bind to X-BD, bringing the AD domain of Gal4 to the promoter and activating the promoter to express the reporter gene. Commonly used reporter genes include *lacZ*, luciferase, and *HIS3*. If the *HIS3* gene is used, it will allow a *his3*⁻ yeast (*his3*⁻ auxotroph) to grow in the absence of histidine if the promoter is activated.

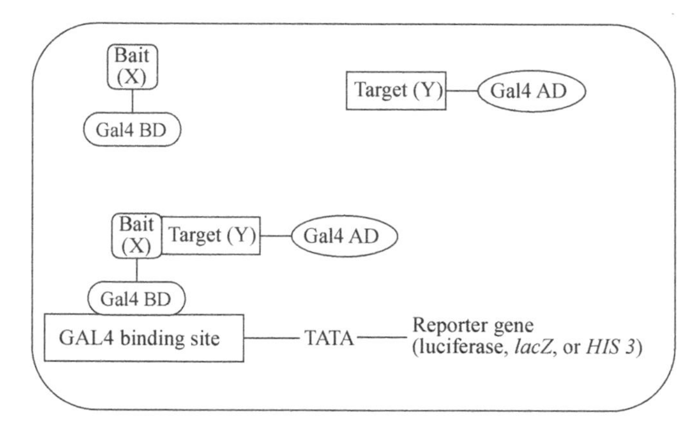

Fig. 17-3: Principle of the yeast two-hybrid system. TATA represents the promoter of the reporter gene.

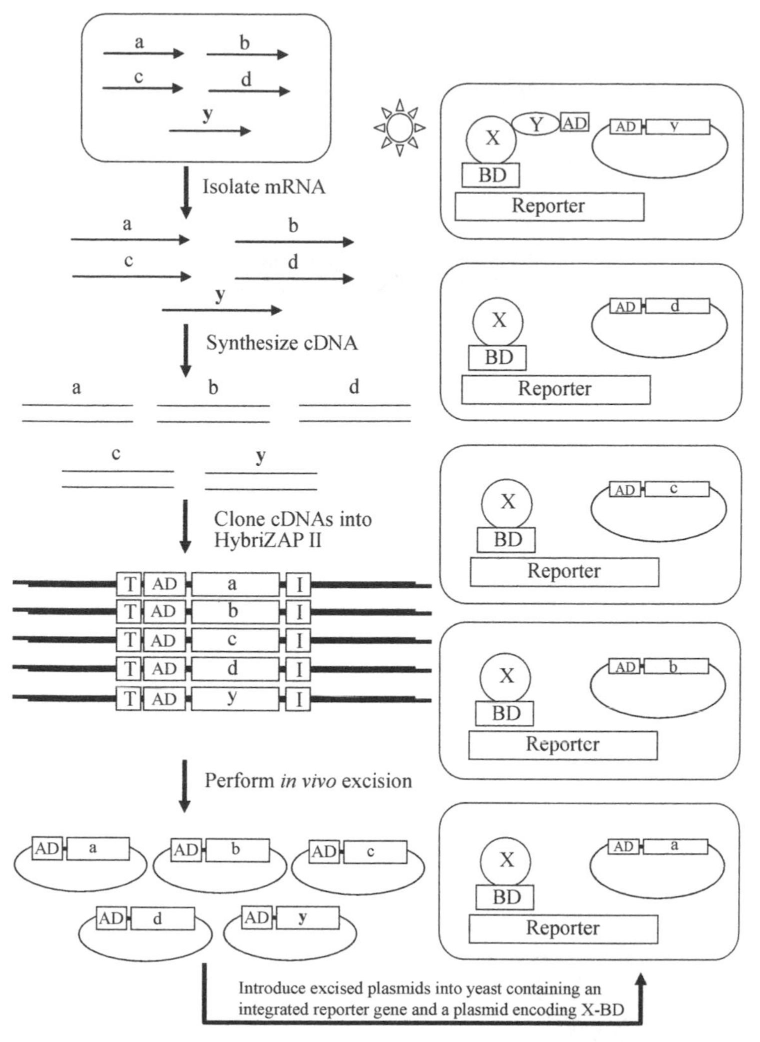

Fig. 17-4: Procedures of the yeast two-hybrid system for isolation of partner proteins.

17.7: Three Hybrid System

The three-hybrid system is used for detection and isolation of the protein that binds to a certain RNA. The target RNA is first fused to the RNA genome of the MS2 bacteriophage. This can be achieved by cloning the gene of the target RNA into a plasmid containing the MS2 genomic RNA gene (Fig. 17-5). When this recombinant plasmid

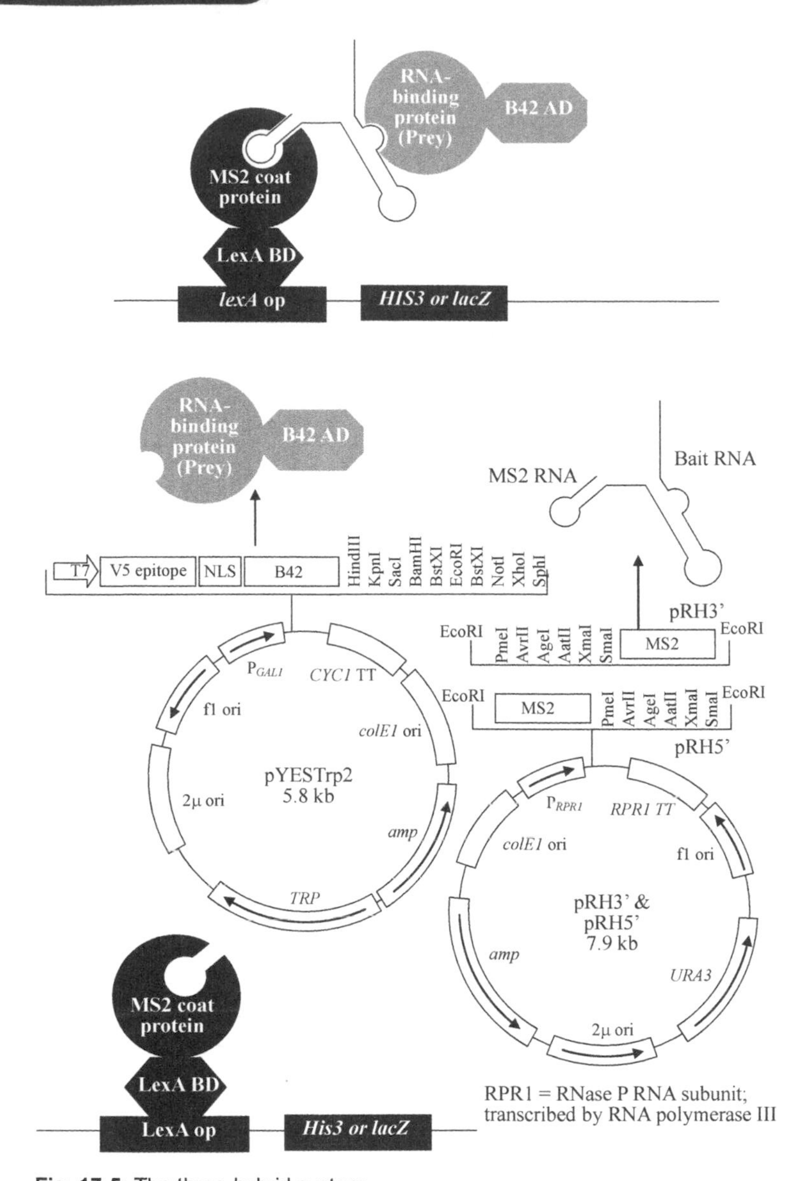

Fig. 17-5: The three-hybrid system.

is introduced into yeast cells, a hybrid RNA consisting of MS2-target RNA is produced. The yeast cells used in this system also contain a plasmid expressing the fusion protein of the MS2 coat protein and LexA binding domain (LexA-BD) and a reporter gene (*lacZ* or *HIS3*) whose expression is controlled by the *lexA* operator. Similar to the two-hybrid system, a plasmid library containing the candidate gene fused to the activation domain of the transcription factor B42 (B42-AD) is first created. B42 is a transcription activator of *E. coli*. When the plasmids are introduced into this yeast strain, some transformants may express the reporter gene. These yeast cells would be the ones containing the plasmid expressing the protein capable of binding to the target RNA. Cloning vectors such as pRH3' or pRH5' can be used to produce the MS2-target RNA fusion. The P_{RPR1} promoter is used to express the fusion gene on these two plasmids. The RPR1 gene encodes the RNA P ribozyme, which is an RNA with enzymatic function. The vector pYESTrp may be used to create the target protein and B42 AD fusion. This plasmid also has a sequence encoding a nuclear localization signal (NLS) which will allow the protein to be transported into the nucleus (Fig. 17-5).

17.8: Primer extension for determination of transcription start site

Once the nucleotide sequence of a gene is determined, the place where translation starts (initiation codon) may be deduced, but the transcription start site cannot be determined by simply examining the nucleotide sequence of the gene. Primer extension is a technique commonly used to determine the transcription start site of a gene. An oligonucleotide primer complementary to the region near the deduced initiation codon is annealed to the mRNAs isolated from cells. Reverse transcription is then performed. This will produce a cDNA corresponding to the region starting from where the primer binds to the 5' end of the target mRNA. As in the example shown in Fig. 17-6, the cDNA synthesized is 17 bases long, indicating that the transcription start site is located 17 bases upstream from the first base of where the primer binds. Once the transcription start site of a gene is determined, it will be easier to identify its promoter as the promoter is usually located within 200 bp upstream from the transcription start site.

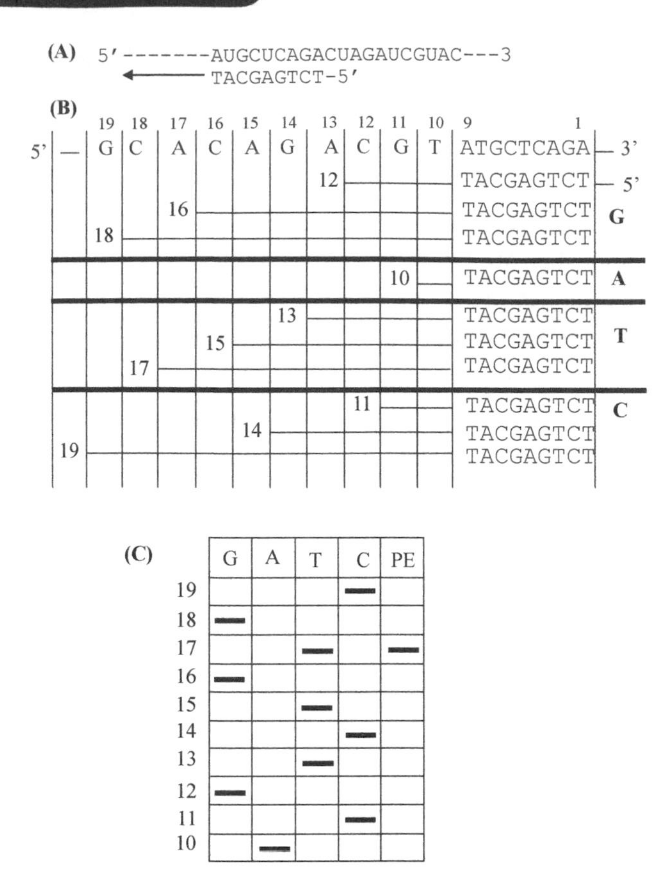

Fig. 17-6: Determination of transcription start site by primer extension. (A). A primer is annealed to the mRNA at the region containing the deduced initiation codon to perform reverse transcription. (B). The DNA fragment containing the region where the transcription initiation may be located is sequenced with the same primer used for the reverse transcription. (C) The primer extended product (PE) is electrophorsed along with the DNA sequencing reaction in the same gel.

17.9: S1 Nuclease Analysis

S1 nuclease can be used to determine the 3' end of a certain mRNA or intron-exon junctions. Since most eukaryotic mRNAs have a poly-A tail, the 3' end of a eukaryotic mRNA can be identified from its nucleotide sequence. To determine the 3' end of a prokaryotic mRNA, a DNA probe corresponding to the region where the transcription may end is labeled at its 3' end. The labeled probe is then annealed to the mRNA. If the 3' end of the mRNA is located within this DNA fragment, a portion of the probe will not anneal to the mRNA and will be single-stranded. S1 nuclease is then used to digest the single-stranded region of this DNA-RNA hybrid, and the exact length of the labeled probe after S1 nuclease digestion is determined. If it is 20 bases long, the 3' end of the mRNA is 20 bases downstream from where the 3' end of the probe binds.

To determine an intron-exon junction, a DNA fragment suspected to contain an intron-exon junction is isolated and labeled at the 5' end of the minus strand. The labeled probe is then annealed to the mRNA. Since the mRNA has no intron, the intron portion in the probe will loop out. S1 nuclease is then used to digest the single-stranded loop region. If the length of the digested probe is 20 bases, the intron-exon junction is located 20 bases downstream from where the 5' end of the probe binds.

17.10: RNase A Protection Assay

The RNase A protection assay (RPA) is a commonly used method to quantify the amount of a certain mRNA. A labeled RNA probe is produced by in vitro transcription and then annealed to mRNAs isolated from cells. The mRNAs that do not hybridize with the probe and the probe molecules that do not bind to the mRNAs are digested with RNase A. The radioactivity derived from the resulting double-stranded RNAs of the probe-mRNA hybrids is then determined, thereby quantifying the target mRNAs.

17.11: Single Strand Conformation Polymorphism

If a double-stranded DNA is denatured, each of the two resulting single-stranded DNAs will form a certain conformation. Since a single base difference between two DNA molecules may result in single-stranded DNAs with different conformations, these single-stranded

DNAs will migrate to different positions during electrophoresis. This phenomenon is called single strand conformation polymorphism (SSCP). SSCP is very commonly used to screen mutations in a certain gene. A portion of the gene suspected to have mutations is amplified with the same primers used to amplify the normal gene with labeled dNTP. The labeled PCR products generated are denatured and electrophoresed in a gel. If a shift in the mobility of DNA bands is observed, the gene contains mutations.

17.12: Heteroduplex Assay

The heteroduplex assay is another method for screening mutations in a certain gene. A portion of approximately 150 bp of both the normal gene and the one suspected to have mutations are amplified separately by PCR using the same set of primers. The resulting PCR products are denatured and mixed together to form hybrid molecules. The hybrids are then electrophoresed in a gel. If mutations exist, the two strands of the hybrids will have imperfect matches, forming a loop in the hybrid structure. Since the size of the hybrid with a loop is larger than that of perfectly matched molecules, this hybrid will migrate slower during electrophoresis.

17.13: Denatured gradient gel electrophoresis

Denatured gradient gel electrophoresis (DGGE) can also be used to screen for mutations in a gene. If a nucleotide of a gene is mutated from A to G, the corresponding base in the other strand would be a C. Since the strength of G-C pairing is stronger than that of A-T pairing, a DNA fragment containing this mutation will be more difficult to denature to become single-stranded DNA. This is the basis of DGGE. A small portion (~150 bp) of the normal gene and the gene suspected to have a mutation is amplified with the same pair of primers by PCR. The two PCR products are then electrophoresed in separate lanes in the same gel which contains an ascending formamide gradient. During electrophoresis, the PCR products will denature when they migrate to where the formamide concentration is sufficient to denature them and slow down their migration. The PCR product with the GC mutation will migrate to a place with a higher concentration of formamide before it is denatured. Therefore, the difference in migration speeds indicates the presence of mutations.

17.14: Random Amplification of Polymorphic DNA or arbitrary primed PCR

Random amplification of polymorphic DNA (RAPD) is a PCR method used for typing microorganisms or cells. It uses only one primer of approximately 10 bases to perform PCR. The primer sequence is randomly chosen and is not specific to any gene. This primer may anneal to several different places in different orientations on the genomic DNA of any cells, generating PCR products of different sizes. If two strains of the same species of an organism differ in their nucleotide sequences, different RAPD patterns will result. The arbitrary primed PCR (AP-PCR) is very similar to RAPD. It also uses only one primer to do PCR. The only difference is that the primers used are known. Some examples of the primers that have been used include the M13 universal primer, M13 reverse primer, T7 primer, and T3 primer.

17.15: Amplified Fragment Length Polymorphism

Amplified fragment length polymorphism (AFLP) is another method for strain typing. The DNAs from two different strains of the same species of an organism may yield different banding patterns when their chromosomal DNAs are cut with the same restriction enzyme. Since numerous bands will result when a chromosomal DNA is cut with a certain restriction enzyme, the digested DNAs will run as a smear instead of distinct bands in a gel. Therefore, the difference in banding patterns cannot be visualized. AFLP is a technique that selectively amplifies some of the DNA fragments, thus making the difference in banding patterns detectable. An example of such an application is shown in Fig. 17-6. A chromosomal DNA is first cut with EcoRI and MseI. PCR is then used to amplify some of the fragments containing an EcoRI site at one end and MseI site at the other end. To be able to amplify these DNA fragments, an adaptor is ligated to each end of the fragments. Since EcoRI digestion generates AATT overhang, the adaptor for the EcoRI end is AATTxxxxxxxx, where xxxxxxxx can be any sequence. The sequence of the other strand of the adaptor is complementary to xxxxxxxx, represented by XXXXXXXX. The sequence of the upper strand of the other primer to be ligated to the MseI end would be TAYYYYYYY. The first two bases, TA, are required for the adaptor to anneal to the TA overhang generated by MseI digestion, and YYYYYY represents any sequence. The sequence of the lower strand of the primer is complementary to YYYYYYY, represented by yyyyyyyy in Fig. 17-7. When these two adaptors are ligated to the DNA fragments, they can be amplified with primers whose sequences are those of the adaptors. In this example, the sequence of the 5' primer would be XXXXXXXXAATT. Since this end is the EcoRI site, the sequence can be extended to become XXXXXXXXAATTC. At this time, decision can be made to selectively amplify a certain population of the PCR products. If the primer XXXXXXXXAATTCG is used as the 5' primer, all the PCR products with a G next to the sequence AATTC will be amplified. Similarly, all PCR products with the sequence GCA next to AATTC will be amplified if an oligonucleotide with the sequence XXXXXXXXAATTC**GCA** is used as the primer. The same strategy can be used for the 3' primer. If an oligonuclrotide with the sequence yyyyyyyyTAA**CTG** is used as the 3' primer, all PCR products with GCA next to AATTC at the 5' end GAC next to TAA at the 3' end will be amplified. With this approach only several, instead of numerous, bands are observed after PCR. If different strains of the organism have different sequences, different AFLP patterns will be observed.

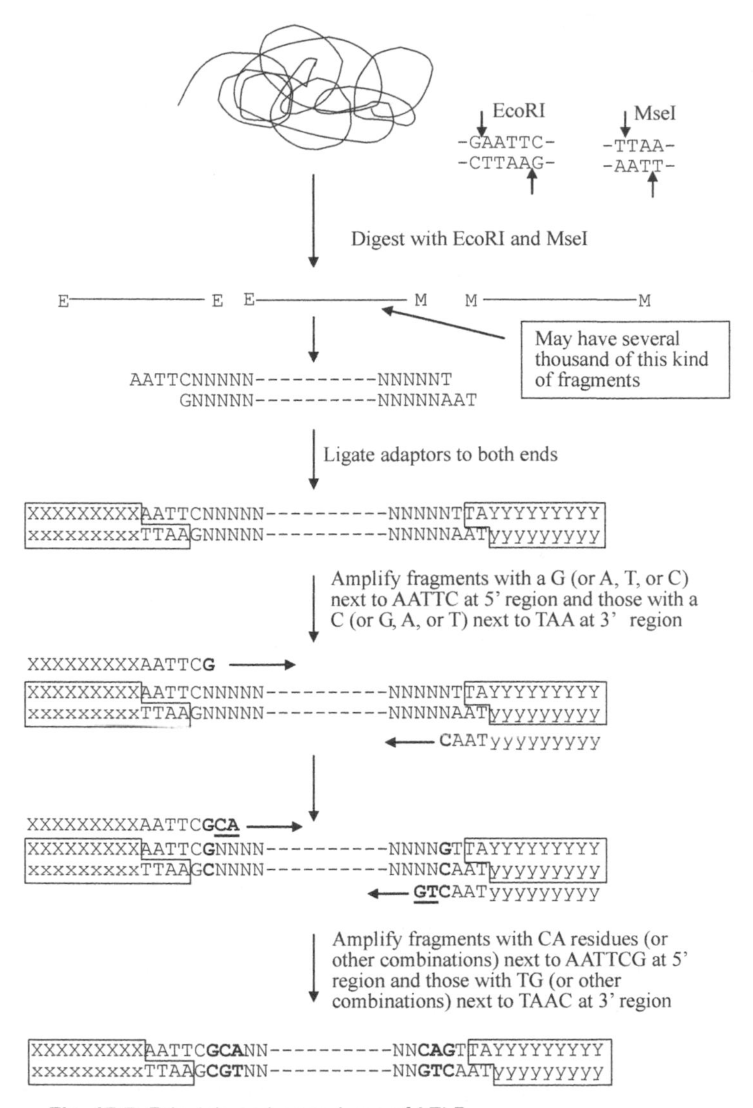

Fig. 17-7: Principle and procedures of AFLP.

17.16: mRNA Differential Display

Gene expression may be changed if cells are treated with certain physical or chemical agents. mRNA differential display is a technique that can be used to detect these changes. The principle of mRNA differential display is very similar to that of RAPD or AP-PCR. The only difference is that mRNA differential display uses two primers (one set) like a regular PCR reaction. The primers used also are not specific to any genes; they are primers which have been shown to be capable of detecting differences in gene expression as a result of a physical or chemical treatment of cells. If the treatment induces production of extra mRNAs, extra PCR products are produced. When these PCR products are sequenced, the identity of genes that are differentially expressed is determined.

17.17: DNA Microarray

DNA microarray allows the study of expression of all genes in a certain type of cell under various conditions. This is possible because the nucleotide sequences of genomes of humans and several animal species and microorganisms have been determined. Synthetic DNA fragments representing different genes of the entire genome are placed on a membrane, glass slide, or matrix similar to that of a computer chip. mRNAs isolated from a certain type of cell are used to produce cDNA or cRNA probes labeled with isotopes or fluorescence. These probes are then hybridized with the microarray. Genes that are expressed in the cells will hybridize with the probes, and the intensity of the hybridization signal will be proportional to the level of gene expression. Since a microarray may contain all the genes in a cell, it is used to study global gene expression, i.e, to determine which genes are expressed in cells under certain conditions, such as in a disease state or after treatment with a drug or physical agent. Some genes may be switched on or off; others may alter their expression levels after the treatment. DNA microarray is currently the most powerful method to study these changes although it is quite costly.

17.18: Serial Analysis of Gene Expression

Serial analysis of gene expression (SAGE) can also be used to study global gene expression in a certain type of cell. In this method, approximately 10 base pairs of each of the genes that are expressed in a cell are sequenced. An example of the SAGE procedure is shown in Fig. 17-8. mRNAs isolated from cells are first converted to double-stranded cDNAs. The cDNAs produced are then digested with NlaIII which produces CATG 3' overhangs. The digested cDNAs are divided into two aliquots. One aliquot is used to ligate an adaptor to each of the NlaIII-digested DNAs. The sequence of the upper strand of the adaptors is 5'-\\\\\\\\\\GGGACATG-3'and that of the lower strand is 3'-////////// CCCT-5', in which the sequences represented by \\\\\\\\\\ (Primer A) and ////////// are complementary. The CATG bases at the end of the adaptor are required for the adaptor to anneal to the ends of NlaIII-digested DNAs. The sequence GGGAC is the recognition sequence of BsmFI which is a Type IIS restriction enzyme. Its cutting site on the upper strand is 10 bases downstream from this sequence, and that of the lower strand is 14 bases downstream from this sequence. Therefore, digestion of the adaptor-ligated cDNAs with BsmFI will cleave a 10-bp fragment from each of the cDNAs present in the reaction.

The same procedure is performed with the other aliquot of the NlaIII digested cDNAs except that a different adaptor is used. In this adaptor, the sequence represented by \\\\\\\\\\ of the adaptor mentioned above is replaced by >>>>>>>>>>, and that represented by //////////// is replaced by <<<<<<<<<<<. All other sequences are the same. After digestion of the adaptor ligated cDNA molecules with BsmFI, the resulting DNA fragments from the two aliquots are mixed and ligated together. The primer with the sequence represented by \\\\\\\\\\ (Primer A) is then used as the 5' primer, and that with the sequence represented by <<<<<<<<<< (Primer B) is used as the 3' primer to amplify the ligated DNA. The resulting PCR products are then digested with NlaIII to generate a 28-bp DNA fragment with CATG overhang at both ends. This 28-bp DNA is composed of two 14-bp fragments from two different cDNA molecules. When the NlaIII digested PCR products are incubated with ligase, many of the 28-bp fragments will be ligated to form a longer DNA fragments. These DNA fragments are then cloned into an appropriate vector, and the insert of the resulting clones are sequenced. Since a 10-bp sequence is sufficient to identify a gene, genes expressed in a certain type of cell can be identified.

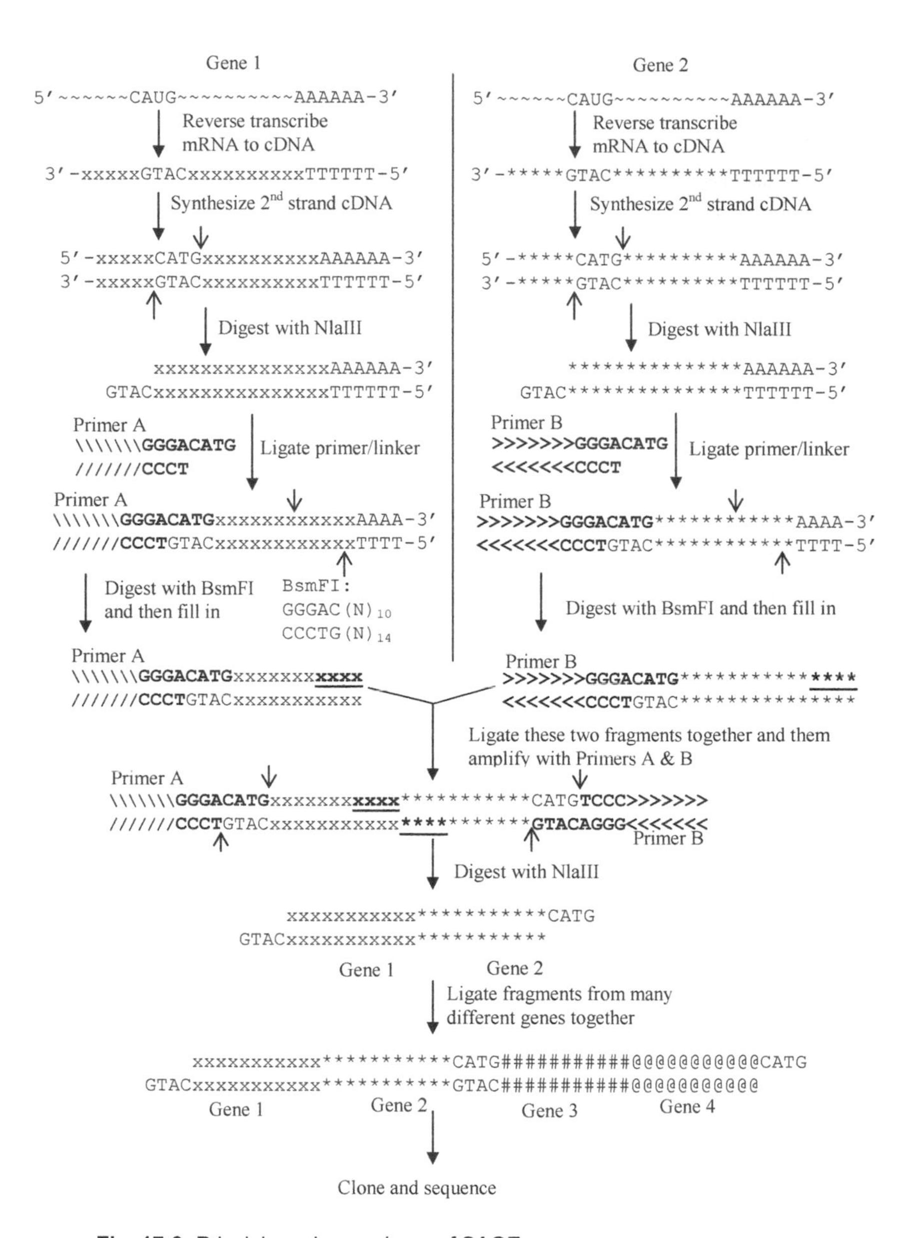

Fig. 17-8: Principle and procedures of SAGE.

17.19: Massively Parallel Signature Sequencing

The principle of massively parallel signature sequencing (MPSS) is similar to that of SAGE. The MPSS technique also sequences a small portion of all the mRNAs in a certain type of cell in order to determine which genes are expressed in the cells. Messenger RNAs isolated from cells are first converted to double-stranded cDNAs. The cDNAs are then linked to microbeads of 5 μm in diameter. Many cDNA molecules are linked to a microbead, and each microbead carries only one type of cDNA. If a gene is expressed at a higher level, more cDNA molecules are linked to the microbead. The cDNA molecules on microbeads are then digested with DpnII, thus allowing the region between the poly-A tail and the first DpnII site upstream from the poly-A tail of each cDNA to remain linked to the microbeads. An adaptor with the upper strand sequence 5'-GATC****GCTGC****-3' and the lower strand sequence 5'-####GCAGC####-3' is then ligated to the cDNAs on the microbeads, where the sequences represented by ***** and ##### are complementary. When these two oligonucleotides are annealed together, an adaptor with a GATC 5' overhang is created. This overhang will anneal to the GATC overhang generated by DpnII digestion. The 5' end of the lower strand of the adaptor is labeled with a fluorescent molecule. Therefore, microbeads carrying adaptor-ligated cDNA molecules can be isolated by flow cytometry. The sequence GCTGA on the adaptor is the recognition sequence of BbvI which is a Type IIS restriction enzyme. The cutting site of BbvI on the upper strand is 12 bases upstream from the sequence GCTGA, and that on the lower strand is 8 bases upstream from this sequence. Therefore, digestion of the adaptor-ligated cDNA molecules will result in a 4-base overhang of unknown sequence.

The BbvII digested microbeads are then injected into a chamber composed of two glass plates with a 5-μm space between them. This allows for a single layer of beads to align in the chamber forming a bead array with each bead containing a different type of cDNA. A total of 1024 different adaptors referred to as encoded adaptors are then injected into the chamber, followed by injection of T4 DNA ligase to ligate the encoded adaptors to the cDNAs. Four different probes, called decoder probes, are then injected into the chamber to identify the first base of the 4-base overhang generated by BbvI digestion. If the first (outmost) base of the overhang is A, the microbead will emit green light when the bead array is irradiated with laser. If it is G, C, or T, it will emit blue, red, or yellow light, respectively. The color of each microbead is recorded, thus determining the first base of the overhang of the cDNAs on each bead. These first base decoder probes are then dissociated from the cDNAs by heating and removed by washing the microbeads with an appropriate buffer. The second set of the decoder probes is then injected into the chamber. Similar to the first set of decoder probes, the beads will be green, blue, red, or yellow if the second base is A, G, C, or T, respectively. After the color of each bead is recorded, the same procedures are repeated for the third and fourth sets of the probes, thus determining the sequence of the 4-base overhang generated by BbvI digestion.

Since the 1024 encoded adaptors that are ligated to the cDNAs also contain the BbvI recognition sequence, another 4 bases next to the fist 4 bases are exposed when the cDNAs on the beads are digested with BbvI again. The same 1024 encoded adaptors are then ligated to BbvI digested cDNA, followed by hybridization with the four different sets of the decoder probes as described above to determine the sequence of the newly exposed 4 bases. If this process is repeated four times, the sequence of 16 bp of each cDNA molecule is determined, and the genes that are expressed in the cell are identified.

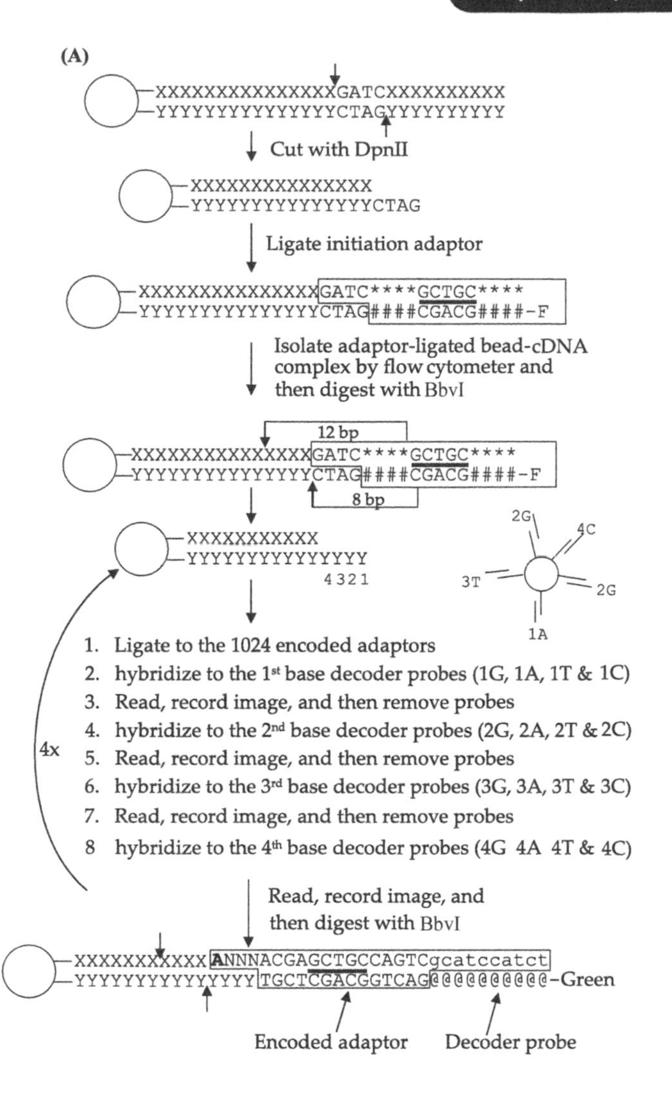
(A)
XXXXXXXXXXXXXXXGATCXXXXXXXXXX
YYYYYYYYYYYYYYYCTAGYYYYYYYYYY
Cut with DpnII
XXXXXXXXXXXXXXX
YYYYYYYYYYYYYYYCTAG
Ligate initiation adaptor
XXXXXXXXXXXXXXXGATC****GCTGC****
YYYYYYYYYYYYYYYCTAG####CGACG####-F
Isolate adaptor-ligated bead-cDNA complex by flow cytometer and then digest with BbvI
12 bp
XXXXXXXXXXXXXXXGATC****GCTGC****
YYYYYYYYYYYYYYYCTAG####CGACG####-F
8 bp
XXXXXXXXXXX
YYYYYYYYYYYYYYY
4321
2G
4C
3T
2G
1A
1. Ligate to the 1024 encoded adaptors
2. hybridize to the 1st base decoder probes (1G, 1A, 1T & 1C)
3. Read, record image, and then remove probes
4. hybridize to the 2nd base decoder probes (2G, 2A, 2T & 2C)
5. Read, record image, and then remove probes
6. hybridize to the 3rd base decoder probes (3G, 3A, 3T & 3C)
7. Read, record image, and then remove probes
8 hybridize to the 4th base decoder probes (4G 4A 4T & 4C)
4x
Read, record image, and then digest with BbvI
XXXXXXXXXXXANNNACGAGCTGCCAGTCgcatccatct
YYYYYYYYYYYYYYYTGCTCGACGGTCAG@@@@@@@@@@-Green
Encoded adaptor
Decoder probe

(B)

1st base encoded adaptors (64 each, 256 total) and decoder probes (4 total)

```
NNNAACGAGCTGCCAGTCcatttaggcg
    TGCTCGACGGTCAG
      1A decoder: gtaaatccgc-Green

NNNGACGAGCTGCCAGTCctgattaccg
    TGCTCGACGGTCAG
      1G decoder: gactaatggc-Blue

NNNCACGAGCTGCCAGTCaccaatacgg
    TGCTCGACGGTCAG
      1C decoder: tggttatgcc-Red

NNNTACGAGCTGCCAGTCcgctttgtag
    TGCTCGACGGTCAG
      1T decoder: gcgaaacatc-Yellow
```

2nd base encoded adaptors (64 each, 256 total) and decoder probes (4 total)

```
NNANACGAGCTGCCAGTCggaacctgaa
    TGCTCGACGGTCAG
      2A decoder: ccttggactt-Green

NNGNACGAGCTGCCAGTCtgtgcgtgat
    TGCTCGACGGTCAG
      2G decoder: acacgcacta-Blue

NNCNACGAGCTGCCAGTCaccgacattc
    TGCTCGACGGTCAG
      2C decoder: tggctgtaag-Red

NNTNACGAGCTGCCAGTCattcctcctc
    TGCTCGACGGTCAG
      2T decoder: taaggaggag-Yellow
```

3rd base encoded adaptors (64 each, 256 total) and decoder probes (4 total)

```
NANNACGAGCTGCCAGTCcgaagaagtc
    TGCTCGACGGTCAG
      3A decoder:gcttcttcag-Green

NGNNACGAGCTGCCAGTCtggtctctct
    TGCTCGACGGTCAG
      3G decoder:accagagaga-Blue

NCNNACGAGCTGCCAGTCtagcggactt
    TGCTCGACGGTCAG
      3C decoder:atcgcctgaa-Red

NTNNACGAGCTGCCAGTCggcgataact
    TGCTCGACGGTCAG
      3T decoder:ccgctattga-Yellow
```

4th base encoded adaptors (64 each, 256 total) and decoder probes (4 total)

```
ANNNACGAGCTGCCAGTCgcatccatct
    TGCTCGACGGTCAG
      4A decoder:cgtaggtaga-Green

GNNNACGAGCTGCCAGTCcaactcgtca
    TGCTCGACGGTCAG
      4G decoder:gttgagcagt-Blue

CNNNACGAGCTGCCAGTCcacagcaaca
    TGCTCGACGGTCAG
      4C decoder:gtgtcgttgt-Red

TNNNACGAGCTGCCAGTCgccagtgtta
    TGCTCGACGGTCAG
      4T decoder:cggtcacaat-Yellow
```

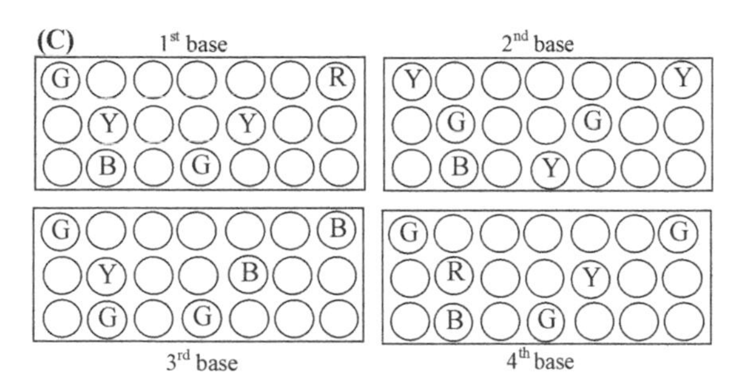

G: Blue, A: Green, T: Yellow, C: Red
(GYGG = ATAA, YGYR = TATC, BBGB =
GGAG, YGBY = TAGT, RYBG = CTGA)

Fig. 17-9: Principle and procedures of MPSS. (A). An adaptor containing BbvI recognition sequence is ligated to DpnII digested cDNA fragments on microbeads. After BbvI digestion, the sequence of the 4-base overhang is determined by ligating encoded adaptors, followed by hybridization with decoder probes. (B). Sequences of encoded adaptors and decoder probes. (C). Bead arrays hybridized with the four sets of decoder probes.

To perform MPSS, cDNA molecules need to be linked to microbeads. To achieve this, a 32-base oligonucleotide (tag) is linked to each cDNA molecule. Each different cDNA molecule is linked to a different tag. A total of 16,777,216 different 32-base tags are made, each with a different sequence. The sequences of these 16,777,216 tags are combinations of the following 8 sequences: TTAC, AATC, TACT, ATCA, ACAT, TCTA, CTTT, and CAAA. A set of oligonucleotides with the sequence ********GGGCCC(N_{32})TTAATTAA######## are first synthesized, where N_{32} represents all possible combinations of these 8 sequences. These oligonucleotides are converted to double-stranded DNA by PCR using oligonucleotides with sequences represented by ******** and ######## as primers. The resulting PCR products are digested with Bsp1201 which recognizes GGGCCC and PacI which recognizes TTAATTAA . The digested products are then cloned between the Bsp1201 and PacI sites on pLCV1, thereby creating a library of 16,777,216 of N_{32} tag clones (Fig. 17-10).

To be able to clone the cDNAs into the N_{32} tag clones, a biotin-conjugated oligonucleotide with the sequence 5'-GACATGCT<u>CGT CTC</u>TGCATTTTTTTTT-3' is used as the primer to perform reverse transcription to convert mRNAs to cDNAs, in which the sequence CGTCTC is the recognition sequence of BsmBI which is also a Type IIS restriction enzyme. Its cutting site on the upper strand is one base 3' to this sequence, and that of the lower strand is five bases downstream from this sequence. Therefore, BsmBI-digested cDNAs will have GCTA 3' overhangs that can be ligated to the BbsI site on pLCV1. The 5' end of the cDNA fragments is ligated to the BamHI site on pLCV1 after BamHI digestion; therefore, the region between poly-A tail and the first DpnII site of each cDNA is cloned into pLCV1 with it 3' end tagged with N_{32} (Fig. 17-11).

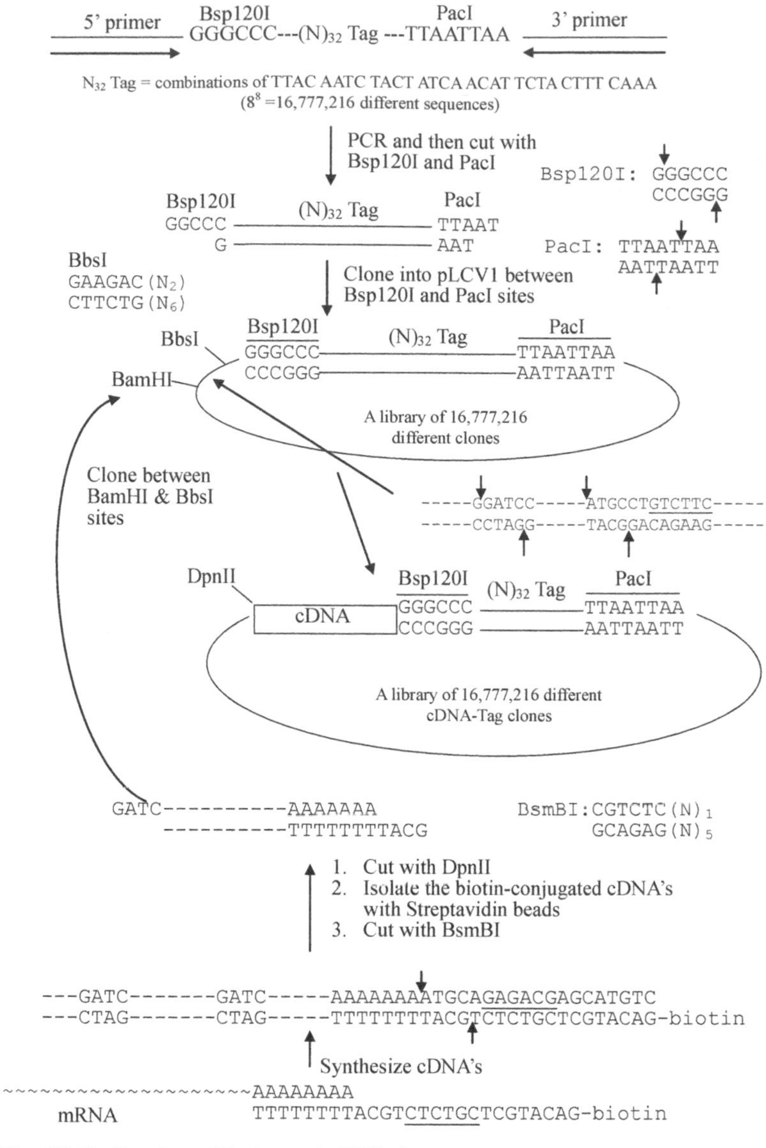

Fig. 17-10: Creation of N_{32}-tagged cDNA clones.

To link the cDNAs to microbeads, the cDNAs in this library of 16,777,216 different clones are amplified with primers that anneal to pLCV1 at the two regions flanking the N_{32}-tagged cDNAs (Fig. 17-11). The PCR products generated are then cut with PacI which generates 3' overhangs. T4 DNA polymerase and dGTP are then added to the

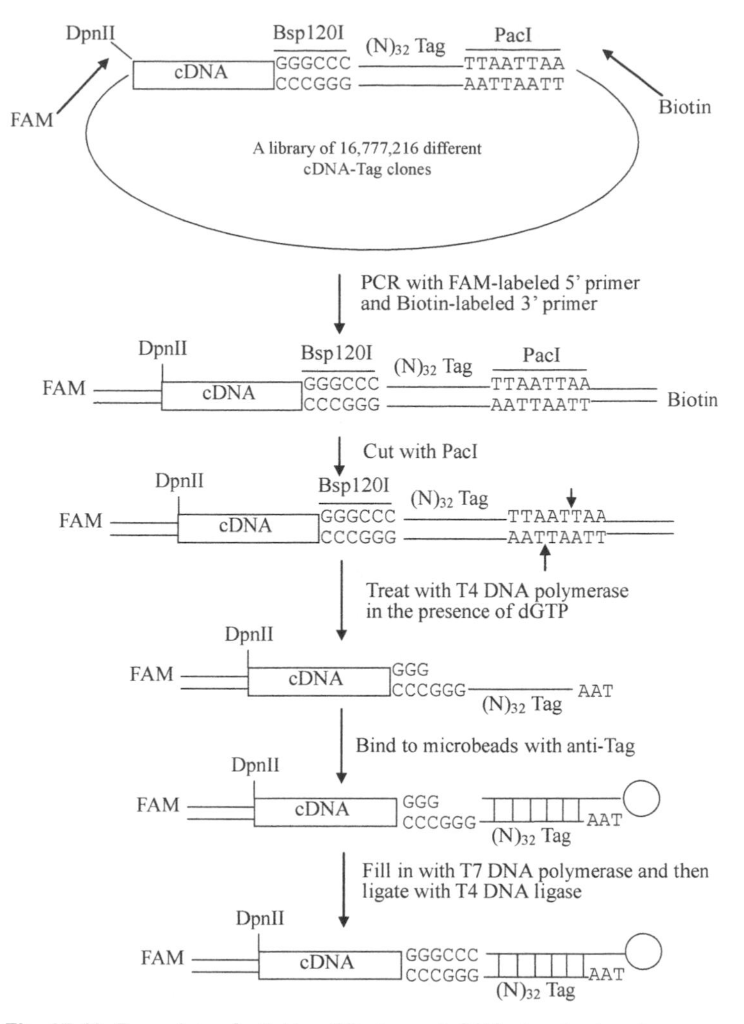

Fig. 17-11: Procedures for linking $(N)_{32}$ tagged cDNAs to microbeads.

reaction. Since T4 DNA polymerase has the single-stranded 3' to 5' exonuclease activity, it will remove nucleotides on the upper strand. Although it also has the 5' to 3' polymerase activity, it will only fill in the three G's next to the 3' end of the cDNA because dATP, dCTP, and dTTP are not present in the reaction. This treatment will generate cDNA molecules linked to single-stranded N_{32} tags whose sequences are complementary to those linked to the microbeads. When these microbeads are mixed with T4 DNA polymerase-treated cDNAs, the cDNAs are linked to microbeads. The gap between the three G's and the tag is then filled in with T7 DNA polymerase and ligated by T4 DNA ligase. Since the 5' primer used to amplify cDNAs from the library is labeled with FAM, beads that are loaded with the cDNAs can be isolated by flow cytometry. The cDNA-loaded beads are then digested with DpnII and subjected to 4 rounds of hybridization reactions described above to determine the sequence of 16 bp from each cDNA molecule.

17.20: Sequencing by Oligonucleotide Ligation and Detection

Sequencing by oligonucleotide ligation and detection (SOLiD™) was developed by Applied Biosystems, Inc. in 2007. It is one of the next (second) generation sequencing methods. To perform SOLiD, the DNA to be sequenced is first converted to single-stranded and fused to an adaptor that is linked to a microbead, 1 μm in diameter. A universal primer is then annealed to the distal portion of the adaptor immediately adjacent to the region to be sequenced. A mixture of 1,024 partially degenerate probes are then added. The probe which anneals next to the universal primer is then ligated to it. Each of the partially degenerate probes is composed of 8 nucleotides with three parts (Fig. 17-12). The last 3 bases at the 3' end of each probe are degenerate with the sequence "nnn". The sequences of the middle two bases (xx) are the 16 different combinations of any two bases, including AA, AC, AG, AT, CA, CC, CG, CT, GA, GC, GG, GT, TA, TC, TG, and TT. The three bases at the 5' end are universal bases (zzz), such as inosine or its derivatives, which can pair with any base. A fluorescent dye is conjugated to the 5' end of each probe. Four different fluorescent dyes with different colors (blue, yellow, green, and red) are used. The fluorescent dye that is conjugated to probes with the middle two bases (xx) being AA, CC, GG, and TT will release blue light when it is irradiated with laser. Green light is emitted if the middle two bases of the probes are AC, CA, GT, or TG. The ones whose middle two bases are AG, CT, GA, or TC will release yellow light, and those that have AT, CG, GC, or TA in the middle will emit red light when irradiated with laser. If the microbead releases green light upon laser irradiation, the sequences of the 4th and 5th bases of the template are AC, CA, GT, or TG. After the results are recorded, the three universal bases (zzz) located at the 5' portion of the probe are removed. If these three bases are ribonucleotides (with a 2' OH group), this portion is RNA which can be digested away with RNaseH. If the linkage between the z and x bases of the probe is a phosphothioate bond, the universal bases can be cleaved with silver ions (e.g. 50 mM $AgNO_3$). Another probe ligation is performed, the color of light emitted from each bead is recorded, and the universal bases on the second probe are again removed. When this process is repeated 5 times, bases 4-5, 9-10, 14-15, 19-20, and 24-25 of the template are sequenced (Fig. 17-12).

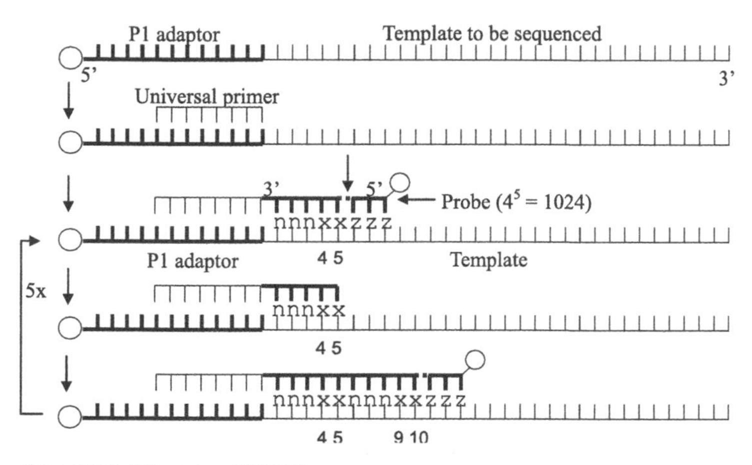

Fig. 17-12: Principle of SOLiD.

The entire system is reset by dissociating and removing all the primers and probes that anneal to the template. A different universal primer is then used to perform the second round of sequencing. This universal primer is called N-1 which anneals to the adaptor at the position one base left of where the first primer binds. Five cycles of the probe ligation reaction are again performed, thus determining the sequences of bases 3-4, 8-9, 13-14, 18-19, and 23-24. After this second round reaction is completed, the system is reset again, and the third round of the reaction is performed with another universal primer, N-2. This primer anneals to the adaptor at the position one base left of where primer N-1 binds. Five cycles of the reaction with primer N-2 are again performed. This process will determine the sequences of bases 2-3, 7-8, 12-13, 17-18, and 22-23. The same procedures are then performed with primer N-3 to determine the sequences of bases 1-2, 6-7, 11-12, 16-17, and 21-22 and with primer N-4 to determine the sequences of bases 0-1, 5-6, 10-11, 15-16, and 20-21. When the results of all 5 rounds of reactions are combined, the sequence of the 25 bases of the template is determined (Fig. 17-13).

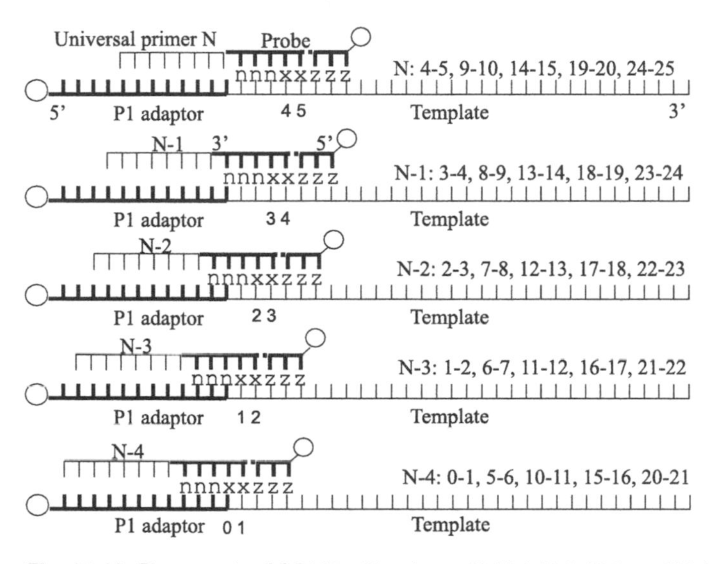

Fig. 17-13: Five rounds of SOLiD with primers N, N-1, N-2, N-3, and N-4 for sequencing of 25 bases of a target DNA.

The sequence determined with primer N-4 then serves as the reference point to identify other bases as shown in Fig. 17-14. If the first reaction with this primer releases green light, the sequence of bases 0 and 1 would be either AC, CA, GT, or TG. Since base number 0 is the last base of the P1 adaptor, its identity is known. If this base is G, base number 1 must be T because among these 4 possibilities only GT has a G at the first position. To identify base number 2, the result of the first reaction performed with primer N-3 is used. If this reaction releases blue light, the sequence of bases 1 and 2 would be AA, CC, GG, or TT. Since base number 1 is already determined to be a T, base number 2 must also be a T. The same principle is used to sequence other bases. If red light is detected in the first reaction performed with primer N-2, the sequence of bases 2 and 3 would be AT, CG, GC, or TA. Since base number 2 is determined to be a T in the reaction with primer N-3, base number 3 has to be an A. Similarly, if the first reaction performed with primer N-1 emits yellow light, the sequence of bases 3 and 4 is either AG, CT, GA, or TC. In this case, base number 4 has to be a G because base number 3 has been determined to be an A. The identity of nucleotide number 5 is determined in the first reaction with primer N. If this reaction releases red light, the sequence of bases 4 and 5 is either AT, CG, GC, or TA. Since base number 4 is already known to be G, base number 5 must be C. All the other nucleotides in the template are determined in the same manner.

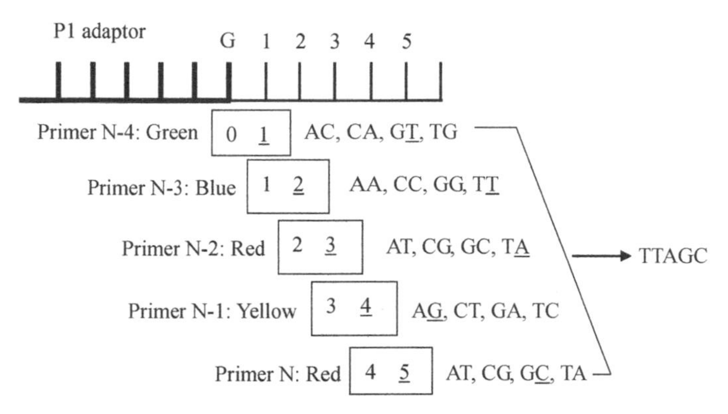

Fig. 17-14: Determination of the sequence of the first 5 bases of the template by SOLiD with primers N-4, N-3, N-2, N-1, and N. The first reaction with these primers releases green blue, red, yellow, and red light, respectively.

The SOLiD system can be used to sequence a specific DNA fragment or the entire genome of certain types of cells or organisms. To sequence a specific DNA fragment, the DNA fragment is isolated. Two different primers, P1 and P2, are ligated to the two ends of the fragment, respectively; the resulting DNA with two different primers at their ends are then amplified by asymmetrical PCR using P1 as the

5' primer and P2 as the 3' primer (Fig. 17-15). In this asymmetrical PCR, the amount of P2 is much greater than that of P1; therefore, the majority of the resulting PCR products are single-stranded and are the minus (lower) strand of the template. These single-stranded DNA molecules are then annealed to the single-stranded adaptor, whose sequence is complementary to that of P1, that is linked to a microbead. Approximately 20,000 molecules of the single-stranded DNA molecules are annealed to a single microbead. The adaptor molecules that are linked to the microbeads are then used as the primer to convert the single-stranded DNA to doubled-stranded. The original single-stranded (minus strand) template is then dissociated from the double-stranded DNA, thus creating a single-stranded (plus strand) DNA covalently linked to the adaptor (Fig. 17-15). The microbeads containing the template DNA are then used for the sequencing reactions described above.

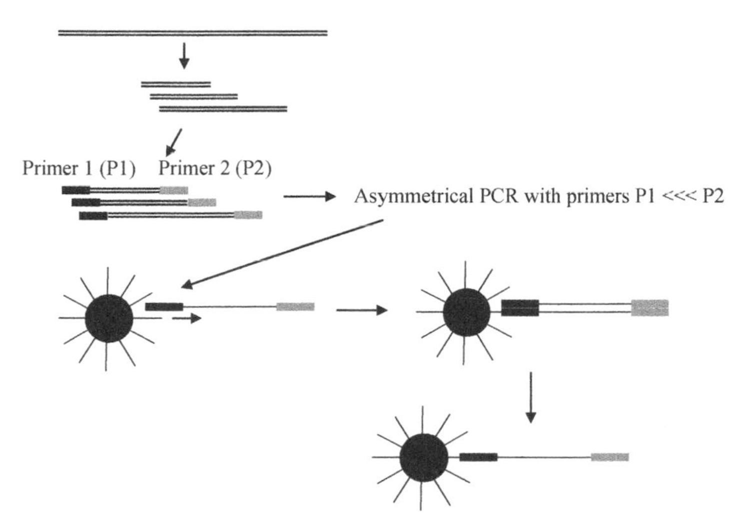

Fig. 17-15: Procedures for linking a single-stranded DNA template to microbeads. The DNA to be sequenced is cut with a certain restriction enzyme. Primers P1 and P2 are then ligated to the ends of each fragment. Asymmetrical PCR is performed to produce the single, minus strand DNA of each fragment. The single-stranded DNA molecules are then annealed to the adaptors that are conjugated to a microbead. The adaptors are used as primers and the single-stranded DNA as the template to synthesize the plus strand DNA which becomes linked to the microbead. The minus strand is then dissociated and removed.

For whole genome sequencing, the genomic DNA is broken into pieces by sonication. An adaptor, referred to as the internal adaptor, is then ligated to both ends of the DNA fragments. The adaptor-ligated DNA fragments are then circularized; this is made possible by making the adaptor capable of self-annealing. The circularized DNA molecules are then digested with EcoP151 which is a Type IIS restriction enzyme. It recognizes the sequence CAGCAG, which is present at each end of the adaptor, and cuts at 25 bp away from this sequence. This digestion will result in DNA fragments containing the internal adaptor flanked by 25 bp DNA on each side (Fig. 17-16). Primers P1 and P2 are then

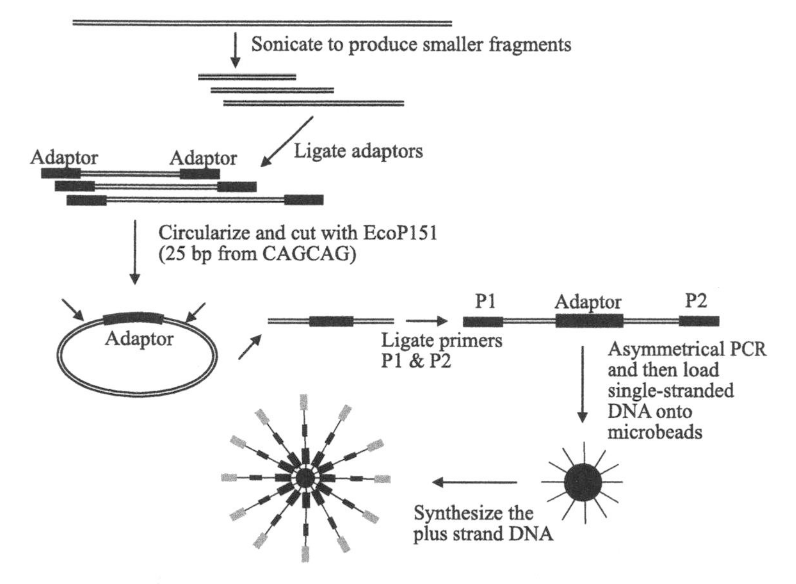

Fig. 17-16: Preparation of fragments for whole genome sequencing by SOLiD. Adaptors are ligated to the ends of small DNA fragments generated by sonication of the genomic DNA. The adaptor-ligated DNA fragments are circularized and then cut with EcoP151, resulting in many DNA fragments each containing the adaptor flanked by 25 bp of genomic DNA on each side. Primers P1 and P2 are ligated to these DNA fragments that are then loaded onto microbeads to produce single, plus strand DNA templates linked to microbeads as described in the legend of Fig. 17-15.

ligated to the ends of these DNA fragments in order to perform the asymmetrical PCR described above. The resulting single-stranded DNA molecules are then loaded onto microbeads in the same manner as that used to sequence a specific DNA fragment. Two different sets of primers are used to perform SOLiD sequencing of the two 25-bp fragments flanking the internal adaptor. Since different microbeads may contain DNA from different portions of the genome, the entire genome can be sequenced if the sequencing results from all microbeads are combined.

The asymmetrical PCR used to produce single-stranded template DNA is performed with a special PCR method called emulsion PCR (Fig. 17-17) in which the PCR is performed in a water-in-oil emulsion. An example of the formulation for making water-in-oil emulsion is 4.5% (v/v) Span 80, 0.4% (v/v) Tween 80, and 0.05% (v/v) Triton X-100 in mineral oil. Span 80 (sobitan monooleate), Tween 80 (polyoxyethylene 20 sorbitan monooleate), and Triton X-100 (octylphenol ethoxylate) are nonionic surfactants. Very small oil droplets are formed when the aqueous PCR reagents are mixed with this formulation. Each oil droplet then serves as a test tube in which the asymmetrical PCR is performed. The concentration of the PCR mixture is diluted so that each aliquot in an oil droplet contains only one microbead and one DNA molecule. Therefore, each of the resulting DNA-loaded microbeads contains only one type of DNA molecule. If two or more different DNA molecules are loaded onto the same microbead, the sequencing results will not be useful and will be rejected by the computer software. It is estimated that a single SOLiD run can sequence approximately 3 billion bases of DNA.

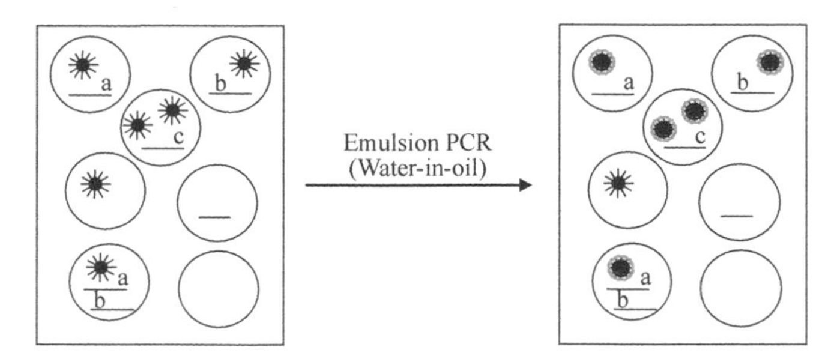

Fig. 17-17: Emulsion PCR for generation of microbeads loaded with single, plus strand template DNA for sequencing by SOLiD. PCR reagents and adaptor-conjugated microbeads are mixed with mineral oil containing nonionic surfactans including Span 80, Tween 80, and Triton X-100 to form oil droplets. The PCR mix is diluted such that each oil droplet contains only one template DNA molecule.

17.21: Total Gene Expression Analysis

Total gene expression analysis (TOGA) is another method for studying global gene expression. Its basic strategy is to divide the expressed genes of a certain type of cell into 256 groups. If a cell expresses 1000 different genes, each group will only have 4 – 5 genes. A portion of the mRNAs in each group is then sequenced. When the results from all 256 groups are combined, all genes that are expressed in the cells are identified. One example of TOGA is shown in Fig. 17-18. mRNAs are isolated from cells and then converted to cDNAs using a biotin-conjugated oligo-dT with the NotI site built in as the primer. The resulting double-stranded cDNA's are then cut with MspI which recognizes the sequence CCGG. In TOGA, the region between the poly-A tail and the first MspI site upstream from the poly-A tail of each cDNA is sequenced. Since the primer used for reverse transcription is conjugated to biotin, the fragments generated by MspI digestion can be isolated by magnet-conjugated streptavidin. The isolated DNA fragments are digested with NotI to remove the primer and then cloned into a vector between ClaI and NotI sites as the ClaI ends are compatible with MspI ends. This process will create a cDNA library with each clone containing the region between poly-A and the first MspI site of each cDNA. Each plasmid of the library is then digested with MspI which will cut out the cDNA flanked by a portion of the plasmid on each side. Since the region located at the 3' side of the cDNA contains the T3 promoter, in vitro transcription is then performed using T3 RNA polymerase and these DNA fragments as templates. The RNAs produced are amplified by RT-PCR using a common 3' primer and four different 5' primers. The 3' primer anneals to the region containing the T3 promoter, while the 5' primers anneal to the region immediately adjacent to the 5' end of the cDNA fragment. Since the sequence of the plasmid is known, the sequence of each 5' primer is the sequence of that of the 3' end of the plasmid DNA located at 5' side of the cDNA and the 1st base of the 5' end of the cDNA. If the last base is G, the primer will amplify all cDNAs whose first base is G. Since the 4 different primers have G, A, T, or C as their last base, the cDNAs can be divided

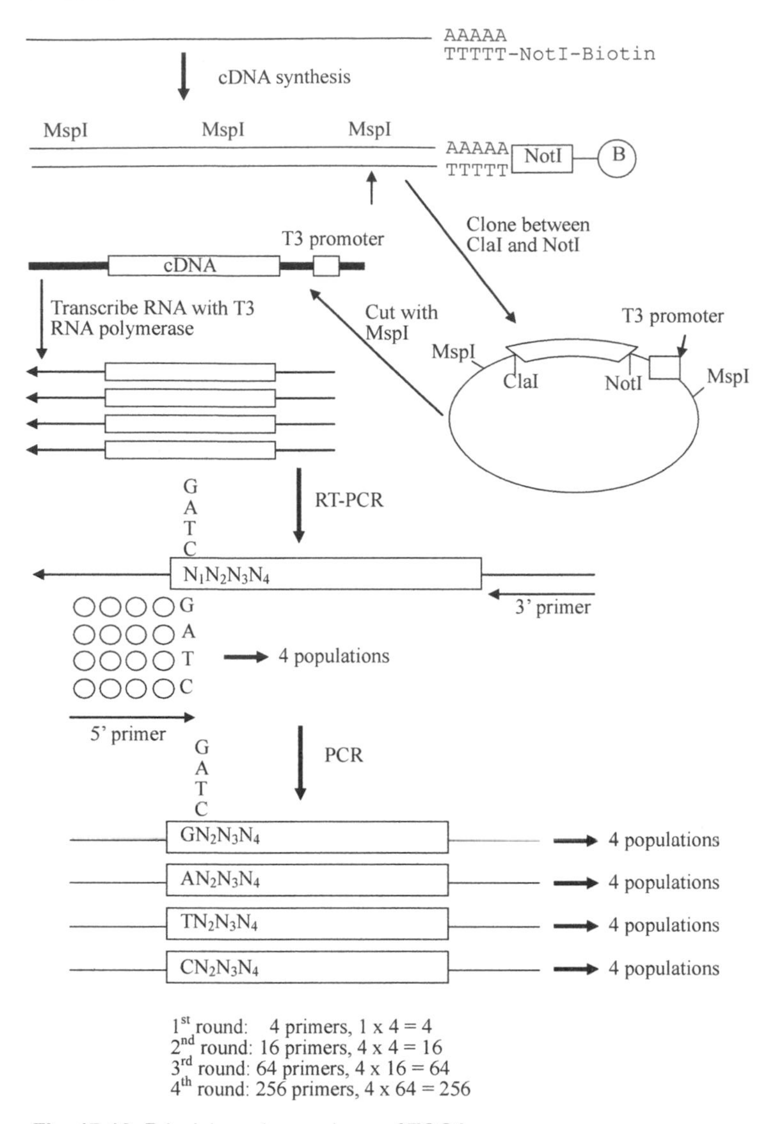

Fig. 17-18: Principle and procedures of TOGA.

into 4 different groups. For the G group, the second base of the cDNAs can also be G, A, T, or C. Therefore, with the use of another 4 primers whose last second base is G, and the last base is G, A. T, or C, the G group can be further divided into another 4 groups. The same procedure is performed with the A, T, and C groups, thus dividing the cDNAs into 16 groups. If this process is repeated 4 times, the cDNAs are divided into 256 groups with each group containing 4 – 5 different cDNAs.

17.22: Representational Difference Analysis

The technique of representational difference analysis (RDA) is used to isolate a portion of the DNA of an unknown etiologic agent. Genomic DNAs are isolated from normal and diseased cells. The two different DNA samples are then digested with the same restriction enzyme. As the example shown in Fig. 17-19, the DNA from normal cells yields a, b, and c fragments, and that from diseased cells yields A, B, C, and D fragments. The extra D fragment is likely derived from the etiologic agent. To isolate the D fragment, a primer is ligated to the 5' ends of A, B, C, and D fragments. These fragments are then denatured and mixed with denatured a, b, and c fragments from normal cells. Since the sequences of the a, b, and c fragments are the same as those of the A, B, and C fragments, A/a, B/b, and C/c hybrids will form, but not D/d hybrid. The upper strand of the D fragment can only pair to its own lower strand to form the D/D molecule. Among these DNA fragments, only the D/D molecules have the primer ligated to the 5' end of both strands, and thus only the D/D fragments can be amplified by PCR with the primer. The PCR products generated are then cloned and sequenced to identify the etiologic agent.

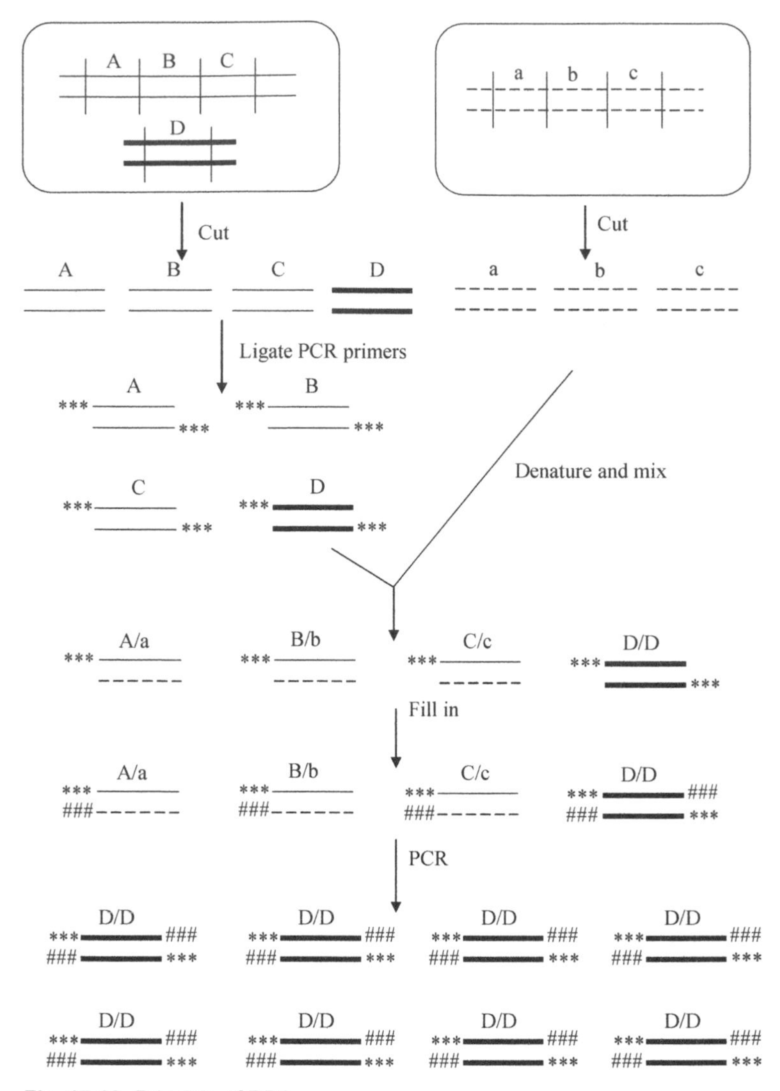

Fig. 17-19: Principle of RDA.

Summary

Nuclear run-on assay is used to determine the rate of transcription. Reporter assay is used to determine the strength of a certain promoter. It can also be used to locate a certain promoter or an enhancer. Some proteins can bind to DNA. The electrophoretic mobility shift assay (EMSA) is the technique most commonly used to study protein-

DNA interactions. To determine the binding sequence of a certain protein, DNase I footprinting assay can be performed. The chromatin immunoprecipitation (ChIP) assay can be used to detect, identify, or isolate DNA fragments to which a certain transcription factor binds. The yeast two-hybrid system is used to investigate protein-protein interactions, and the three-hybrid system is used to identify proteins that bind to RNA. To determine the start site of transcription, primer extension is performed. S1 nuclease analysis can be used to determine the 3' end of mRNAs and the intron-exon junctions. RNase A protection assay is used to determine the levels of certain mRNAs in a cell. Single-strand conformation polymorphism (SSCP), heterduplex assay, and denatured gradient gel electrophoresis (DGGE) are used to screen for mutations in a certain gene. For strain typing, random amplification of polymorphic DNA (RAPD), arbitrary-primed PCR (AP-PCR), and amplified fragment length polymorphism (AFLP) can be used. Messenger RNA differential display is used to identify genes that are differentially expressed after treatment of cells with a certain physical or chemical agent. DNA microarray allows for the study of global gene expression. It can also be used to investigate differential gene expression. Serial analysis of gene expression (SAGE), massively parallel signature sequencing (MPSS), sequencing by oligonucleotide ligation and detection (SOLiD), and total gene expression analysis (TOGA) can be used to identify genes that are expressed in certain types of cells. Representational difference analysis (RDA) can be used to identify unknown etiologic agents.

Sample Questions

1. Reporter assay is usually used to study: a) translation efficiency, b) transcription elongation, c) promoter strength, d) binding of protein to DNA.
2. Electrophoresis mobility shift assay (EMSA) is used to investigate: a) translation efficiency, b) transcription elongation, c) promoter strength, d) binding of protein to DNA.
3. DNase I footprinting is used to determine: a) transcription start site, b) translation start site, c) the sequence of a DNA-binding protein, d) promoter strength.
4. Chromatin immunoprecipitation (ChIP) is used to: a) isolate a DNA-binding protein, b) isolate a DNA fragment to which a certain protein binds, c) determine the structure of

a chromosome, d) investigate the function of a promoter.

5. Two-hybrid system is used: a) to create fusion protein for expression, b) to investigate protein-protein interaction, c) to detect and isolate proteins which bind to a certain protein, d) to make RNA-DNA hybrid.
6. Which of the following is used to determine the initiation site of transcription: a) primer extension, b) PCR, c) in vitro transcription, d) in vitro translation.
7. S1 nuclease analysis is used to: a) determine the 3' end of mRNA, b) determine intron-exon junction, c) quantify a certain mRNA, d) determine RNA secondary structure.
8. RNase A protection assay is used to: a) protect RNA from degradation, b) quantify a certain mRNA, c) make RNA-DNA hybrid, d) determine RNA structure.
9. Single strand conformation polymorphism (SSCP) is used to: a) detect mutation in DNA, b) determine the secondary structure of DNA, c) identify a point mutation in DNA, d) determine the secondary structure of RNA.
10. Heteroduplex assay is used to: a) detect mutation in DNA, b) make RNA-DNA hybrid, c) identify a point mutation in DNA, d) determine the secondary structure of RNA.
11. Denatured gradient gel electrophoresis (DGGE) is used to: a) detect mutation in DNA, b) determine DNA sequence, c) identify a point mutation in DNA, d) determine RNA sequence.
12. Amplified fragment length polymorphism (AFLP) is used to: a) amplify RNA, b) perform strain typing, c) investigate protein polymorphism, d) determine DNA sequence.
13. mRNA differential display is used to: a) determine the abundance of a certain RNA, b) study difference in gene expression between two different types of cells, c) study alternative mRNA splicing, d) determine difference in mRNA structure.
14. DNA microarrays yield which of the following information: a) identity of genes that are expressed in a certain cell, b) transcription initiation site, c) the relative level of expression of genes, d) replication origin.
15. SAGE (serial analysis of gene expression) is used to: a) study the level of expression of a certain gene, b) identify genes that are expressed in a certain type of cells, c) determine transcription initiation site, d) study interactions between genes.
16. MPSS (massively parallel signature sequencing) is used to: a) sequence a cloned DNA fragment, b) sequence the entire gene that is expressed in a certain cell, c) identify genes that are expressed in a certain type of cells, d) sequence a small portion of all the genes that are expressed in a certain type of cells.

Suggested Readings

1. Alam, J. and Cook, J. L. (1990). Reporter genes: application to the study of mammalian gene transcription. Anal. Biochem. 188: 245-254.
2. Albrecht, G., Vermaas, E., Williams, S. R., Moon, K., Burcham, T., Pallas M., DuBridge, R. B., Kirchner, J., Fearon, K., Mao, J-I., and Corcora, K. (2000). Gene expression analysis by massively parallel signature sequencing (MPSS) on microbead arrays. Nature Biotech. 18: 630-634.
3. Brenner, S., Williams, S. R., Vermaas, E. H., Storck, T., Moon, K., McCollum, C., Mao, J-I., Luo, S., Kirchner, J. J., Eletr, S., DuBridge, R. B., Burcham, T., and Albrecht, G. (2000). In vitro cloning of complex mixtures of DNA on microbeads: physical

separation of differentially expressed cDNAs. Proc. Natl. Acad. Sci. USA 97: 1665-1670.

4. Brenowitz, M., Senear, D. F., Shea, M. A., and Ackers, G. K. (1986) "Footprint" titrations yield valid thermodynamic isotherms. Proc. Natl. Acad. Sci. USA 83: 8462-8466.

5. Cullen, B. R. (1987). Use of eukaryotic expression technology in the functional analysis of cloned genes. Meth. Enzymol. 152: 684-704.

6. Dedon, P. C., Soults, J. A., Allis, C. D., and Gorovsky, M. A. (1991). A simplified formaldehyde fixation and immunoprecipitation technique for studying protein-DNA interactions. Anal. Biochem. 197: 83-90.

7. Fields, S. and Song, O. (1989). A novel genetic system to detect protein-protein interactions. Nature 340: 245-246.

8. Fried, M. and Crothers, D. M. (1981). Equilibrium and kinetics of *lac* repressor-operator interactions by polyacrylamide gel electrophoresis. Nucleic Acids Res. 9: 6505-6525.

9. Ganguly, A., Rock, M. J., and Prockop, D. J. (1993). Conformation sensitive gel electrophoresis for rapid detection of single-base differences in double-stranded PCR products and DNA fragments: evidence for solvent-induced bends in DNA heteroduplexes. Proc. Natl. Acad, Sci. USA 90: 10325-10329.

10. Hovig, E., Smith-Sorensen, B., Brogger, A., and Borresen, A.-L. (1991). Constant denaturant gel electrophoresis, a modification of denaturing gradient gel electrophoresis, in mutation detection. Mutation Research 262: 63-71.

11. Kanazawa, A., O'Dell, M., Hellens, R. P., Hitchin, E., and Metzlaff, M. (2000). Mini-scale method for nuclear run-on transcription assay in plants. Plant. Mol. Bio. Reporter 18: 377-383.

12. Lemon, B. and Tjian, R. (2000). Orchestrated responses: a symphony of transcription factors for gene control. Genes Dev. 14: 2551-2569.

13. Liang, P. and Pardee, A. B. (1992). Differential display of eukaryotic messenger RNA by means of the polymerase chain reaction. Science. 257: 967-971.

14. Lisitsyn, N., Lisitsyn, N. and Wigler, M. (1993). Cloning the differences between two complex genomes. Science 259: 946-951.

15. Mitchell, P. and Tjian, R. (1989). Transcriptional regulation in mammalian cells by sequence-specific DNA-binding proteins. Science 245: 371-378.

16. Myers, R. M., Maniatis, T., and L. S. Lerman. (1987). Detection and localization of single base pair changes by denaturing gradient gel electrophoresis. Meth. Enzymol. 155: 501-527.

17. Orita, M., Iwahana, H., Kanazana, H., Hayashi, K., and Sekiya, T. (1989). Detection of polymorphisms of human DNA by gel electrophoresis as single-strand conformation polymorphisms. Proc. Natl. Acad. Sci. USA 86: 2766-2770.

18. Pabo, C. T. and Saue, R. T. (1992). Transcription factors: structural families and principles of DNA recognition. Annu. Rev. Biochem. 61: 1053-1095.

19. Prost, E. and Moore, D. D. (1986). CAT vectors for analysis of eukaryotic promoters and enhancers. Gene 45: 107-111.

20. Ptashne, M. (1988). How eukaryotic transcriptional activators work. Nature 335: 683-689.

21. SenGupta, D. J., Zhang, B., Kraemer, B., Pochart, P., Fields, S., and Wickens, M.

(1996). A three-hybrid system to detect RNA-protein interactions in vivo. Proc. Natl. Acad. Sci USA 93: 8496-8501.

22. Sawadago, M. and Roeder, R. G. (1985). Interaction of a gene-specific transcription factor with the adenovirus major late promoter upstream of the TATA box region. Cell 43: 165-175.
23. Sutcliffe, J. G., Foye, P. E., Erlander, M. G., Hilbush, B. S., Bodzin, L. J., Durham, J. T., and Hasel, K. W. (2000). TOGA: an automated parsing technology for analyzing expression of nearly all genes. Proc. Natl. Acad. Sci. USA 97: 1976-1981.
24. Velculescu, V. E., Zhang, L., Vogelstein, B., and Kinzler, K. W. (1995). Serial analysis of gene expression. Science 270: 484-487.
25. Vos, P., Hoqers, R., Bleeker, M., Reijans, M., van de Lee, T., Hornes, M., Frijters, A., Pot, J., Peleman, J., Kuiper, M., and Zabeau, M. (1995). AFLP: a new technique for DNA fingerprinting. Nucleic Acids Res. 11: 4407-4414.
26. Williams, J. G. K., Kubelik, A. R., Livak, K. J., Rafalski, J. A., and Tingey, S. V. (1990). DNA polymorphisms amplified by arbitrary primers are useful as genetic markers. Nucleic Acids Res. 18: 6531-6535.
27. Williams, R., Peisajovich, S. G., Miller, O. J., Magdssi, S., Tawfik, D. S., and Griffiths, A. D. (2006). Amplification of complex gene libraries by emulsion PCR. Nature Methods. 3: 545-550.
28. Wood, K. V. (1990). Firefly luciferase: a new tool for molecular biologists. Promega Notes: 28: 1-3.
29. Young, R. A. (2000). Biomedical discovery with DNA arrays. Cell 102: 9-15.

Glossary

Accession number – The unique identifier assigned to a new DNA or protein sequence submitted to a major database.

Actin – A family of proteins, very abundant in eukaryotic cells (8 - 14% total cell protein) and one of the major components of the actomyosin motor and the cortical microfilament meshwork. Since ß-actin mRNA is usually produced at an equal level in nucleated cells, it is often used to normalize RNA sample levels.

Activator – A protein that turns on or up-regulates gene expression by binding to transcription control sites.

Adaptor – A double stranded, synthetic oligonucleotide with sequence of a certain restriction enzyme. By ligation of the adaptor to a DNA fragment, sticky ends can be generated to aid in ligation of the fragment into a vector.

Adenovirus – A group of DNA viruses that cause respiratory disease, including one form of the common cold. Adenoviruses can also be genetically modified and used in gene therapy to certain diseases.

Adenylate cyclase – An enzyme that uses ATP as a substrate to generate cyclic AMP, in which 5' and 3' positions of the sugar ring are connected via a phosphate group.

Agarose – A purified agar, the lower its sulfate content, the more highly purified it is. When agarose is heated and cooled, it forms a gel matrix.

Aliquot – A representative sample of a fixed proportion or to divide into portions. Often samples are aliquoted into small portions to avoid repeated freeze/thaws or contamination.

Allele – One of several alternate forms of a gene occupying a given locus on a chromosome or plasmid.

Alternative splicing – The process by which alternative exons within a single RNA are combined during the splicing process, resulting in the generation of mRNAs encoding different protein sequences. Alternative splicing may be tissue- or disease-specific.

Amber codon – One of the three codons that cause termination of protein synthesis. Its sequence is UAG.

Amber, ochre, opal suppressor – Gene mutations that code for tRNAs that recognize stop codons as sense codons. An amino acid is placed in the growing polypeptide chain at the stop codon instead of terminating the peptide.

Aminoacyl-tRNA – A tRNA linked to an amino acid. The COOH group of the amino acid is linked to the 3'- or 2'-OH group of the terminal base of the tRNA.

Ampicillin – An antibiotic frequently used in the selection of bacterial transformants containing a cloning vector encoding resistance to this antibiotic, generally because of the presence of an ampicillin resistance gene derived from a transposable element. While this antibiotic is considered bacteriocidal in liquid culture, causing sensitive cells to lyse, it is bacteriostatic in solid media, allowing sensitive cells to exist as viable sphereoplasts with damaged cell walls until the ampicillin in the media degrades, allowing continued cell growth. The enzyme encoded by the ampicillin resistance gene present in many cloning vectors inactivates this antibiotic.

Amplicon – Any small DNA fragment which can be replicated.

Amplified Fragment Length Polymorphism (AFLP) – A method for discriminating similar DNA among organisms or sequences based on polymorphisms that alter the number or length of restriction fragments after PCR amplification.

Anion – An ion with a negative charge. Anions migrate to the anode of an electrophoretic gel.

Annealing – Hydrogen bonding of bases between complementary single strands of DNA.

Anode – The positive (+) electrode. Negatively charged compounds move toward the anode. The anode is connected to the power supply of a gel system with the red voltage lead.

Antibiotic – A natural, synthetic or semi-synthetic product, that inhibits bacterial growth.

Antibody – An immunoglobulin molecule that reacts specifically with another (usually foreign) molecule, the antigen.

Anticodon –A three-nucleotide sequence of a transfer RNA (tRNA) that is complementary to the three-nucleotide codon of an mRNA.

Antigen (immunogen) – A substance that causes production of an antibody directed against itself.

Antiparallel – Having the opposite polarity (e.g. the two strands of a DNA molecule).

Antisense – In general, the complementary strand to a coding sequence of DNA or mRNA. Antisense RNA hybridizes with mRNA. The antisense strand acts as the template in the synthesis of mRNA or cDNA second strand. Also referred to as the minus strand (-strand).

Antitermination – A prokaryotic mechanism to reverse premature termination of RNA synthesis during transcription. Antitermination occurs when the RNA polymerase ignores the termination signal. In the lambda phage, proteins N and Q are anti-terminators. The N protein prevents termination in early operons of lambda, but not in other phage or bacterial operons.

AP site – A position in a double-stranded DNA sequence that is missing an A or G base (apurinic) or a C or T base (apyrimidinic).

ARE – Adenylate/uridylate-rich element, an element located in the 3' untranslated region.

Asymmetric PCR – Preferential PCR amplification of one strand of DNA by lowering the concentration of one primer.

att sites – The loci on a lambda phage and the bacterial chromosome at which recombination integrates the phage into, or excises it from, the bacterial chromosome.

Autoradiography – A process to detect radioactively labeled molecules based on their ability to create an image on photographic or X-ray film. The energy they emit exposes the film.

Autosplicing (self-splicing) – The ability of an intron to excise itself from an RNA by a catalytic action that depends only on the sequence of RNA in the intron.

Avidin – A glycoprotein which binds to biotin with very high affinity.

B-form DNA – A right-handed double helix with ten base pairs per turn of the helix. This is the form found under physiological conditions and whose structure was proposed by Crick and Watson.

BAC – Bacterial Artificial Chromosome, a cloning vector capable of carrying between 100 and 300 kb of sequence. Because of its large cloning potential, BAC is often used as an intermediate in large-scale genomic sequencing projects.

Bacteriophage (phage) – A virus that infects bacteria. Phages consist of a protein coat or capsid enclosing the genetic material, DNA or RNA, that is injected into the bacterium after binding with the proper receptor.

Base – Commonly used when referring to a nucleotide; specifically, it is the purine or pyrimidine residue portion of a nucleoside or nucleotide.

Base mispairing – A coupling between two bases that does not conform to the Watson-Crick rule, e.g., adenine with cytosine, thymine with guanine.

Base pair – One pair of hydrogen bonded, complementary nucleotides within a duplex strand of nucleic acid. Under Watson Crick rules, these pairs consist of one pyrimidine and one purine: i.e., C G, A T (DNA) or A U (RNA). However, "noncanonical" base pairs (e.g., G U) can occur in RNA secondary structure.

Basic local alignment search tool (BLAST) – A nucleic acid and protein sequence comparison program based on the creation of a matrix of similarity scores for all possible pairs of residues, defining the high-scoring segments, and statistically evaluating the significance of the results. Reference: Altschul, SF, W Gish, W Miller, EW Myers, DJ Lipman. Basic local alignment search tool. 1990. J Mol Biol 215: 403-410.

Biotin – An essential coenzyme for carboxylation reactions. Often used in labeling reactions as a component of avidin-biotin systems.

Biotinylation – The covalent linkage of biotin to DNA and proteins. Biotin may be detected with avidin or streptavidin linked to an enzymatic moiety or fluorescent molecule.

Blotting – The general term for the transfer of protein, RNA, or DNA molecules from a relatively thick acrylamide or agarose gel on to a membrane (usually nitrocellulose with a nylon backbone in the case of nucleic acid or polyvinylidene difluoride in the case of protein) by capillary action or an electric field, preserving the two dimensional spatial arrangement. Once on the membrane, the molecules are immobilized by crosslinking to the substrate, typically by baking or by ultra-violet irradiation. Selected species are detected at high sensitivity by hybridization (in the case of DNA & RNA), or antibody labeling (in the case of protein).

Blue/White selection – A method of colony selection. Wild type colonies without insert are blue (meaning that they are ß-gal$^+$), while colonies with inset are white (colorless) (meaning that they are ß-gal$^-$) because the insert disrupts *lacZ* gene engineered into the cloning vector.

Blunt end – The end of double stranded DNA that has been cut at the same position on both strands by a restriction endonuclease such that no 3' or 5' overhanging nucleotides are produced. The ligation of blunt-ended fragments is much less efficient than that of "sticky ends". However, any blunt-ended fragment can be inserted into a vector with blunt ends.

bp – Abbreviation for base pairs. The length of a fragment of double-stranded DNA is measured in bp.

Buffer – A solution that acts to minimize the change in concentration of a specific chemical species in solution against addition or depletion of this species. pH buffers are weak acids or weak bases in aqueous solution. The working range is given by pKa +/- 1.

CAAT box – The regulatory sequence upstream from some eukaryotic structural genes.

cAMP – An AMP molecule where the phosphate is linked at the 3' and 5' positions. It acts in the regulation of many cascades of signal transduction and transcription.

CAP – Catabolic activator protein or cAMP receptor protein (CRP). Acts to initiate transcription of certain genes in *E. coli.*

Cap – The linkage of a methylated guanosine to the 5' end of an hnRNA to prevent degradation. Capping is one step in the processing of pre-mRNA to become mature mRNA. It is added post-transcriptionally, and is not encoded in the DNA.

Capillary electrophoresis – The separation of nucleic acids based on length through a matrix held in a tube of small diameter using an electric field. The cathode and the capillary are introduced into the nucleic acid containing solution and the fragments are drawn into the capillary based on their ability to move through the matrix, with smaller fragments moving faster.

Capsid – A protein coat that encloses the genome of many viruses.

cat (chloramphenicol acetyltransferase) – the enzyme that inactivates the antibiotic choramphenicol by acetylation. Widely used as a reporter gene. A cat assay measures the efficiency and action of a gene promoter through its ability to produce cat.

Cathode – The negative (-) electrode. Positively charged proteins move toward the cathode. The cathode is connected to the power supply in a gel system with the black voltage lead.

cDNA – Complementary DNA. Viral reverse transcriptase can be used to synthesize a strand of DNA that is complementary to an RNA (e.g. an isolated mRNA). After second strand synthesis to create a ds-cDNA, the cDNA can be used as a probe to locate the gene, cloned in the double-stranded form, or used in other ways. cDNA is common template for PCR and construction of libraries.

cDNA library – A collection of DNA clones generated from mRNA sequences. This type of library contains only protein-coding DNA (genes) and does not include any non-coding DNA.

Centromere – A constricted region of a chromosome that includes the site of attachment to the mitotic or meiotic spindle.

Chelator – A molecule that binds metal ions.

Chemiluminescence – Visible light produced by chemical action and not accompanied by heat. Luciferase or other luminescent systems can be used in visualizing hybridized or bound molecules conjugated with the enzyme. This is an alternative to radioactive labeling and detection of molecules.

Chip (oligonucleotide array) – In biotechnology, a component of a device for screening genomic or cDNA for mutations, polymorphisms, or gene expression. A chip is a small (a few centimeters on each side) standardized glass or other solid surface on which thousands of oligodeoxynucleotide probes have been synthesized or robotically deposited in a predetermined array, so that automated recording of fluorescence from each of the spots may score successful hybridizations. A chip may be designed for the detection of all known genes of a species (human, mouse, yeast), or selected specific sequences.

Chloramphenicol – A protein synthesis-inhibiting antibiotic sometimes used in the selection of bacterial transformants containing a cloning vector encoding resistance to this antibiotic, generally because of the presence of a chloramphenicol resistance gene. Resistance to this antibiotic is usually accomplished by acetylation of the chloramphenicol.

Chromatin – a complex of DNA and protein in the nucleus of a eukaryotic cell. This complex was recognized originally because it readily stains for DNA.

Cis – a prefix meaning on the same side, or on the near side; it is the opposite of trans.

Cis acting – A genetic element (such as a promoter or other regulatory locus) with an effect like promotion or suppression on targets (such as genes) located on the same chromosome.

Cistron – A genetic element which is roughly equivalent to the sequence of DNA that codes for one polypeptide chain.

Clear plaque – A locus that contains lysed bacterial cells and appears as a cleared area on a bacterial lawn.

Clone – A large number of cells, viruses, or molecules which are identical and which are derived from a single ancestral cell, virus or molecule. To clone something is to produce identical copies of it. It can be the process of isolating single cells or viruses and letting them proliferate (a "biological clone"), or the process of isolating and replicating a piece of DNA by recombinant DNA techniques (a “molecular clone”).

Cloning – The synthesis of multiple copies of a target DNA by joining of a DNA fragment (insert) with a vector followed by replication in a host cell such as *E. coli*. Cloning usually involves the following: preparation of insert and vector, ligation, transformation or infection, and identification and verification of recombinant clones.

Coding region – The genomic sequence between the start and stop codons.

Coding strand (sense strand) – The DNA strand that has the same sequence as the mRNA and is related by the genetic code to the protein sequence that it represents.

Codon – A combination of three consecutive nucleotides that codes for a single amino acid or termination signal. The code is degenerate, i.e. each amino acid is coded for by more than one codon (except Met and Trp).

Codon bias – The tendency for an organism to use certain codons more than others to encode a particular amino acid.

Cohesive end – The same as sticky end

Colony – The progeny of a single micro-organism grown in culture on solid media and visible as a spot on a plate.

Colony forming unit (cfu) – An individual cell that is able to clone itself into an entire colony of identical cells.

Colony hybridization – A technique for screening bacterial colonies for those that contain a desired polynucleotide sequence using a labeled probe with a sequence complementary to part of the desired sequence thus labeling clones with that sequence.

Compatible ends – Two blunt DNA ends or two sticky ends with complimentary overhangs. Two compatible ends can be joined together without any gap or mismatch.

Competent *E. coli* – Bacterial cells chemically or otherwise modified to take up exogenous DNA and thus be transformed. Competence can arise naturally in some bacteria. A similar state can be induced in *E. coli* by treatment with chemicals such as calcium chloride.

Complementary – In nucleic acid chemistry, this is the relationship between two polynucleotides that can combine in an antiparallel double helix; the bases of each polynucleotide are in a hydrogen-bonded inter-strand pair with a complementary base, A to T (or U) and C to G.

Complementation – Return of the normal phenotype to a mutated host cell by the introduction of foreign DNA to correct the defect caused by the mutation.

Concatamer – Two or more of the same linear DNA molecule covalently linked end-to-end.

Concatenate – To link together two circular molecules, as in a chain.

Conjugation – Mating between two bacterial cells during which a part of the chromosome or other genetic element is transferred from one to the other.

Consensus sequence – A DNA, RNA or protein sequence in which each nucleotide or amino acid represents the most commonly found unit at that particular position when many homologous sequences are aligned.

Conserved sequence – A sequence which is similar or even identical across multiple species, indicating that this sequence has been maintained by evolution despite speciation. In other words a conserved sequence has a high degree of sequence identity to other sequences between species. The TATA promoter sequence is an example of a highly conserved DNA sequence, being found in most eukaryotes.

Constitutive – A process that occurs all the time, unchanged by any form of stimulus or external condition.

Constitutive expression – Expression of a gene which requires only the interaction of RNA polymerase with the promoter. These genes are transcribed at all times. Genes which are constitutively expressed are mostly housekeeping genes.

Contig – The term comes from a shortening of the word contiguous. A contig is a group of clones that represent an overlapping region of the genome. It is used to construct the final sequence of the entire region.

Copy number – The number of copies of a plasmid maintained in a bacterium (relative to the number of copies of the origin of the bacterial chromosome).

Core enzyme – The smallest aggregate of an enzyme's subunits that has enzymatic activity. In the case of *E. coli* RNA polymerase, the core enzyme consists of those subunits needed for elongation. It does not include the sigma factor.

Core sequence – The segment of DNA that is common to the attachment sites on both the phage lambda and bacterial genomes. It is the location of the recombination event that allows phage lambda to integrate.

Cos – The short form name of the cohesive site of λ phage, the only sequence required for λ phage packaging. After cleavage of the *cos* site by a λ enzyme, there are complementary 3' overhangs to facilitate concatamerization and recircularization after packaging. The *cos* sequence contains two portions: Initiator–recognized by λ to begin packaging and terminator to cut λ concatamer so that only one copy of the λ genome is packaged. This sequence is often engineered into a plasmid to create a cosmid. These vectors allow cloning of 40-50 kb inserts.

Cosmid – A plasmid cloning vector into which phage λ *cos* initiation and termination sequences have been inserted; as a result, the plasmid DNA and any insert contained *cos* can be packaged as a λ particle. Cosmids are often used for construction of genomic libraries because of their ability to carry relatively long pieces of insert DNA, compared with plasmids.

Cot value – The product of DNA concentration and the time for its reassociation. 1/2 Cot is the Cot value required for 50% completion of reassociation. This varies directly with the length of DNA to be reassociated.

CpG island – A short stretch of DNA in which the frequency of the CG sequence is higher than other regions. The "p" simply indicates that a phosphodiester bond connects "C and G". CpG islands are often located around the promoters of genes frequently expressed in a cell. At these locations, the CG sequence is not methylated. By contrast, the CG sequences in inactive genes are usually methylated to suppress their expression.

CRP activator – A positive regulator protein activated by cyclic AMP. It is needed for RNA polymerase to initiate transcription of many operons of *E. coli*.

Cytoplasmic domain – The part of a transmembrane protein that is exposed to the cytosol.

Dalton – A unit of molecular mass approximately equal to that of a hydrogen atom or 1.66 x 10-24 grams. Alternatively, there is approximately 1 mole of daltons per gram. Abbreviated Da.

Daughter strand – The newly synthesized DNA strand.

Deacylated tRNA – A form of tRNA that has no amino acid or polypeptide chain attached because it has completed its role in protein synthesis and is ready to be released from the ribosome. Also called an uncharged tRNA.

Deamination – The removal of an amino group. Accidental deamination may change the cytosine to uracil, or the methylated cytosine to thymine.

Degeneracy – In reference to proteins, the genetic code is degenerate in that multiple different codons in mRNA can specify the same amino acid in an encoded protein. In reference to nucleic acids, probes or primers whose design is based on protein sequence must include all molecules containing each of the possible combination of nucleotides since the mapping of codons to amino acids is many to one. The total number of different oligonucleotides in the resulting mixture is known as the degeneracy of the probe or primer.

Degenerate PCR – A form of PCR where primers are chosen to amplify unknown DNA sequences based on similar sequences.

Deletion – Removal of a portion of a gene (single base or fragment). In genetics it is defined as the loss of one or more nucleotides from a chromosome.

Denaturation – Conversion of DNA or RNA from the double stranded state to single stranded state; separation of the strands is most often accomplished by heating. The temperature required to bring denaturation is a function of the unique sequence of the molecule, i.e., GC base pairs confer 4°C of thermal stability each, whereas AT base pairs confer only 2°C thermal stability each.

Denaturing gradient gel electrophoresis – A method for detection of single base substitutions in DNA fragments. Putative mutant and normal double-stranded DNA fragments are applied along one edge of a slab gel that contains a denaturing agent (e.g. urea and formamide) in a gradient perpendicular to the direction of electrophoresis. At a low concentration of the denaturant the fragments are more mobile, and at high concentration they are denatured and therefore less mobile. A single-base change in DNA can result in a difference in this mobility.

Density gradient – A column of liquid in which the density varies continually with position, usually as a consequence of variation of a solute. Density gradients are widely used for centrifugal and gravity induced separations of cells, organelles and macromolecules. The separations may exploit density or size differences between particles.

DEPC – Diethylpyrocarbonate, a chemical which is used to inactivate RNases.

Depurination – The cleavage of N-glycosidic bonds of DNA to form apurinic DNA.

Derepression – In biochemistry, a repressor inhibits the activity of an operator gene. By inactivating the repressor, the operator gene becomes active again. This effect is called derepression.

Detergent – An agent that reduces the surface tension of a liquid. Detergents are often used to disrupt lipid bilayers, including cellular plasma membranes; they can also be used to solubilize proteins.

Dialysis – A technique for the separation of macromolecules from smaller molecules by placing them within a semi-permeable membrane. Dialysis tubing comes in different sizes. This tubing is placed in a large volume of water or buffer. Only the low-molecular-mass diffusible molecules cross the membrane and pass into the larger volume; the macromolecules are confined to their original space.

Dicer – an RNase III nuclease that catalyzes the digestion of RNA into siRNAs.

Dideoxy sequencing – A chain-termination method for sequencing DNA that utilizes dideoxynucleotides. Also called the Sanger method

Dimer – A protein complex composed of two proteins. In a homodimer, the two proteins are identical, whereas a heterodimer contains two different proteins.

Direct repeat – Identical or related sequences present in two or more copies in the same orientation in the same molecule of DNA; they do not have to be adjacent.

Divergence – The percent difference in nucleotide sequence between two related DNA sequences or in amino acid sequences between two related proteins.

DNA (Deoxyribonucleic acid) is a nucleic acid that contains the genetic instructions used in the development and functioning of all known living organisms and some viruses. The main role of DNA molecules is the long-term storage of information. DNA is often compared to a set of blueprints, a recipe, or a code, since it contains the instructions needed to construct other components of cells, such as proteins and RNA molecules. The DNA segments that carry this genetic information are called genes, but other DNA sequences have structural purposes, or are involved in regulating the use of this genetic information. Chemically, DNA consists of two long polymers of simple units called nucleotides, with backbones made of sugars and phosphate groups joined by ester bonds. These two strands run in opposite directions to each other and are therefore anti-parallel. Attached to each sugar is one of four types of molecules called bases. It is the sequence of these four bases along the backbone that encodes information.

DNase – Short form of Deoxyribonuclease, a class of enzymes which digest DNA. The most common is DNase I, an endonuclease which digests both single and double-stranded DNA.

DNA-binding proteins – Proteins which interact with DNA, typically to pack or modify the DNA, e.g., histones, or to regulate gene expression, e.g., transcription factors. Among those proteins that recognize specific DNA sequences, there are a number of characteristic conserved "motifs" believed to be essential for specificity.

DNA fingerprint – The unique pattern of bands on an electrophoresis gel of DNA fragments produced after digestion with a certain enzyme. These pattern differences correspond to DNA polymorphisms or mutations.

DNA fingerprinting – A technique for analyzing the differences between individuals of the fragments generated by using restriction enzymes to cleave regions that contain short repeated sequences or by PCR. The lengths of the repeated regions are unique to every individual, and as a result the presence of a particular subset in any two individuals can be used to define their common inheritance (e.g., a parent-child relationship).

DNA ligase – The enzyme that makes a bond between an adjacent 3'-OH and 5'-phosphate end where there is a nick in one strand of duplex DNA.

DNA polymerase – An enzyme that catalyzes the synthesis of new DNA (based on the sequence of a template DNA strand; a DNA-dependent, DNA polymerase). DNA polymerase is also involved in repair of DNA. Retroviruses possess a unique DNA polymerase (reverse transcriptase) that uses an RNA template (an RNA-dependent, DNA polymerase).

DNA replicase – A DNA-synthesizing enzyme required specifically for replication

DNA replication – The process in which a copy of a DNA molecule is made from a double stranded DNA template molecule, and thus the genetic information it contains is duplicated. The parental double stranded DNA molecule is replicated semi-conservatively, i.e. each copy contains one of the original strands paired with a newly synthesized strand that is complementary in terms of AT and GC base pairing. Replication is a complex process, involving many enzymes.

DNA synthesis – The linking together of nucleotides (as deoxyribonucleotide triphosphates) to form DNA. In vivo, most synthesis is DNA replication, but incorporation of precursors also occurs in repair. In the special case of retroviruses, DNA synthesis is directed by an RNA template (see reverse transcriptase). DNA is synthesized in the 5' to 3' direction, but read the template sequence 3' to 5'. DNA synthesis forms phosphodiester linkage between the α phosphate of the nucleotide to be added and the 3' carbon of the last nucleotide of the polymer with the release of PPi.

Domain – A specific portion of a peptide or nucleic acid which has a particular function.

Dot blot – A method for detecting a specific protein or message. A DNA, RNA, or protein containing solution is denatured and then spotted directly onto either a nitrocellulose membrane or a nylon membrane. It is hybridized with a probe or a specific antibody. There is no separation of the target DNA or RNA based on size by electrophoresis as in Southern or Northern blots. Dot blot is a good method for screening a large number of samples.

Double-strand breaks (DSB) – Breaks that occur on both strands of a DNA duplex opposite one another. Genetic recombination is initiated by such breaks. DSBs are a potentially lethal form of DNA damage. Ionizing radiation and a host of drugs can induce

DSBs directly or indirectly, which determines their therapeutic value in cancer treatment

Downstream – A term which identifies sequences proceeding farther in the direction of expression; for example, the coding region is downstream from the initiation codon, toward the 3' end of an mRNA molecule.

dsDNA – Double stranded DNA

EMSA – Electrophoretic Mobility Shift Assays

Editing – In DNA replication, the 3'-to-5' exonuclease activity of a polymerase that removes incorrectly paired bases, also referred to as proofreading.

Electroelution – Extraction of nucleic acid from an agarose gel by excision of the nucleic acid containing portion of the gel, packaging of the slice in dialysis tubing with electrophoresis buffer, and placing it back into the electrophoresis chamber to drive the nucleic acid out of the gel.

Electrophoresis – The separation of molecules using an electrical current, usually applied to a gel matrix. The gel matrix has a sieving effect which allows molecules to be separated on the basis of size. The electric field separates molecules on the basis of charge.

Electroporation – A method for introducing foreign nucleic acid into bacterial or eukaryotic cells. It uses a brief, high voltage DC charge to render the cells permeable to the nucleic acid.

Elongation – The stage in a macromolecular synthesis reaction (replication, transcription, or translation) when the nucleotide or polypeptide chain is extended by the addition of individual subunits.

Elongation factors – The proteins that associate with ribosomes cyclically during the addition of each amino acid to the polypeptide chain.

Eluate – The liquid which is produced during chromatography.

Elution – The removal, by a suitable solvent of one material from another that is insoluble in that solvent, as in column chromatography.

End labeling – The addition of a radioactive or otherwise labeled group to the 3' or 5'-end of a DNA molecule. 5'-end labeling involves replacement of the 5' phosphate of the DNA for the labeled γ-phosphate of a donor nucleotide utilizing a kinase enzyme. 3'-end labeling involves the addition of labeled nucleotide(s) to the 3' end of the DNA molecule utilizing a terminal transferase enzyme

Endonuclease – One of a large group of enzymes that cleave nucleic acids at positions within the chain, but not at the 3' or 5' end of the polymer. These enzymes can be specific for single or double stranded DNA. Bacterial restriction endonucleases are crucial in recombinant DNA technology for their ability to cleave double-stranded DNA at highly specific sites.

Enhanceosome – A complex of transcription factors that assembles cooperatively at an enhancer.

Enhancer – A eukaryotic transcriptional control element. A DNA sequence which acts at some distance to enhance the activity of a specific promoter sequence. Unlike promoter sequences, the position and orientation of the enhancer sequence is generally not important to its activity.

Epigenetics – Changes in phenotype (appearance) or gene expression caused by mechanisms other than changes in the underlying DNA sequence. Non-genetic factors cause the organism's genes to behave (or "express themselves") differently. The best example of epigenetic changes in eukaryotic biology is the process of cellular differentiation. During morphogenesis, totipotent stem cells become the various pluripotent cell lines of the embryo which in turn become fully differentiated cells.

Episome – A piece of hereditary material that can exist as free, autonomously replicating DNA or can be attached to and integrated into the chromosome of the cell, in which case it replicates along with the chromosome.

Epitope – A molecular region on the surface of an antigen capable of eliciting an immune response and reacting with the specific antibody produced by such a response.

Error rate – The frequency of mistakes in nucleotide incorporations for a DNA polymerase.

Escherichia coli (E. coli) – A common Gram-negative bacterium useful for cloning experiments. Hundreds of strains of *E. coli* exist; many are present in the intestinal tract.

Ethidium Bromide – A dye which binds to DNA. Ethidium bromide intercalates between bases of a DNA strand and fluoresce under UV light so that DNA can be visualized. Ethidium bromide staining is commonly used to visualize RNA or DNA in agarose gels placed on UV light boxes. Ethidium bromide is also used in cesium chloride gradients during large scale plasmid isolation to separate supercoiled circular DNA from linear and relaxed or linear DNA, based on the amount of ethidium bromide each form can incorporate. Since this agent will induce strand breaks in DNA when exposed to UV light, care must be taken in its handling and use with DNA.

Eukaryote – A cell or organism having a visible nucleus or nuclei.

Excision – Release of phage or episome or other sequence from the host chromosome as an autonomous DNA molecule.

Exon – A segment of DNA that specifies the genetic code for a protein, as distinguished from an intervening sequence (intron). Exons are the sequences represented in the mature mRNA. Exons are usually joined together in mRNA in the same order as their organization in DNA, but alternate splicing can take place to include, exclude or rearrange some exons. Introns tend to undergo more frequent mutation, but exon sequences are conserved, since it is more likely that mutations will have negative consequences for the organism.

Exonuclease – A class of enzymes involved in DNA repair; cleave nucleotides from the 5' or 3' end of a DNA strand. Exonucleases can be single or double strand DNA specific.

Express – To "express" a gene is to cause it to function. A gene which encodes a protein will, when expressed, be transcribed and translated to produce that protein. A gene which encodes an RNA rather than a protein (for example, a rRNA gene) will produce that RNA when expressed.

Expression – This term is usually used to refer to the entire process of producing a protein from a gene, which includes transcription, translation, post-translational modification and possibly transport reactions.

Expression cloning – A method of gene identification and cloning based on transfection of a large number of cells with cDNAs in an expression vectors (e.g. a cDNA library). The proteins coded for by these cDNAs are expressed in the host cells and then the colonies are screened for a functional property of the target gene product (e.g. binding of a radiolabeled hormone to identify receptors, or induction of transforming activity for putative oncogenes or with an antibody to the desire protein, etc.).

Expression vector – A plasmid or phage designed for production of a polypeptide from inserted foreign DNA under specific controls. The vector may provide a promoter, transcriptional start site, ribosomal binding sequence, and initiation codon.

Exteins – The sequences that remain in the mature protein that is produced by processing a precursor via protein splicing.

F factor – Bacterial sex or fertility plasmid. Bacteria containing this plasmid are considered male.

Fidelity – The accuracy of DNA polymerase in copying DNA. Fidelity is the inverse of the error rate.

Filamentous bacteriophage – A type of ssDNA bacteriophage. Its morphology is long and thin, like a filament. M13 is a filamentous bacteriophage.

FISH (fluorescence in situ hybridization) – A technique of directly mapping the position of a gene or DNA clone within a genome. The DNA probe is labeled with a fluorophore, and the hybridization sites are visualized as spots of light by epifluorescence. Frequently, several probes can be used at one time, to mark specific chromosomes with different colored fluorophores ('chromosome painting'). Also used to detect mRNAs in tissue sections as in other in situ hybridizations, but the method of detection is by fluorescence instead of enzymatic or radioactive label.

Fluorescein – Dye that fluoresces green when exposed to blue or ultraviolet light.

Fluorescence – The emission of electromagnetic radiation following absorption of shorter wavelength light.

Fluorescence Resonance Energy Transfer (FRET) – The excitation of a donor fluorescent dye is transferred to a receptor dye, leading to the fluorescence of the acceptor dye instead of the donor dye. The transfer is possible only if the two dyes are in close proximity. Applications include receptor-ligand interactions and real-time PCR assays.

Fluorophore (fluor) – A chemical group responsible for the fluorescence of a compound or macromolecule.

Footprinting – A technique for identifying the site on DNA bound by some protein by virtue of the protection of bonds in this region against attack by nucleases

Frame-shift mutation – An insertion or deletion of a number of bases not divisible by three in an open reading frame in a DNA sequence. Such mutations result in a change in the amino acids coded from that point on and may create premature stop codons.

Fully methylated – A site that is a palindromic sequence and is methylated on both strands of DNA.

Fusion protein – A product of recombinant DNA in which the foreign gene product is juxtaposed ("fused") to either the carboxyl-terminal or amino-terminal portion of a polypeptide encoded by the vector itself. Use of fusion proteins often facilitates expression of otherwise lethal products and the purification of recombinant proteins.

Gap – A space introduced into a sequence within a multiple sequence alignment that allows similar or conserved sequence regions to be aligned in columns.

GC box – A common pol II promoter element consisting of the sequence GGGCGG.

Gel shift – Also called EMSA or gel retardation. A method by which the interaction of a DNA with a protein is detected. The mobility of the nucleic acid is monitored in an agarose gel in the presence and absence of the protein: if the protein binds to the nucleic acid, the complex migrates more slowly in the gel. By virtue of a second shift in mobility caused by the binding of a specific antibody to the nucleic acid-protein complex, the identity of a protein can be determined.

Gene – Originally defined as the physical unit of heredity, it is now probably best defined as the unit of inheritance that occupies a specific locus on a chromosome. It might also be defined as the set of DNA sequences that are required to produce a single polypeptide.

Gene expression – The conversion of the information encoded in a gene first into messenger RNA, and then to a protein.

Gene therapy – Treatment of a disease, which is caused by the malfunction or the absence of a gene, by stably or transiently transfecting the cells with a normal copy of the gene.

Genetic code – The correspondence between a sequence of three nucleic acid bases (triplet) in DNA (or mRNA) and amino acids in proteins. A triplet acts as a "codeword" (codon) for one amino acid, or a start or stop codon. There are 64 different codons, but only 20 commonly used amino acids; therefore, the code is degenerate, with more than one codon specifying a particular amino acid.

Genetic recombination – Formation of new combinations of alleles in offspring (viruses, cells, or organisms) as a result of exchange of DNA sequences between molecules. It occurs naturally, as in crossing over between homologous chromosomes in meiosis or experimentally, as a result of genetic engineering techniques.

Genome – The total genetic information content of an organism.

Genomic library – A collection of DNA clones, derived from restriction fragments that have been cloned in vectors. The cloned DNA includes all, or part, of the genetic material of an organism, both the coding and non-coding regions.

Genotype – The genetic constitution of an organism; determined by its nucleic acid sequence. The genome of an organism is its DNA; its genotype is the sum of the information possessed in that DNA.

Glycoslyation – The process of adding sugar units. The sugar is added covalently to nitrogen or oxygen atoms present in the side chains of certain amino acids of proteins. Glycosylation may have a profound effect on the folding, stability and antigenicity of proteins.

Green Fluorescent Protein (GFP) – The green fluorescent protein is a protein from jellyfish. GFP converts a blue light (activation light) to a green light (emission light). It has been widely used as a reporter protein because of its ability to label living cells. Variants of GFP have been engineered to have different activation and emission light spectrum, allowing more than one fluorescent protein reporters to be used in one living cell.

Gyrase – The prokaryotic topoisomerase II that utilizes ATP to generate negative supercoils of DNA.

Haploid – A set of chromosomes including one copy of each autosomal gene and one sex chromosome. This is referred to 1N.

Haplotype – The combination of alleles in some defined region of DNA. It is the genotype of a particular locus.

Helicase – A class of enzymes catalyzing the ATP-dependent unwinding of double-stranded DNA. They unwind DNA ahead of an advancing replication fork or during transcription.

Helper phage M13K07 – An M13 bacteriophage co-infected with an M13 or phagemid vector to direct replication and packaging of the vector. The helper phage is mutated so that it will preferentially package the vector DNA.

Heteroduplex DNA – Generated by base pairing between complementary single strands derived from different parental duplex molecules; heteroduplex DNA molecules occur during genetic recombination in vivo and during hybridization of different but related DNA strands in vitro. Because the sequences of the two strands differ, the molecule is not perfectly base paired.

Heterogenous (hn) RNA – pre-mRNA transcripts of nuclear genes prior to processing. It has wide size distribution and low stability.

Heterozygote – An individual with two different alleles at some particular locus.

Hfr – A bacterium that has an integrated F plasmid within its chromosome. Hfr stands for high frequency recombination, referring to the fact that chromosomal genes are transferred from an Hfr cell to an F- cell much more frequently than from an F+ cell.

hnRNA – heterogenous nuclear RNA. A class of RNA found in the nucleus but not the nucleolus, with a very wide range of sizes, 2 - 40 kilobases. It represents the primary transcripts of RNA polymerase II and includes precursors of all mRNAs from which introns are removed by splicing.

Holliday structure – An intermediate structure in homologous recombination, for which the two duplexes of DNA are connected by the genetic material exchanged between two of the four strands, one from each duplex.

Holoenzyme – Functional enzyme consisting of all units, differentiated from the core enzyme which lacks one or more subunits, but still performs certain, but not all, functions.

Homologous recombination – The exchange of sequence between two related but different DNA (or RNA) molecules, resulting in a new chimeric molecule. There must be a region of homology in the recombination partners.

Homology – In comparing DNA sequences, it is the degree to which two sequences have the same nucleotide sequence. Homology is usually expressed as a percent of total nucleotides compared. In comparing two protein sequences, degree to which two peptides have similar amino acids at the same position, i.e. a basic, acidic, or neutral amino acid.

Homozygote – An individual with the identical alleles at a particular locus.

Host strain (bacterial) – The bacterium used to harbor a plasmid. Typical host strains include DH5α (general purpose *E. coli* strain), JM109 (suitable for growing M13 phages), and XL1-Blue (general-purpose, good for blue/white lacZ screening).

Hot start – The process of inhibiting DNA polymerase in a PCR reaction until the block temperature is beyond the point where mispriming may occur. May be accomplished in several ways, i.e. chemical or antibody mediated suppression of activity.

Housekeeping (constitutively expressed) genes – Genes (theoretically) expressed at constant levels in all cells. They provide the basic functions of all cell types. Commonly probed for in Northern blots to normalize the signal of target RNAs.

***hsd* system** – Host specificity for DNA. This system, at work in some bacteria, acts to protect the cell from bacteriophage infection by recognizing and cleaving foreign DNA.

Hybridization – Formation of H-bonding between complementary single stranded fragments of nucleic acid. This is a way to assess sequence identity and detect specific sequences. The hybridization can be carried out in solution or with one component immobilized on a gel or, most commonly, nitrocellulose membrane. Degree of hybridization can be manipulated by altering the stringency of the incubation and washes. Hybridizations can be done with all combinations: DNA-DNA, DNA-RNA, or RNA-RNA.

Hydrophilic – "Water loving", molecules or portions of molecules that are capable of forming hydrogen bonds, and can do so with water. They readily disperse in an aqueous environment.

Hydrophobic – "Water hating", molecules or portions of molecules which do not interact with water. Molecules like hydrocarbons are only slightly soluble in water. When they do dissolve, these elements will aggregate to eliminate water.

Inducer – A small molecule that induces (provokes) transcription of a gene through its interaction with transcriptional regulating protein(s). In *E. coli*, lactose acts as an inducer of the *lac* operon by binding the *lac* repressor, removing it from the *lac* promoter.

Induction – Production of a certain enzyme in cells only when the substrate for the enzyme is present. An example of induction is seen in the transcription and translation of the *lac* operon in *E. coli* only when lactose is the only carbon source in the cell's environment.

Infection – In molecular biology, introduction of viral nucleic acid into a permissive cell via binding of the virus to a receptor and injection of its nucleic acid.

Initiation codon – A special codon (usually AUG) used to start synthesis of a protein

Initiation factors – The set of catalytic proteins required, in addition to mRNA and ribosomes, for protein synthesis to begin. These proteins associate with the small subunit of the ribosome.

Insert – Common name for any fragment ligated into a cloning site of a vector.

In situ hybridization – Technique for revealing patterns of gene transcription in a section of tissue or other preparation of cells affixed to a microscope slide using labeled DNA or RNA probe to hybridize to complementary mRNA sequences (cellular DNA is double stranded).

Intein – The part that is removed from a protein by protein splicing.

Intercistronic region – The region between the termination codon of one gene and the initiation codon of the next gene

Intergenic – DNA sequence of unknown function located upstream, downstream, or between genes.

Integrase – a site specific enzyme which directs specific DNA-protein interactions between the enzyme and DNA for recombination.

Integration – Insertion of viral or other DNA into the host cell genome. With viral DNA integration, the site of integration may be specific or random.

Interleukin (IL) – A cytokine that acts as a signal between different leukocyte populations.

Intron (intervening sequence) – A segment of DNA that is transcribed, but removed from the primary transcript by splicing. Introns are mainly found in eukaryotic DNA. New evidence shows that introns are also involved in transcriptional control of some genes, and in some cases are catalytic, as in Type I self-splicing introns (ribozymes). Genes may contain as many as 80 introns.

Inverted repeats – Two copies of the same or related sequence of DNA repeated in the opposite orientation on the same molecule. Adjacent inverted repeats constitute a palindrome.

Inverted terminal repeat – Inverted repeats at the ends of some transposable elements or viruses.

IPTG (Isopropyl- ß-D-thiogalactopyranoside) – An irreversible binder of the LacI repressor; this binding removes the repressor from the operator of the *lac* operon so that transcription can proceed. Used in blue/white selection for cloning and in control of expression of protein from sequences inserted into expression vectors.

Isoschizomer – A restriction enzyme that recognizes the same sequence as another restriction endonuclease. It may 1) recognize the same sequence and cut in the same place, 2) recognize the same sequence and cut in a different place, or 3) recognize the same sequence, but have its action modified in another way.

Isozyme (isoenzyme) – Any of the chemically distinct forms of an enzyme that perform the same biochemical function on the same substrate.

Junk DNA – Genomic DNA that serves, as yet, no known function

kb – the abbreviation for 1000 base pairs of DNA or 1000 bases of RNA.

Kinase – An enzyme that catalyzes the addition of a phosphate group to a protein or nucleic acid.

Kinetic PCR – Another name for Real-time PCR.

Klenow fragment – The large C-terminal fragment of the bacterial DNA polymerase I (76 kD) that remains after treatment with subtilisin. It has 5' to 3' polymerase activity, and 3' to 5' exonuclease activity, but no 5' to 3' exonuclease activity.

Knock down – A technique for reducing gene expression without mutating the gene in the genome. Frequently used for gene knock-down is siRNA.

Knock in – A technique for adding a gene into the genome.

Knock out – The excision or inactivation of a gene in an intact organism or even an animal. The inactivation is often carried out by a method involving homologous recombination.

Kozak Translation Initiation Consensus – A consensus sequence (A/GCCATGG) for mammalian translation initiation site.

Labeling – A process in which nucleic acids or proteins are tagged with a radioactive or non-radioactive marker.

***Lac* operon (lactose operon)** – Group of adjacent and coordinately controlled genes involved in the metabolism of lactose in *E. coli*. The *lac* operon was the first example of a group of genes under the control of an operator region to which a lactose repressor binds. *lacZ* codes for ß galactosidase, *lacY* for the permease, *lacA* for the transacetylase.

Lag phase – The first phase of bacterial growth. It is a period of slow growth when the cells are adapting to the high-nutrient environment. The lag phase has high biosynthesis rates, as proteins necessary for rapid growth are produced

Lagging strand – The daughter strand synthesized from the minus strand of DNA during replication. The template strand is read 3' to 5', but the advancing replication fork reveals more 3' sequence after a primer has already annealed to the open strand; therefore, the daughter strand is synthesized discontinuously in the form of short fragments (Okazaki fragments) that are later ligated.

Lambda bacteriophage (λ phage) – Bacterial DNA virus, first isolated from *E. coli*. It has a lytic cycle and a lysogenic cycle. Studies on the control of these alternative cycles have been very important for our understanding of the regulation of gene transcription. It is used as a cloning vector, accommodating fragments of DNA up to 15 kilobase pairs long.

Lariat – An intermediate structure in RNA splicing where the ends of adjacent exons are brought in close proximity with the intervening sequence (intron) forced into a looped formation. Formation of the lariat structure is mediated by snRNPs and exon and intron sequences.

LB medium – Originally Lysogeny Broth. It is now more commonly referred to as Luria-Bertani medium (agar or broth). It is used for the cultivation and maintenance of *E. coli* in recombinant DNA work.

Leading strand – The daughter strand synthesized from the plus strand of DNA during replication. DNA is synthesized continuously in the 5' to 3' direction as the advancing replication fork reveals more template, so that more nucleotides can be continuously added to the 3' of the growing strand.

Library – A set of cloned fragments which together represent all the genetic material (genome), or all the transcriptionally active genes (cDNA) of an organism or tissue.

Ligase – A major class of enzymes that link two DNA fragments together. Ligases also covalently close nicks in double stranded DNA. The ends to be ligated must contain a 3' -OH and 5' phosphosate terminus.

Linker – A blunt-ended, double stranded, synthetic oligonucleotide with a certain restriction site. To aid in ligation of blunt ended fragments, linkers are ligated to the fragment. After digestion of the linker-ligated fragment with the proper restriction enzyme, the sticky ends can more easily be ligated into a linearized vector with compatible ends. The advantage to this technique is that the relatively inefficient blunt-ended ligation can be favored by greatly increasing the number of available ends.

Locus – Position on a chromosome occupied by a gene(s) of a particular trait.

Log phase – The second of the three phases of bacterial growth; it follows the lag phase and precedes the stationary phase. In logarithmic phase, the cells are growing exponentially, doubling in number with each generation.

Long Terminal Repeat (LTR) – Direct repeats at the ends of retroviral genomes.

Loop – A single stranded region at the end of a hairpin or other ds region in an RNA or ssDNA, or between inverted repeats in DNA.

Luciferase assay – Firefly luciferase protein catalyzes luciferin oxidation and light is generated in the reaction. The luciferase protein is frequently used as a report for measuring promoter activity or transfection efficiency.

Lysis – The bursting of infected bacteria at the end of the lytic cycle, when phage progeny are released. In more general terms, the chemical, enzymatic, or mechanical destruction of the plasma membrane of a cell.

Lysogen – A bacterial cell which contains stable, prophage DNA in its chromosome. The lysogen functions as a normal cell as long as the viral DNA remains in the lysogenic state.

Lysogeny – Ability of a phage to stably incorporate into a bacterial genome. For example, after circularization, the λ chromosome integrates into a specific site on the host chromosome. In lysogeny, most of the integrated viral genes (prophage) are inactivated, and the virus can maintain a long-term, non-lethal relationship with its host.

Lysozyme – A glycosidase that hydrolyzes the bond between N-acetyl-muramic acid and N-acetyl-glucosamine, thus cleaving an important polymer of the cell wall of many bacteria. Present in tears, saliva, and in the lysosomes of phagocytic cells, it is an important anti-bacterial defense, particularly against gram-positive bacteria.

Lytic infection – The cycle of infection of a cell by a virus or bacteriophage in which mature virus or phage particles are released from the host cell by enzymatic lysis of the host cell.

M 13 – A commonly used bacteriophage that infects *E. coli*, but does not kill them. The growth rate of the infected cell slows as host resources are used to synthesize phage progeny. The viral DNA is replicated as single stranded DNA, a property used in M13 cloning vectors for the production of single stranded DNA for sequencing.

Maxam and Gilbert method – A DNA sequencing method developed in the 1970's based on base-specific chemical cleavage reactions (to cut sequences at each A, A or G, C, or C and T). This system uses 5'-end labeled, dsDNA. It was the first method developed for DNA sequencing, but is seldom used today.

Megabase – A unit of length for DNA fragments equal to 1 million nucleotides.

Melting temperature – The temperature at which dsDNA denatures. Parameters which determine melting temperature are sequence, length, and G:C content. Melting temperature can be estimated by summing [(A+T) x 2] + [(G+C) x 4].

Methylation – The process by which methyl groups are added to certain nucleotides in genomic DNA. This may affect gene expression, as methylated DNA is not easily transcribed. DNA methylation is thought to play an important developmental role in sequentially restricting the transcribable genes available to distinct cell lineages. In bacteria, methylation plays an important role in the restriction systems, as type I restriction enzymes cannot cut sequences with certain specific methylations.

Methylation Dependent Restriction System (MDRS) – A system used in bacteria to detect and cleave foreign DNA that is methylated at specific sequences.

Methyltransferase – An enzyme that adds a methyl group to a substrate, which can be a small molecule, a protein, or a nucleic acid.

microRNA (miRNA) – A class of molecules consisting of a non-coding sequence of approximately 22 nucleotides. miRNA controls gene expression at the post-transcriptional level

Microsatellite – A simple sequence repeat (SSR). It might be a homopolymer ('...TTTTTTT...'), a dinucleotide repeat ('....CACACACACACACA.....'), trinucleotide repeat ('....AGTAGTAGTAGTAGT...') etc. Due to polymerase slip (a.k.a. polymerase chatter) during DNA replication, these repeat sequences may become altered with copies of the repeat unit created or removed. Consequently, the exact number of repeat units may differ between individuals.

Minicell – A bacterial cell without a nucleus produced by asymmetric cell division enclosing only cytoplasm in one of the daughters. Minicells are useful in protein expression and purification.

Minisatellite – DNAs consisting of ~10 copies of a short repeating sequence. The length of the repeating unit is measured in 10s of base pairs. The number of repeats varies between individual genomes.

Minus strand DNA – The single-stranded DNA sequence that is complementary to the viral RNA genome or the plus strand of a DNA.

Mismatch – Base pairs that do not conform to AT and GC rules.

Mismatch repair – Repair that corrects bases that do not pair properly. The process preferentially corrects the sequence of the daughter strand, sometimes on the basis of their states of methylation.

Missense mutation – A change in a single codon that causes the replacement of one amino acid by another in a protein sequence

Molecular Biology – The branch of biology that deals with biological phenomena at the molecular level through the study of DNA and RNA, proteins, and other macromolecules involved in genetic information and cell function.

Monocistronic gene – A genetic unit that gives a single polypeptide chain when translated. Virtually all eukaryote mRNAs code for a single protein, but many bacterial mRNAs are polycistronic, especially those transcribed from operons.

Motif – A recurring pattern of short sequence of DNA, RNA, or protein, that usually serves as a recognition, binding, or active site. The same motif can be found in a variety of types of organisms.

mRNA – messenger RNA. The mature, processed product of DNA transcription used for translation of a gene into a protein. mRNA is a copy of the genetic information in DNA. It constitutes 1-5% of total RNA. mRNA encodes virtually all of the polypeptides synthesized by the cell. Eukaryotic mRNA is capped at the 5' terminus, polyadenylated at the 3' end, and spliced to remove intervening sequences.

mtDNA (mitochondrial DNA) – Double-stranded DNA contained within mitochondria that encodes mitochondrial tRNAs, rRNAs, and proteins. Several copies of the mitochondrial genome are found in each organelle.

Multiple cloning site – An artificially constructed region within a cloning vector containing a number of closely spaced restriction endonuclease cleavage sites, typically each site is unique in the vector.

Mutagens – Chemicals or physical means which induce changes in DNA sequence, directly or indirectly

Mutation – Change. Usually restricted to change in the DNA sequence of an organism, which may arise in a variety of different ways.

Mutation frequency – Rate of occurrence of a particular mutation in a population.

Mutation rate – The number of mutations per gene per generation.

Nascent protein – A protein that has not yet completed its synthesis; the polypeptide chain is still attached to the ribosome via a tRNA.

Nascent RNA – A ribonucleotide chain that is still being synthesized.

Native gel electrophoresis – An electrophoresis gel run under conditions which do not denature proteins (i.e., in the absence of SDS, urea, 2-mercaptoethanol, or other denaturant).

Negative supercoiling – The twisting of closed dsDNA in a direction opposite of the turns of the helix itself. This twisting acts to relieve torsional strain in the closed molecule.

Neoschizomers – A subset of isoschizomers that share the same recognition sequence, but cleave the DNA at a different site within that sequence.

Nested PCR – A very sensitive method for amplification of DNA. This method takes part or all of the product of a single PCR reaction (after 30 35 cycles), and subjects it to a new round of PCR using a different set of PCR primers which anneal within the region flanked by the original primer pair.

N-formyl-methionyl-tRNA (tRNAfMet) – The aminoacyl-tRNA that initiates bacterial protein synthesis. The amino group of the methionine is formylated.

Nick – A point in a double stranded DNA molecule where there is no phosphodiester bond between adjacent nucleotides of one strand, typically through damage or enzymatic action.

Nick translation – A technique used to label DNA which uses the sequence 5' to a nick as the primer. DNase I creates nicks in the dsDNA as a starting point. *E.coli* DNA polymerase I adds a nucleotide, copying the complementary strand, to the free 3' OH group at a nick; at the same time its exonuclease activity removes the first nucleotide 5' of the nick. This sequence of events continues until the strand is replaced. Addition of a labeled nucleotide to the reaction mixture results in a radioactively labeled product.

Non-coding DNA – DNA that does not code for part of a polypeptide chain or RNA. This includes introns and pseudo-genes. In eukaryotes, the majority of the DNA is non-coding.

Nonsense mutation – Any change in DNA that replaces a codon specifying an amino acid with a termination codon (UAG, UGA, or UAA).

Nonsense suppressor – A mutant tRNA which recognizes a termination codon as a sense condon.

Nontranscribed spacer – The region between transcription units in a tandem gene cluster.

Northern blot – A technique for transferring RNA to a nitrocellulose substrate from an agarose gel. Once immobilized, with preservation of the spatial arrangement of the nucleic acid, the RNA can be hybridized to a labeled complementary nucleic acid probe.

Nuclear run on (run off) – A technique used to determine RNA synthesis rate. Nuclei from cells expressing the gene are incubated with a radiolabel to visualize any transcripts that were being made at the time of harvest. All transcripts are incubated with an immobilized DNA complementary to the target RNA and the relative signal strength correlates the level of transcription of the target gene.

Nuclease – An enzyme capable of cleaving the phosphodiester bonds of nucleic acids.

Nucleic acids – Linear polymers of nucleotides, linked by phosphodiester bonds. In deoxyribonucleic acid (DNA), the sugar group is 2'-deoxyribose, and the bases of the nucleotides adenine, guanine, thymine and cytosine are bound to the sugar with an N-glycosidic linkage. Ribonucleic acid (RNA) has ribose as the sugar, and uracil replaces thymine. DNA functions as a stable repository of genetic information in the form of base sequence. RNA has a similar function in some viruses but more usually serves as an informational intermediate (mRNA), a transporter of amino acids (tRNA), or a component of the translational machinery (rRNA) in a structural capacity..

Nucleoside – A ribose or deoxyribose sugar with an N-glycosidically linked base (purine or pyrimidine).

Nucleotide – A molecule composed of an organic base, sugar and phosphate group, which constitutes the "building block" of nucleic acids (DNA and RNA).

Nucleus (plural, nuclei) – An organelle containing chromosomes. It is the site of cellular DNA replication and RNA synthesis.

Nuclide – An isotope.

Null mutation – A change that completely eliminates the function of a gene.

Ochre codon – The codon UAA, one of the three that causes termination of protein synthesis. This is the most frequently used stop codon in *E. coli* genes.

Ochre suppressor – A mutant tRNA that inserts an amino acid at the site of the ochre termination codon and allows translation to continue.

Okazaki fragments – The short stretches of 1000-2000 bases produced during discontinuous replication of the lagging strand of DNA. They are later joined into a continuous strand by ligase.

Oligonucleotide – Linear sequence of a few (up to 50) nucleotides joined by phosphodiester bonds. Above this length, the term polynucleotide is used. Usually single stranded.

Oncogene – One of a number of genes that cause the transformation of normal cells into malignant cells, especially a viral gene that transforms a host cell into a tumor cell.

Opal codon – The codon UGA, one of the three that causes termination of protein synthesis.

Open complex – The stage of initiation of transcription when RNA polymerase causes the two strands of DNA to separate to form the "transcription bubble."

Open reading frame – A series of triplets coding for amino acids without any termination codons and can potentially translate as a polypeptide chain.

Operator – The site on DNA at which a repressor protein binds to prevent transcription from initiating at the adjacent promoter.

Operon – A unit of bacterial gene expression and regulation, including structural genes and control elements in DNA recognized by regulator gene products.

Origin (*ori*) – A sequence of DNA at which replication is initiated. Bacterial chromosomes and plasmids have origins of replication. Different plasmids with the same origins of replication cannot co-exist in the same bacterium.

Overhang – A terminus of a dsDNA molecule which has one or more unpaired nucleotides in one of the two strands. Cleavage of DNA with many restriction endonucleases leaves such overhangs. Unpaired nucleotides at the 5' end of a dsDNA would be a 5' overhang; a 3' overhang would have unpaired nucleotides at the 3' end.

P site – The site in the ribosome that is occupied by peptidyl-tRNA, the tRNA carrying the nascent polypeptide chain, still paired with the codon to which it bound in the A site.

PAGE – Poly Acrylamide Gel Electrophoresis – The most widely used form of electrophoresis in the vertical format and uses polyacrylamide as the separation medium. PAGE gels are made from an acrylamide matrix crosslinked by TEMED, and the crosslinking is catalyzed by ammonium persulfate. Typically, it is used for smaller molecular weight DNA and proteins. Can be carried out under denaturing or non-denaturing conditions.

Palindromic sequence – Nucleic acid sequence that is identical to its complementary strand when each is read in the 5' → 3' direction (CCTAGG). Palindromic sequences are often recognition sites for Type II restriction endonucleases.

Parental strand – The DNA that is being replicated.

Peptide – Two or more amino acids covalently linked by a bond formed between the carboxy terminus of one amino acid and the amino terminus of the next—this link is called a peptide bond. A peptide may be a portion of a protein, a protein subunit, or a complete and functional protein by itself.

Peptide bond – A covalent bond between two amino acids, in which the carboxyl group of one amino acid and the amino group of an adjacent amino acid are linked.

Peptidyl transferase – The activity of the large ribosomal subunit that synthesizes a peptide bond when an amino acid is added to a growing polypeptide chain. The actual catalytic activity is a property of the rRNA.

Phage – See bacteriophage.

Phagemid – Plasmid vector containing the replication origin of M13 or f1 phage. This allows the vector to be replicated as a single stranded DNA and packaged as a viral particle. Phagemids are useful in sequencing.

Phenotype – Observable physical or biochemical characteristics of an organism, as determined by both genetic makeup and environmental influences.

Phosphatase – An enzyme that removes a phosphate from a nucleic acid or protein.

Phosphodiester bond – The covalent bond between the 3' hydroxyl of the sugar ring in one nucleotide and the 5' phosphate group of the sugar ring of the downstream nucleotide residue within a nucleic acid.

Phosphorylation – The creation of a phosphate derivative of an organic molecule. This is usually achieved by transferring a phosphate group from ATP. A specific kinase enzyme catalyzes this reaction.

Pilus – A surface appendage on a bacterium that allows the bacterium to attach to other bacterial cells. It appears as a short, thin, flexible rod. During conjugation, it is used to transfer DNA from one bacterium to another

Plaque – An area of clearing in a lawn of bacteria, representing bacterial cells that have been infected by a bacteriophage.

Plaque forming unit – An infectious virus particle capable of forming a plaque on an *E. coli* lawn. For a virus like bacteriophage λ, the number of viable viral particles, established by counting the number of plaques formed by serial dilution of the phage stock, is the titer, as expressed in plaque forming units/ unit volume. For example, a cDNA library might have a titer of 50,000 pfu/ml.

Plaque lift – A technique that uses a membrane to "lift" bacteriophage clones growing on a plate. Screening of the membrane identifies the intended clones.

Plasmid – An extrachromosomal genetic element in bacteria that has an origin of replication. The size of plasmids can vary from 500 bp to several hundred kb. Plasmids are circular and usually supercoiled.

Plus strand DNA – The strand of DNA that contains the same sequence as the mRNA, ie., the coding strand. Can also refer to the strand of DNA which lies in the direction of DNA synthesis from an M13 bacteriophage origin of replicaton.

Ploidy – The number of copies of the chromosome set present in a cell; a haploid has one copy, a diploid has two copies and so forth.

Point mutation – A single nucleotide substitution within a gene. Point mutations do not lead to a shift in reading frames, but may cause a single amino acid substitution, or the introduction of a stop codon. A single base change in the sequence of the ß –globin gene is responsible for sickle cell anemia.

Poly(A)-binding protein – The protein that binds to the 3' stretch of poly(A) on a eukaryotic mRNA

Poly A$^+$ tail – A lengthy (10-200) adenylic acid polymer which is covalently linked to the 3' end of newly synthesized hnRNA molecules in the nucleus. This is part of the processing of the transcript.

Polyadenylation – The addition of a sequence of polyadenylic acid to the 3' end of eukaryotic RNA after transcription.

Polycistronic gene – One genetic unit that codes for several proteins. Polycistronic mRNA includes coding regions representing more than one gene, and can include coding sequences read on both strands. Most polycistronic transcripts occur in prokaryotic organisms.

Polymerase – An enzyme that links individual nucleotides together into a long strand, using another strand as a template. There are two general types of polymerase — DNA polymerases (which synthesize DNA) and RNA polymerase (which makes RNA). Within these two classes, there are numerous sub-types of polymerase, depending on what type of nucleic acid can function as template and what type of nucleic acid is formed. A DNA-dependant DNA polymerase will copy one DNA strand starting from a primer, and the product will be the complementary DNA strand. A DNA-dependant RNA polymerase will use DNA as a template to synthesize an RNA strand.

Polymerase chain reaction (PCR) – The first practical system for in vitro replication of DNA. Two synthetic oligonucleotide primers, which are complementary to 18-22 bp regions of the target DNA (one for each strand) to be amplified, are added to a DNA solution that includes the target sequence, in the presence of excess deoxynucleotides and a thermostable DNA polymerase. Through a series of temperature cycles, the target DNA is repeatedly denatured, annealed to the primers and a daughter strand is extended from the primers. The daughter strands themselves act as templates for subsequent cycles. Fragments bounded by both primers are amplified exponentially, rather than linearly. Millions of copies of products from a single template are possible in 30-40 cycles.

Polyribosome – A series or cluster of ribosomes on a single mRNA that translate the sequence at the same time. Also called a polysome.

Polypeptide – Chains of amino acids joined by peptide bonds. Distinction between peptides, oligopeptides and polypeptides are arbitrary by length; an oligopeptide is usually 20 to 50 residues in length. Polypeptide refers to any chain of amino acids, not necessarily functional or structural proteins.

Post-translational processing – The reactions that alter a protein's structure, such as phosphorylation, glycosylation, or proteolytic cleavage.

Pribnow box – A sequence about 10 nucleotides upstream from the place where transcription starts. The consensus sequence is 5'-TATAAT-3'.

Primary transcript – RNA transcript immediately after transcription in the nucleus before RNA splicing or polyadenylation to form the mature mRNA.

Primase – A type of RNA polymerase that synthesizes short segments of RNA to be used as primers for DNA replication

Primer – A short DNA or RNA sequence that will anneal to ssDNA or RNA specifically and provides a free 3'-OH end at which a DNA polymerase starts to synthesize a DNA chain.

Primer dimer – The 3'-end of one primer may pair with the 5'-end of the same primer or a different primer. The primer-primer pairs may be extended to form primer dimers. Primer dimers should be avoided by better primer design.

Primer extension – an enzymatic process using DNA polymerase to add nucleotides to an oligonucleotide primer based on the complementary strand of DNA or RNA to which the primer is specifically annealed. The many uses of this technique include probe production and determination of mRNA initiation point.

Probe – A DNA or RNA molecule which has been radioactively labeled or otherwise marked (e.g. biotin), and used to detect or identify a target DNA or RNA sequence. The term may also refer to antibody used in Western blots.

Processivity – The extent to which an RNA or DNA polymerase adheres to a template before dissociating. Processivity is determined by the average length (in kilobases) of the newly synthesized nucleic acid strands.

Processing – The post-translational modification of a polypeptide or post-transcriptional modification of a primary RNA transcript.

Progeny – offspring or descendents

Prokaryotic – Organisms which lack nuclei like bacteria.

Promoter – A sequence of DNA 5' to the coding region of a gene. It promotes and regulates the transcription of that gene. Generally consists of an RNA polymerase binding site and sites for binding of various transcription and regulatory molecules (DNA, protein, hormone, inducer, etc.) that affect the initiation or fidelity of transcription.

Proofreading – A mechanism for correcting errors in protein or nucleic acid synthesis. This involves a checking of the individual units after they have been added to the chain.

Prophage – A viral genome covalently integrated in the chromosome of host cell. Usually quiescent.

Proteomics – Analysis of protein regulation, expression, structure, post-translational modification, interactions, and function.

Proto-oncogene – A gene present in a normal cell that usually carries out normal cellular function, but can become oncogenic. The prefix "c" indicates a cellular gene like *c-myc*.

Provirus – A duplex DNA sequence in the eukaryotic chromosome corresponding to the genome of an RNA retrovirus.

Pseudogene – An inactive but stable component of the genome derived by mutation of an ancestral, active gene.

Pulsed field gel electrophoresis (PFGE) – A method for separation of large (>50 kb) pieces of DNA based on size, including complete chromosomes and genomes, by rapidly alternating the direction of electrophoretic migration in agarose gels.

Purine – Adenine and guanine bases of DNA and RNA. Each contains a two ring structure with 5 carbons and 5 nitrogens.

Pseudogene – Non-functional DNA sequences that are very similar to the sequences of known genes. Some probably result from gene duplications that become non functional because of the loss of promoters, accumulation of stop codons, mutations that prevent correct processing etc.

Pyrimidine – Cytosine, thymine, and uracil bases of nucleic acids. Cytosine and thymine are components of DNA, while cytosine and uracil are the pyrimidines of RNA. Each consists of a single ring structure of 4-5 carbons and 2-3 nitrogens.

Pyrimidine dimmer – A dimer that forms when ultraviolet irradiation generates a covalent link between two adjacent pyrimidine bases in DNA. It blocks DNA replication and transcription.

Quencher – A quencher molecule absorbs the fluorescent emission of reporter molecule when in close vicinity, i.e., the reporter is not fluorescent when the quencher is close by and becomes fluorescent when the quencher is removed.

Random primer labeling (oligolabeling) – A method of generating labeled DNA molecules based on the sequence of a dsDNA template. Random hexamers (oligonucleotides of six units, the mix contains all combinations of the four nucleotides) are annealed to the denatured template at the sites specific for each hexamer. Not all hexamers will find a complementary sequence, but some will. Klenow fragment is used to synthesize a new strand of DNA using the annealed hexamer as a primer. If radioactively or otherwise labeled nucleotides are used in this reaction, the new strand will incorporate the label. Probes generated in this fashion contain many different probe sequences, determined by where a primer has annealed to the template and how long the extended strand becomes. Gel electrophoresis of this probe mixture shows a smear, not a band, to reflect this heterogeneity of probe sequences.

Radiolabel – A radioactive ligand incorporated into a small molecule or biomolecule that enables detection and quantification of the molecule.

Rapid amplification of DNA ends – (RACE; 3' RACE; 5' RACE) Techniques, based on the polymerase chain reaction, for amplifying either the 5' end (5' RACE) or 3' end (3' RACE) of a cDNA molecule, given that some of the sequence in the middle is already known. The two procedures differ slightly; in 3' RACE, first strand cDNA is prepared by reverse transcription with an oligo dT primer (to match the poly-A tail), from an mRNA population believed to contain the target. PCR then proceeds with a gene-specific, forward primer and an oligo-dT reverse primer. 5' RACE is an example of anchored PCR; the first-strand cDNA population is tailed with a known sequence, either by homopolymer tailing (eg. with dA) or by ligation of a known sequence. PCR then proceeds as before with a primer specific for the gene, and one specific for the added tail.

Rare cutter – A type of restriction enzyme that cuts at sites that occur infrequently within chromosomes.

rDNA – DNA that codes for ribosomal RNA.

Reading frame – mRNA is translated as a series of base triplets to specify the amino acids in a protein chain. Three reading frames exist in any mRNA, in most cases only one reading frame will produce a functional protein.

Readthrough – The transcription of nucleic acid sequence beyond its normal termination sequence.

Real time PCR – Amplification of a nucleic acid sequence using the polymerase chain reaction and primers complementary to the 5' and 3' ends of the target sequence with analysis of the production of amplicons as the reaction proceeds. Tracking of the amplification of the target sequence is based on the fluorescence emitted by the amplicons; either intercalated nonspecifically into any dsDNA, or using a probe complementary to some internal portion of the target sequence. The increase in fluorescence correlates linearly with an increase in the number of amplification products.

Rec A – The product of the *rec*A gene of *E. coli*; a protein with the ability to both activate proteases and exchange ssDNA. The protease-activating activity controls the SOS response; exchange of ssDNA is crucial for recombination events.

Receptor – A molecule which spans a membrane and, when bound by a signal molecule (for example a hormone), transmits information across that membrane.

Recombinant – An organism or cell in which genetic recombination has taken place.

Recombinant DNA – Spliced DNA formed from two or more different sources that have been cleaved by restriction enzymes and joined by ligases.

Recombineering (recombinogenic engineering) – A genetic and molecular biology technique based on homologous recombination systems in *E. coli* to modify DNA. The procedure is widely used in the generation of target vectors for making conditional mouse knockout. The term was first coined in 2001 by Ellis et al. (Proc. Natl. Acad. Sci, 98:6742-6746). Recombineering is based on homologous recombination in *E.coli* mediated by phage proteins, either RecE/RecT from Rac prophage or Red alpha/beta/gamma from bacteriophage lambda (Zhang et al, Nature Genetics, 1998, 20, 123-8; Muyrers et al, Nucleic Acids Res., 1999, 27, 1555-7). The lambda Red recombination system is now most commonly used. These homologous recombination systems mediate the efficient recombination of a target fragment (with homology sequences as short as 30 bps) into the DNA construct. The sequence homologies (or arms) flanking the desired modifications are homologous to regions 5' and 3' to the region to be modified.

Redundancy – The concept that two or more gene products or functional pathways fulfill the same function, so that no single one of them is essential.

Relaxed DNA – Circular DNA that is not supercoiled. Relaxed DNA is less compact than supercoiled DNA and migrates more slowly during gel electrophoresis.

Release factor (RF) – A factor required to terminate protein synthesis to cause release of the completed polypeptide chain and the ribosome from mRNA.

Renaturation – The conversion of denatured protein or DNA to its native configuration. DNA and protein are denatured by heat, destroying secondary and tertiary structure. DNA can reassume both secondary and tertiary structures, including base pairing, by cooling of the environment, without input of additional energy.

Repair – When a DNA strand is damaged, it is excised and replaced by the synthesis of a new stretch through repair synthesis. This can also take place by recombination reactions, when the duplex region containing the damaged strand is replaced by an undamaged region from another copy of the genome.

Repetitive DNA – Nucleotide sequences in DNA that are present in the genome as numerous copies. Originally identified by the value on the Cot curve derived from kinetic studies of DNA renaturation. These sequences are not thought to code for polypeptides. One class of repetitive DNA, termed highly repetitive DNA, is found as short sequences, 5-100 nucleotides, repeated thousands of times in a single long stretch. It typically comprises 3-10% of the genomic DNA and is predominantly satellite DNA. Another class, which comprises 25-40% of the DNA and termed moderately repetitive DNA, usually consists of sequences about 150-300 nucleotides in length dispersed evenly throughout the genome, and includes Alu sequences and transposons.

Replication – Production of an identical DNA from a template DNA using host cell proteins and processes.

Replication fork – The point at which strands of parental duplex DNA are separated so that replication can proceed. A complex of proteins including DNA polymerase is found there.

Response element – A portion of a gene which must be present in order for that gene to respond to some hormone or other stimulus. Response elements are binding sites for transcription factors. Certain transcription factors are activated by stimuli such as hormones or heat shock. A gene may respond to the presence of that hormone because the gene has in its promoter region a binding site for hormone-activated transcription factor. Example: the glucocorticoid response element (GRE).

Reporter gene – A gene that encodes an easily assayed product (eg. CAT or ß-galactosidase) that is coupled to the upstream regulatory sequence of another gene. Once transfected into cells, the reporter gene can be used to determine factors that activate response elements in the upstream region of the gene of interest.

Repression – Inhibition of transcription by the binding of a repressor protein to a specific site on DNA (or mRNA).

Repressor protein – A protein that binds to an operator on DNA or RNA to prevent transcription or translation, respectively.

Restriction – To restrict DNA means to cut it with a restriction enzyme

Restriction endonuclease – An enzyme that recognizes specific short sequences of DNA and cleave the duplex, sometimes at the target site, sometimes elsewhere, depending on type.

Restriction enzymes (restriction endonucleases) – Enzymes that cut dsDNA at specific sites. In bacteria, their function is to destroy foreign DNA, such as that of bacteriophages. Types I, II, and III have different patterns of recognition, digestion, and modification. Type II enzymes are the most often used in molecular biology. Restriction enzymes are generally named according to the bacterium from which they were first isolated (first letter of genus name and the first two letters of the species name). Multiple enzymes from a single bacterial strain are numbered according to the protein fraction of that contains the activity.

Restriction fragments – Fragments of dsDNA generated by digesting dsDNA with one or more restriction endonucleases.

Restriction fragment length polymorphism (RFLP) – A technique that allows relationships to be established by comparing the characteristic polymorphic patterns that are obtained when certain regions of genomic DNA, amplified by PCR if necessary, are cut with certain restriction enzymes.

Restriction map – Representation of a DNA molecule showing the positions of restriction endonuclease recognition sites as numbered from an arbitrary position on a circular DNA or from the 5' terminus of a linear DNA.

Restriction site – The specific sequence in DNA which a restriction enzyme recognizes and cuts.

Retrovirus – An RNA virus that propagates via conversion into duplex DNA before integration into the host genome. Retroviruses undergo mutation at a faster rate than DNA viruses, often making it difficult to develop adequate chemotherapeutic agents.

Retroviral vector – Modified retroviruses used in the genetic modification of cells as a means of introducing foreign DNA into the genome or knocking out host genes.

Reverse transcriptase – RNA directed DNA polymerase. This enzyme, first discovered in retroviruses, can construct a ssDNA molecule from a single stranded RNA, creating an RNA:DNA hybrid.

Reverse transcription – The synthesis of DNA on a template of RNA. This is accomplished by using a reverse transcriptase enzyme and a primer complementary to some portion of the RNA, often the poly-A^+ tail.

RF – Replicative form. Commonly, the form of a bacteriophage genome in which it can be replicated by the host machinery.

Rho dependent terminator – Terminator of prokaryotic transcription which requires the Rho protein to physically pull the nascent RNA from its contact with the DNA and transcription machinery.

Rho factor – A protein involved in assisting bacterial RNA polymerase to terminate transcription at rho dependent sites.

Rho independent terminator – Sequences of DNA that cause prokaryotic RNA polymerase to terminate transcription in the absence of the rho factor. The sequence, located at the 3' end of genes, usually has inverted GC rich and AT rich portions capable of forming a hairpin structure.

Ribonucleases – The class of enzymes that cleave RNA. They may be specific for single-stranded or double-stranded RNA, and may be either endonucleases or exonucleases.

Ribonucleotide – A nucleotide that contains ribose as its sugar and is a component of RNA.

Riboprobe – A strand of RNA synthesized in vitro (usually radiolabeled) and used as a probe for hybridization reactions. An RNA probe can be synthesized at very high specific activity and can be used to detect DNA or RNA targets.

Ribosome – A complex cellular particle that is involved in the translation of messenger RNAs to make proteins. Ribosomes consist of ribosomal RNAs and several proteins.

Ribotyping – A technique where bacterial genomic DNAs are digested and separated by gel electrophoresis. Universal probes that target specific conserved domains of ribosomal RNA coding sequences are used to detect the banding patterns.

Ribozyme – An RNA molecule that has the ability to catalyze the cleavage or formation of covalent bonds in RNA or DNA strands at specific sites.

RNA (Ribonucleic acid) – Polymers of ribonucleotides. All RNA species are synthesized by transcription of DNA sequences, but may involve post transcriptional modification.

RNAi – RNA interference. A mechanism by which small double-stranded RNAs interfere with expression of any mRNA having a similar sequence. These small RNAs are known as 'siRNA', for short interfering RNAs. The mode of action for siRNA includes dissociation of its strands, hybridization to the target RNA, and then fragmentation of the target. Importantly, the remnants of the target molecule may also act as an siRNA itself; thus the effect of a small amount of starting siRNA is amplified and can have long-lasting effects on the recipient cell. The RNAi effect has been exploited in numerous research projects to deplete the cell of specific messages, thus examining the role of those messages by their absence.

RNase – A class of enzymes whose substrate is RNA. May be specific for single stranded, double stranded or hybrid RNA. Some have specific endonuclease or exonuclease activity.

RNA ligase – An enzyme that functions in tRNA splicing to make a phosphodiester bond between the two exon sequences that are generated by cleavage of the intron.

RNase A protection assay – A sensitive method to determine the amount of a specific mRNA present in a complex mixture of mRNA. A radioactive RNA probe (in excess) is allowed to hybridize with a sample of mRNA (for example, total mRNA isolated from tissue), after which the mixture is digested with RNase A. Only the probe which is hybridized to the specific mRNA will escape the nuclease treatment, and can be detected on a gel. The amount of radioactivity which was protected from nuclease is proportional to the amount of mRNA to which it hybridized.

RNA polymerases – Enzymes that polymerize ribonucleotides in accordance with the information present in DNA. Prokaryotes have a single enzyme for the three RNA types. Eukaryotes have three RNA polymerases. RNA polymerase I synthesizes the 18s, 5.8s, and 28s rRNA's. RNA polymerase II acts primarily in transcription of protein encoding genes as well as snRNA's and miRNA's. RNA polymerase III synthesizes tRNA's and the 5s rRNA.

RNA splicing – The process of excising introns from RNA and connecting the exons into a continuous mRNA.

Rolling circle replication – A mode of replication most often found in filamentous bacteriophages. The replication fork proceeds in one direction around a circular template for an indefinite number of revolutions. The newly synthesized DNA strand in each revolution displaces the strand synthesized in the previous revolution, giving a tail containing a linear series of sequences complementary to the circular template of the strand.

rRNA – Ribosomal RNA. Several RNAs that form part of the structure of ribosomes. rRNA brings mRNA and tRNA together during translation. About 80-85% of total RNA is rRNA

RT-PCR (reverse transcriptase-polymerase chain reaction; reverse transcription-PCR) – PCR in which the starting template is RNA, implying the need for an initial reverse transcription step to make a DNA template. Some thermostable polymerases have appreciable reverse transcriptase activity; however, it is more common to perform an explicit reverse transcription, inactivate the reverse transcriptase or purify the product, and proceed to a separate conventional PCR.

Sanger sequencing – A DNA sequencing method introduced in the 1970's based on enzymatic incorporation of chain-terminating nucleotide analogs; also called 'dideoxy sequencing.' It makes use of the ability of DNA polymerase to faithfully synthesize a complementary copy of a DNA template using a short DNA fragment as a primer and to randomly incorporate an analog of dNTP - a ddNTP which has a 3'-H instead of 3'-OH group. Since 3'-H is not a substrate for chain elongation, the growing DNA strand is terminated once the analog is incorporated. During sequence analysis, four separate reactions are carried out. Each reaction is supplied with all four dNTPs and a small amount of one of the four ddNTPs. Each reaction forms a nested series of fragments i.e. the "G" tube will contain populations of fragments that are terminated at every G. Incorporation of a trace in the reaction allows for detection of fragments after separation by denaturing PAGE.

Satellite colonies – The small colonies that surround a big antibiotic resistant colony. They survive because other cells are doing the work of destroying the antibiotic in their immediate vicinity on the plate.

Satellite DNA – DNA, usually containing highly repetitive sequences, that has a base composition and density sufficiently different from that of normal DNA so that it sediments as a distinct band in cesium chloride density gradients.

Secondary structure – Structures produced in polypeptide or nucleotide (single or double strand) chains involving interactions between amino acids or nucleotides within a chain. Especially helical structures.

Sense strand (+ strand or coding strand) – A strand of DNA or RNA whose sequence contains information to produce a protein. In dsDNA, it is the strand that has the same sequence (except for containing T instead of U) as the mRNA.

Sequence – A primary structure of a DNA molecule, in terms of the order of bases it contains. As a verb, to sequence is to determine the structure of a piece of DNA by determining the identity and order of the nucleotides it contains.

Sequence tagged site (STS) – Short, tagged tracts of DNA sequence that are used as landmarks in genome mapping. In most instances, 200 to 500 base pairs of sequence define a sequence tagged site that is operationally unique in the human genome (i.e., can be specifically detected by the polymerase chain reaction in the presence of all other genomic sequences).

Serial Analysis of Gene Expression (SAGE) – A technique for profiling gene expression in cells. SAGE characterizes a short segment of DNA, called a SAGE tag, in each expressed gene. cDNA is digested with the restriction enzyme Nla III, ligated to a linker, then digested with BsmF1, an enzyme that cleaves 10-14 bases from the Nla III cut. SAGE tags are then concatenated, cloned, and sequenced.

Shine Dalgarno sequence (Ribosomal binding sequence) – In prokaryotic organisms, all or part of the specific sequence AGGAGG located on mRNA just upstream of an AUG start codon. This sequence is complementary to the sequence at the 3' end of 16S rRNA and is involved in binding of the ribosome to mRNA.

Shotgun cloning – Cloning of all fragments generated by digestion of a large molecule of DNA in a single ligation reaction, so that transformation of the products will give a population of different clones containing all the digestion products.

Shuttle vector – Cloning vector that can replicate in cells of more than one organism, e.g. *E. coli* and yeast. This combination allows DNA from yeast to be replicated in *E. coli* and tested directly for complementation in yeast. Shuttle vectors are constructed so that they have replication origins of the various hosts.

Sigma factor –The subunit of bacterial RNA polymerase needed for initiation; it is responsible for selection of promoters.

Single Base Primer Extension Assay (SBE) – A technique for detecting known SNP site. A primer that anneals immediately adjacent to the SNP is extended by one base using a fluorescently labeled ddNTP. No further extension is possible because ddNTP is used.

Single Strand Conformation Polymorphism (SSCP) – A technique to identify changes in target DNA sequences from different individuals. Secondary structure of ssDNA is determined by sequence; therefore, changes in sequence (even a single base change) will alter the secondary structure. After gel electrophoresis of ssDNA, changes in sequence between isolates is identified by a change in the rate of movement through the gel.

Single stranded DNA (ssDNA) – DNA that consists of only one chain of nucleotides rather than the two base pairing strands found in the double helix form. Single stranded DNA can be produced experimentally by high thermal or basic conditions. Reverse transcription forms a DNA:RNA hybrid which becomes ssDNA with digestion of the RNA.

Silencer – A short sequence of DNA that can inactivate expression of a gene in its vicinity.

Silent mutation – A mutation in a gene that causes no detectable change in the properties of the gene product. It is a mutation that does not change the sequence of a protein because it produces synonymous codons.

Single Nucleotide Polymorphism (SNP) – A single nucleotide position in a genomic DNA sequence that varies from one individual to another.

Single-Stranded DNA Binding protein (SSB) – SSB binds to ssDNA. Binding of SSB to ssDNA serves several functions: prevent DNA secondary structure from forming, prevent primase from priming on ssDNA, enhance DNA synthesis rate and error repair by DNA polymerase III holoenzyme, enhance duplex DNA unwinding by DnaB helicase. SSB is also essential for DNA repair and recombination.

siRNA – short interfering RNAs.

Site-directed mutagenesis –The production of a mutation, usually a point mutation or an insertion, into a particular location in a cloned DNA fragment. This fragment may be used to knock out a gene in the organism of interest by homologous recombination.

snRNA (small nuclear RNA) – One of many small RNA species confined to the nucleus; several of them are involved in splicing or other RNA processing reactions.

Southern blot – A technique where DNA is separated by electrophoresis (usually on agarose gel), and then transferred to a paper like membrane (usually nylon-backed nitrocellulose) by capillary action or an electric field, preserving the spatial arrangement. Once on the membrane, the molecules are immobilized by baking or ultraviolet irradiation. The DNA is denatured to single strands so that it can be hybridized with a specific probe. Southern blots were named after their developer, Dr. Ed Southern, a molecular biologist in Edinburg.

Southwestern blot – A technique for identifying the protein that binds to a nucleic acid on a matrix. This is similar to what is done in Southern, Northern and Western blots.

Spacer DNA – DNA sequence between genes. In bacteria, they are only a few nucleotides long. In eukaryotes, they can be extensive and include repetitive DNA, comprising the majority of the DNA of the genome. The term is used particularly for the DNA located between the many tandemly-repeated copies of the ribosomal RNA genes.

Specific activity – A measure of the relative abundance of radioactive molecules in a labeled sample and is reported as radioactivity per unit mass (or per unit number) – for example, cpm/μmol. High specific activity allows for detection of low copy targets, since each molecule hybridized contains several radioactively labeled monomers.

Splice – The removal of introns and joining of exons in RNA; introns are spliced out. Splicing is carried out by recognition of signals at the 5' and 3' ends of the intervening sequence, assembly of the spliceosome using snRNPs and the formation of a lariat structure with cleavage of two phoshodiesterase bonds and ligation of the exons.

Splicesome – A complex of small nuclear RNA and protein particles (snRNP or "snurps") that participate in hnRNA splicing.

ssDNA (single stranded DNA) – DNA that consists of only one chain of nucleotides rather than the two base pairing strands found in DNA in the double helix form. Single stranded DNA can be produced experimentally by high thermal or basic conditions. Reverse transcription forms a DNA:RNA hybrid which becomes ssDNA with digestion of the RNA. ssDNA is important in antisense therapies and in *in situ* hybridization.

Stable transfection (stable expression) – A clone of cells in which the transgene has been physically incorporated into the genome, rather than remaining episomal. It thus provides a stable, long term expression.

Standard curve – A graph prepared by plotting the migration of a set of known macromolecules (DNA, RNA, or protein) and used to estimate the size of unknown macromolecules by comparing their migration to that of the known macromolecules.

Star activity – Recognition and cutting action of a restriction endonuclease at a site other than its regular recognition sequence as a result of digestion in the wrong buffer conditions.

Start codon (or initiation codon) – The codon at which translation of a polypeptide chain begins. This is usually the first AUG triplet in the mRNA molecule from the 5' end, where the ribosome binds to the cap and begins to scan in a 5' - 3' direction. However, the surrounding sequence context is important and may lead to the first AUG being bypassed by the scanning ribosome in favor of an alternative, downstream AUG. Occasionally other codons like UUG may serve as initiation codons.

Startpoint – The position on DNA corresponding to the first base incorporated into RNA

Stationary phase – The final phase of bacterial growth. The cells reduce their metabolic activity and consume non-essential cellular proteins due to depleted nutrients.

Stem – The base-paired segment of a hairpin structure in RNA or single stranded DNA

Stem-Loop – A feature of RNA secondary structure, in which two complementary, inverted sequences which are separated by a short intervening sequence within a single strand of RNA base pair to form a "stem" with a "loop" at one end. Similar to hairpins, but these usually have very small loops and longer stems.

Sticky end – DNA fragments with 3' or 5' overhang ends generated by restriction enzymes that can readily anneal with the ends of other fragments with compatible sequences; therefore, they are called "sticky ends."

Stop codon – One of three triplets (UAG, UAA, or UGA) that causes protein synthesis to terminate. They are also known historically as nonsense codons. The UAA codon is called ochre, the UGA codon is called opal, and the UAG codon is called amber.

Stringency – A condition of salt concentration and temperature which controls the degree to which non-homologous fragments of ssDNA can remain hybridized to a target DNA. High temperature and low salt = high stringency.

Subcloning – Excision of a fragment of DNA from one vector and ligating it into another vector.

Subtraction library – A library prepared by subtracting mRNA's that are not wanted.

Subtractive hybridization – A method to identify DNA/RNA present in one sample but not in others.

Supercoiling – In circular DNA or closed loops of DNA, twisting of the DNA about its own axis changes the number of turns of the double helix. Negative supercoiling in circular DNA relives torsional strain induced by the turning of the helix. Supercoiling can be detected by agarose electrophoresis because supercoiled DNA migrates faster than relaxed DNA.

Supernatant – The liquid fraction of a sample after centrifugation.

Suppressor tRNA – A transfer RNA that recognizes a termination codon in an RNA and adds an amino acid residue instead of terminating the chain, generating a readthrough protein.

SYBR Green – A fluorescent dye which binds to double-stranded DNA. Used for quantitation of double-stranded DNA such as PCR products.

TA cloning – A specific type of cloning for PCR products that takes advantage of the tendency of Taq DNA polymerase to add a dA at the 3' end of newly synthesized DNA strands, thus leaving a single base 3' overhang. Vectors are prepared with a single base dT 3' overhang, allowing ligation of sticky ends without restriction endonuclease digestion of either the PCR product or the vector.

TAMRA – Tetramethylrhodamine (TAMRA). A fluorescent dye the rhodamine group.

TATA box – The binding site for general transcription factors of eukaryotic genes. It is usually located at about 25 bp upstream from the transcription start site.

Tandem repeat – Short, identical or near identical sequences repeated in the same (direct) or opposite (indirect) direction.

TaqMan probe – A specifically designed probe used for real-time PCR. At one end of the TaqMan probe is a fluorescent reporter; at the other end is a quencher. When binds to template DNA, the reporter fluorescence is quenched by the quencher due to fluorescence resonance energy transfer (FRED). PCR amplification frees fluorescent reporter from the template and generates fluorescence.

Taq DNA polymerase – The DNA polymerase which is stable at high temperatures; isolated from the thermophilic bacterium *Thermus aquaticus.* Very useful in PCR reactions which must cycle repetitively through high temperatures during the denaturation step.

Telomere – The terminal portion of a linear chromosome. These portions are not replicated via a replication fork. They are replicated by a telomerase, which carries an RNA primer to initiate synthesis.

Template – The parent strand of DNA or RNA which is used as the basis to synthesize a complementary nucleic acid strand. The sequence of bases in the new strand is determined by the parent strand.

-10 sequence – The consensus sequence TATAAT centered about 10 bp before the start point of a bacterial gene. It is involved in the initial melting of DNA and binding of by RNA polymerase.

Tetramer – A protein composed of four peptide subunits linked by electrostatic or covalent bonds.

Terminal protein – A protein that allows replication of a linear genome to start at the very end. It attaches to the 5' end of the genome through a covalent bond and contains a nucleotide that serves as a primer.

Terminase – An enzyme involved in λ DNA replication. Terminase makes 12-nucleotide staggered cuts in the two DNA strands at a site named *cos*, thereby generating the "sticky ends" that allow an infecting molecule to create concatamers in the host cell.

Terminator – A sequence of DNA, present at the end of the transcript, that can form a stem-and-loop structure and causes RNA polymerase to terminate transcription.

Thermocycler – A computer controlled heat block that will cool or warm to a series of user designated temperatures, each for a specific length of time, and then repeat for a user designated number of cycles.

Tertiary structure – A third level of structure due to the interactions between amino acid residues or nucleotides which are not closely positioned within the primary structure of the molecule.

-35 sequence – The consensus sequence centered about -35 bp before the start point of a bacterial gene. It is involved in initial recognition by RNA polymerase.

Tissue-specific expression – Gene function which is restricted to a particular tissue or a cell type. Tissue specific expression is usually the result of an enhancer which is activated only in the proper cell type.

Tm – See also melting temperature. The midpoint of the temperature range over which DNA is denatured by heat; more specifically, the temperature at which a ds nucleic acid molecule is 50% melted into single strands. It is dependent on the number and proportion of G-C base pairs as well as the ionic conditions.

Topoisomerase – An enzyme that can change the degree of supercoiling in DNA by cutting one or both strands. Type I topoisomerases cut only one strand of DNA, while Type II topoisomerases cut both strands of DNA. Both enzymes increase the degree of negative supercoiling in DNA.

Trans – A prefix denoting across, through, or beyond; it is the opposite of cis.

***Trans*-acting** – The action of a regulatory factor that binds to a DNA element and activates or silences the expression of the gene containing the DNA element.

Transactivation – Stimulation of transcription by a transcription factor binding to DNA.

Transcript – An RNA molecule synthesized by transcriptional machinery complementary to the sequence of a DNA template.

Transcription – Synthesis of RNA by RNA polymerases using a DNA template. The nucleotide at which transcription starts is designated +1. Transcription of a eukaryotic gene yields an hnRNA with 5' and 3' untranslated regions, exons, and introns.

Transcription factor – Protein required for effective recognition by RNA polymerases of specific stimulatory sequences in promoters.

Transcription unit – The distance between sites of initiation and termination by RNA polymerase. Prokaryotic transcription units may include more than one gene.

Transduction – The transfer of a gene from one bacterium or cell to another by a bacteriophage or virus, respectively..

Transfection – Introduction of foreign DNA into eukaryotic cells. It is usually accomplished using DNA precipitated with calcium ions or packaged by liposomes, but a variety of other methods like electroporation can be used.

Transfer region – A segment on the F plasmid required for bacterial conjugation.

Transformation – Introduction of foreign DNA into bacteria. Transformation can be recognized by changes in growth characteristics, such as resistance to a certain antibiotic.

Transgenic – A cell that has had genes from another cell put into its genome through recombinant DNA techniques.

Transient transfection – Introduction of foreign DNA into eukaryotic cells for short-term studies such as 2-3 days.

Transition – A nucleotide substitution in which one pyrimidine is replaced by the other pyrimidine, or one purine is replaced by the other purine (e.g., A is changed to G, or C is changed to T) (contrast with Transversion).

Translocation – Breakage and removal of a large segment of DNA from one chromosome, followed by insertion of the segment to a different chromosome

Translation – Production of a protein based on the genetic code. Ribosomes perform the synthesis and tRNAs bring the correct amino acids into close proximity to form peptide bonds by reading the information on mRNA.

Transposition – The movement of DNA from one location to another location on the same or a different DNA molecule within a cell.

Transposon – A DNA sequence able to excise and insert itself at a new location in the genome without a homologous recombination event. Many transposons carry antibiotic resistance determinants and have insertion sequences at both ends.

Transversion – A nucleotide substitution in which a purine replaces a pyrimidine, or vice versa (e.g., A is changed to T, or T is changed to G) (see Transition).

Trimer – A protein composed of three peptide subunits linked by electrostatic or covalent bonds.

tRNA – Transfer RNA. Active in assembly of polypeptide chains, tRNA translates mRNA base sequences into protein amino acid sequences. tRNAs have a unique secondary structure and contain many modified nucleotides. The anticodon of tRNA H-bonds to the codon of mRNA to put amino acid in position to be added to the growing polypeptide chain. About 10-20% of total RNA is tRNA.

Triplet – A three-nucleotide sequence; a codon.

Turbid plaque – A speckled clearing in a lawn of bacteria representing an area of cells infected with a bacteriophage. Only some of the cells in the plaque have phage in the lytic phase, so only some of the cells are lysed. The rest have phage in the lysogenic state; thus they remain speckled.

Two hybrid assay – An assay that detects interaction between two proteins by means of their ability to bring together a DNA-binding domain and a transcription-activating domain. The assay is performed in yeast using a reporter gene that responds to the interaction.

UNG – Uracil N-glycosylase. When employed in conjunction with a uracil containing PCR reaction, this enzyme is used to degrade PCR products of previous reactions to reduce the possibility of false positive results.

Ultraviolet light (UV) – Light in the wavelength range of 280 to 320 nanometers that is used to illuminate DNA and RNA samples bound with fluorescent dyes, such as ethidium bromide, allowing visualization of the nucleic acids.

Unit – A measurement of the function of an enzyme for a given volume or mass. A unit of activity for a restriction enzyme is typically defined as the amount of enzyme required to digest (or "restrict") one microgram of reference DNA in one hour at optimal temperature.

Universal primer – A primer with its sequence complementary to a region flanking the cloning sites. When an insert DNA is cloned into the multiple cloning site of a vector, any insert DNA can be sequenced using this primer.

Up mutation – A mutation in a promoter that increases the rate of transcription

Upstream – A sequence of DNA or RNA opposite of the direction of transcription or translation. Upstream also refers to early events in any process that involves sequential reaction.

Variable number tandem repeat (VNTR) – A minisatellite DNA; the repetition of a 35-80 bp sequence, to a size of up to 2 kb, that is characteristic of an individual. VNTRs are used in forensic science to identify, or exclude, suspects of a crime.

Vector – A commonly used term for a genetic element that can be used to transfer DNA sequences from one organism to another. This can be plasmid, phagemid, cosmid, YAC, or virus.

Volume excluders – For restriction enzymes, buffer components such as glycerol and polyethylene glycol that reduce the amount of water at the enzyme interface.

Voltage – An electrical unit of measure used during electrophoresis. It is the electromotive force applied to drive charged molecules through a gel matrix in order to separate the molecules by size and/or charge during electrophoresis. Proper voltage can be critical to the resolution of different species in a mixed sample. A very low voltage may allow diffusion of samples and loss of band resolution, while a very high voltage can distort sample mobility. Excessive power (Watts = voltage times current) can generate excessive heat, and distort sample mobility or melt an agarose gel matrix.

Western blotting (protein blotting) – Electrophoretic transfer of PAGE-separated proteins onto a nitrocellulose or polyvinylidene diflouride (PVDF) membrane. The membrane becomes an exact positional representation of the original gel. The transferred proteins are readily accessible for probing with antibodies or staining for visualization.

Wild type – The native or predominant genetic constitution before mutations, usually referring to the genetic constitution that normally exists in nature.

Wobble position – The third base position within a codon, which can often (but not always) be altered to another nucleotide without changing the encoded amino acid (see Degeneracy).

YAC (yeast artificial chromosome) – A vector system that allows extremely large segments of DNA to be cloned. Useful in chromosome mapping; contiguous YACs covering the whole *Drosophila* genome and certain human chromosomes are available.

Z-DNA – The left-handed helical form of DNA. Z-DNA has been found for sequences containing alternating C and G bases.

Zinc finger – A protein structural motif common in DNA binding proteins. Four Cys residues are found for each "finger" and one finger can bind a molecule of zinc. A typical configuration is: CysXxxXxxCys--(intervening 12 or so aa's)--CysXxxXxxCys.

Zoo blot – A Northern blot of mRNA from multiple organisms.

Index

A

Q

R

S

T

U

V

W

X

Y

Z

Lightning Source UK Ltd.
Milton Keynes UK
UKOW06f0617300815

257755UK00001B/44/P